Signals and Communication Technology

The series "Signals and Communications Technology" is devoted to fundamentals and applications of modern methods of signal processing and cutting-edge communication technologies. The main topics are information and signal theory, acoustical signal processing, image processing and multimedia systems, mobile and wireless communications, and computer and communication networks. Volumes in the series address researchers in academia and industrial R&D departments. The series is application-oriented. The level of presentation of each individual volume, however, depends on the subject and can range from practical to scientific.

More information about this series at http://www.springer.com/series/4748

Joachim Speidel

Introduction to Digital Communications

Joachim Speidel
Institute of Telecommunications
University of Stuttgart
Stuttgart, Baden-Wurttemberg
Germany

ISSN 1860-4862 ISSN 1860-4870 (electronic)
Signals and Communication Technology
ISBN 978-3-030-13123-4 ISBN 978-3-030-00548-1 (eBook)
https://doi.org/10.1007/978-3-030-00548-1

This Springer imprint is published by the registered company Springer Nature Switzerland AG
The registered company address is: Gewerbestrasse 11, 6330 Cham, Switzerland

Preface

Digital communication has found an increasing interest in the past 70 years starting with the telephone network on copper wires, the development of the optical transmission, and the emerging Internet based on wire-line and wireless transmission technologies. Today, the trend to serve an increasing number of mobile users and also machines with information through digital networks is unbroken.

The new book *Introduction to Digital Communications* is aiming at graduate students, scientists, and engineers, who are interested in getting an introduction to modern digital communications. The main focus is on the fundamentals of the physical layer from the perspective of the theory of linear time-invariant as well as time-variant systems. The book draws a bow from single input single output to multiple input multiple output systems with an emphasis on wireless transmission over time-variant channels. The main concern lies in an accurate mathematical description, wherein the findings and lemmas are proven in detail. Various chapters are enriched by numerical examples and also illustrated with results from computer simulations provided by the open platform "webdemo" of the Institute of Telecommunications at the University of Stuttgart, http://www.inue.uni-stuttgart.de.

Organization of the Book

The book covers three main parts and the fourth part with two Appendices.

Part I

It deals with the principles of digital transmission, which are important for wire-line as well as wireless communications. It describes the main building blocks for single input single output (SISO) systems. The concept of quadrature amplitude modulation is introduced. An important part is the design of the overall system for minimal intersymbol interference with Nyquist's first criterion. The introduction of the equivalent baseband system allows the concise definition of the link between the transmitter input and the receiver output as a "black box" without details of the modulation, the spectral signal shaping, and the channel. For the receive signal, several detection methods are described in detail, such as threshold decision, maximum likelihood, and maximum a posterior detection. Also, the difference between symbol-by-symbol and sequence detection is addressed, and the maximum likelihood sequence estimator is described as an example. With an adequate model of the noise at the receiver, the symbol error probability is calculated.

The following chapters in Part I are devoted to the wireless transmission. The main difference is the wireless channel, which changes its characteristic with time. Therefore, the theory of linear time-variant systems is introduced to describe the building blocks of the system with time-variant impulse responses and delay spread functions. As not all students and engineers are frequently involved with this topic, the book contains an own Part II devoted to the theory of linear time-variant systems. Selected points are briefly reported for Part I; hence, the reader is not required to study Part II beforehand. However, for a deeper understanding, the reader should get involved in Part II. The introduction of the equivalent baseband system, which is then time-variant, follows. With this model, the increase of the output signal bandwidth at the receiver compared to the transmit signal is shown as an example. The multipath channel model is described in detail. As the wireless transmission link is multifaceted, a statistical characterization of the channel is helpful. To this end, various channel models are reviewed, such as the Rayleigh and Nakagami-m fading as well as the model according to Clarke and Jakes.

Part II

It is devoted to the theory of linear time-variant systems. In many cases, this topic is just touched upon during the education of graduate students in Electrical Engineering and Computer Science. Therefore, this dedicated Part II is provided. The input–output relation given by the time-variant convolution is addressed in detail and the mathematical properties are derived. We outline the relation with the well-known (time-invariant) convolution used by engineers in most applications. The time-variant impulse response and the delay spread function turn out to be the proper system descriptions in the time domain. Also, the system functions in the frequency domain are presented, such as the time-variant transfer function and the Doppler spread function. For the statistical description of randomly changing time-variant systems, autocorrelation functions as well as power spectral densities of the system functions are studied.

Part III

It deals with multiple input multiple output (MIMO) systems. First, the input–output relation is derived using matrix notation. We discuss the principle MIMO channel models, such as the time-variant finite impulse response and the i.i.d. Gaussian model. Furthermore, spatial correlations at the transmitter and the receiver are incorporated leading to the Kronecker model. Linear and nonlinear MIMO receivers are investigated in detail, such as the zero-forcing, the minimum mean squared error, and the maximum likelihood receiver. An important question is how many bits per channel use can be transmitted over an MMO channel. This issue is studied together with the maximization of the channel capacity. Next, the principles of spatial prefiltering and space-time encoding are investigated to improve transmission quality and to increase the data rate. In the last chapter, we leave the single-user transmission and consider the MIMO principle for a multitude of users in a network. Various multi-user MIMO schemes for the uplink and downlink are discussed, which can reduce the interference when the users transmit their signals in the same time slots and frequency bands.

Part IV

In **Appendix A**, a summary of the characterization of random variables and stochastic processes is given.

Appendix B provides an overview on the most important lemmas of linear algebra required for the understanding of some topics of this book.

Acknowledgement

The author is indebted to Prof. Stephan ten Brink for providing the facilities of the Institute of Telecommunications at the University of Stuttgart, to Mrs. Agnes Schoen-Abiry for drawing most of the figures, and to Maximilian Arnold for some good advice about LYX.

Stuttgart, Germany Prof. Dr.-Ing. Joachim Speidel

Actual notes can be found at:
http://www.inue.uni-stuttgart.de/institut/mitarbeiter/joachim_speidel_01.en.html

Contents

Part I Digital Communications Over Single Input Single Output Channels

1   Transmission System with Quadrature Amplitude Modulation....   3
   1.1   Introduction   3
   1.2   The Transmitter   4
   1.3   Signal Constellation Diagrams   8
   1.4   Transmission Channel   9
   1.5   Receiver   10
   1.6   Equivalent Baseband System Model   13
   1.7   Intersymbol Interference   14
   References   14

2   Removal of Intersymbol Interference   17
   2.1   Nyquist's First Criterion in the Time Domain   17
   2.2   Nyquist's First Criterion in the Frequency Domain   18
   2.3   Raised Cosine Nyquist Lowpass Filter   20
   References   22

3   Characterization of the Noise at the Receiver   23
   3.1   Channel Noise $n_C(t)$ 23
   3.2   Noise After Demodulation and Lowpass Filtering   25
   3.3   Noise After Sampling   26
   3.4   Summary   28

4   Detection Methods   31
   4.1   Receive Signal Under Detection   31
   4.2   Maximum Likelihood Symbol-by-Symbol Detection   31
      4.2.1   Maximum Likelihood Detection   31
      4.2.2   Threshold Detection   32
      4.2.3   Symbol Error Probability for Threshold Detection   33

4.3   Maximum A-Posterior Symbol-by-Symbol Detection   36
4.4   Maximum Likelihood Sequence Detection   38
    4.4.1   System Model   38
    4.4.2   State Space Trellis Diagram   39
    4.4.3   Maximum Likelihood Sequence Detection   40
    4.4.4   Solution Using the Viterbi Algorithm   42
    4.4.5   Viterbi Equalizer   45
References   45

5   Digital Transmission over Wireless, Time-Variant Channels   47
5.1   Transmission System with Time-Variant Channel   47
5.2   Overall Time-Variant Impulse Response.   49
5.3   Overall Delay Spread Function   51
5.4   Overall Doppler Spread Function.   52
    5.4.1   Fourier Transform of the Overall Delay Spread
         Function   52
    5.4.2   Principal System Model Parameters   54
5.5   Equivalent Time-Variant Baseband System and Receiver
    Output Signal   57
    5.5.1   Equivalent Time-Variant Baseband System   57
    5.5.2   Receiver Output Signal   58

6   Basic Parameters of Wireless Channels   63
6.1   Path Loss   63
6.2   Shadowing   64
References   65

7   Wireless System with Multipath Propagation   67
7.1   Multipath Model of Time-Invariant Channel   67
7.2   Multipath Model of Time-Variant Channel   69
    7.2.1   Delay Spread Function of the Time-Variant Multipath
         Channel   69
    7.2.2   Delay Spread Function of the Equivalent
         Time-Variant Multipath Baseband System   70
    7.2.3   Doppler Spread Function of the Equivalent
         Time-Variant Multipath Baseband System   71
    7.2.4   Receiver Output Signal $q_R(t)$ 72
    7.2.5   Fourier Spectrum $Q_R(f_t)$ of the Receiver Output
         Signal   72
7.3   Multipath Channel and Mobile Receiver   73
    7.3.1   System Model   73
    7.3.2   Doppler Shift   74

  7.3.3 Delay Spread Function 75
  7.3.4 Receiver Output Signal $q_R(t)$ with Doppler Shift 75
7.4 Frequency Selective Fading of Multipath Channel 77
References 78

8 Statistical Description of Wireless Multipath Channel 81
8.1 Complex Gaussian Multipath Model 81
8.2 Channel Model with Rayleigh Fading 82
8.3 Channel Model with Rician Fading 83
8.4 Channel Model with Nakagami-m Fading. 83
8.5 Channel Model of Clarke and Jakes. 84
References 88

Part II Theory of Linear Time-Variant Systems

9 Introduction and Some History 91
References 93

**10 System Theoretic Approach for the Impulse Response
of Linear Time-Variant Systems** 95
10.1 Continuous-Time, Time-Variant Impulse Response 95
10.2 Modified Time-Variant Impulse Response—the Delay
Spread Function 96
10.3 Discrete-Time, Time-Variant System 98
  10.3.1 Discrete-Time Delay Spread Function 98
  10.3.2 Transition to Continuous-Time Delay Spread
  Function 100

11 Properties of Time-Variant Convolution 103
11.1 Relation Between Time-Variant and Time-Invariant
Convolution 103
11.2 Properties 104
11.3 Summary 108
11.4 Examples 110

12 System Functions and Fourier Transform 113
12.1 Time-Variant Transfer Function 113
12.2 Delay Doppler Spread Function 114
12.3 Doppler Spread Function 115
12.4 Spectrum of the Output Signal 115
12.5 Cascades of Time-Variant and Time-Invariant Systems 116
  12.5.1 Cascade of Time-Invariant $g_1(\tau)$ and Time-Variant
  System $g_2(t, \tau)$ 116

12.5.2 Cascade of Time-Variant $g_1(t,\tau)$ and Time-Invariant
        system $g_2(\tau)$ 117
12.5.3 Cascade of Two Time-Variant Systems $g_1(t,\tau)$
        and $g_2(t,\tau)$ 118
12.6 Summary 120

13 Applications 121

**14 Interrelation Between Time-Variant and Two-Dimensional
    Convolution** 125
14.1 Input–Output Relation 125
14.2 Fourier Spectrum of the Output Signal 126

15 Randomly Changing Time-Variant Systems 127
15.1 Prerequisites 127
15.2 Correlation Functions of Randomly Changing Time-Variant
      Systems 128
15.3 Wide Sense Stationary Time-Variant Systems 130
      15.3.1 Wide Sense Stationarity 130
      15.3.2 Autocorrelation Functions and Power Spectral
              Densities 130
15.4 Time-Variant Systems with Uncorrelated Scattering 132
      15.4.1 Delay Cross Power Spectral Density of $g(t,\tau)$ 133
      15.4.2 Autocorrelation Function of Time-Variant
              Transfer Function 133
15.5 Wide Sense Stationary Processes with Uncorrelated
      Scattering 134
      15.5.1 Delay Cross Power Spectral Density of $g(t,\tau)$ 134
      15.5.2 Doppler Power Spectrum 134
      15.5.3 Autocorrelation Function of Time-Variant Transfer
              Function 134
15.6 Simplified Parameters for Time-Variant Systems 135
      15.6.1 Coherence Bandwidth 135
      15.6.2 Coherence Time 136
References 137

Part III Multiple Input Multiple Output Wireless Transmission

16 Background 141
References 142

17 Principles of Multiple Input Multiple Output Transmission 143
17.1 Introduction 143
17.2 MIMO Transmission System with Quadrature Amplitude
      Modulation 143

17.2.1 System Model 143
17.2.2 Input–Output Relation of MIMO System
with Time-Variant Channel...................... 146
17.3 Deterministic Models for Wireless MIMO Channels 148
17.3.1 Uniform Linear and Uniform Circular Antenna
Arrays...................................... 148
17.3.2 Finite Impulse Response Channel Model 149
17.3.3 Spatial Channel Models 150
17.3.4 Spectral Properties of the Channel Model 150
17.4 Statistical Models for MIMO Channels 153
17.4.1 I.I.D. Gaussian MIMO Channel Model 153
17.4.2 Covariance Matrix of the MIMO Channel 154
17.4.3 MIMO Channel Model with Correlation 156
17.4.4 MIMO Channel Model with Transmit and Receive
Correlation (Kronecker Model) 158
17.4.5 Exponential Covariance Matrix Model............. 162
References 162

18 Principles of Linear MIMO Receivers 165
18.1 Introduction 165
18.2 Operation Modes for MIMO Systems 166
18.3 Zero-Forcing Receiver for Equal Number of Transmit
and Receive Antennas............................... 168
18.4 Zero-Forcing Receiver for Unequal Number of Transmit
and Receive Antennas............................... 169
18.4.1 Receiver with More Antennas than Transmitter,
$N > M$ 169
18.4.2 Receiver with Less Antennas than Transmitter,
$N < M$ 174
18.5 Signal-to-Noise Ratio of Linear Receivers 177
18.5.1 Signal-to-Noise Ratio with Zero-Forcing Receiver 178
18.5.2 Normalization of the Channel Matrix H............ 179
18.6 Minimum Mean Squared Error receiver 181
18.6.1 Prerequisites.............................. 181
18.6.2 Receiver Matrix 182
18.7 Linear Combiner for single Input Multiple Output System 185
18.7.1 Principle of Linear Combining and the
Signal-to-Noise Ratio 185
18.7.2 MMSE Receiver for SIMO System (Maximum
Ratio Combiner)............................ 186
18.7.3 Equal Gain Combiner........................ 188
18.8 Decision of Receiver Output Signal 190
References 191

19   Principles of Nonlinear MIMO Receivers   193
    19.1   Maximum Likelihood MIMO Receiver   193
    19.2   Receiver with Ordered Successive Interference
           Cancellation   196
    19.3   Comparison of Different Receivers.   199
    References   201

20   MIMO System Decomposition into Eigenmodes   203
    20.1   MIMO System Transformation Using Singular Value
           Decomposition   203
    20.2   Implementation of the MIMO Eigenmode Decomposition   206

21   Channel Capacity of Single-User Transmission Systems   209
    21.1   Channel Capacity of SISO System.   209
           21.1.1   AWGN Channel with Real Signals and Noise   209
           21.1.2   AWGN Channel with Complex Signals and Noise   211
    21.2   Channel Capacity of MIMO Systems with Statistically
           Independent Transmit Signals and Noise   213
           21.2.1   Prerequisites.   213
           21.2.2   Instantaneous MIMO Channel Capacity   215
           21.2.3   Alternative Formulas for the MIMO Channel
                  Capacity   219
    21.3   MIMO Channel Capacity for Correlated Transmit Signals   220
    21.4   Channel Capacity for Correlated MIMO Channel   221
    21.5   Maximizing MIMO System Capacity Using the Water
           Filling Algorithm   222
           21.5.1   Prefilter for Transmit Power Allocation   222
           21.5.2   Computation of the Optimal Power Allocation
                  Coefficients a_i   224
           21.5.3   Graphical Interpretation of the Water Filling
                  Solution.   227
           21.5.4   Iterative Solution and Example   228
    21.6   Capacity of a Stochastic MIMO Channel   230
           21.6.1   Ergodic Channel Capacity.   231
           21.6.2   Outage Capacity.   231
    References   231

22   MIMO Systems with Precoding   233
    22.1   Principle of MIMO Precoding   233
    22.2   Zero-Forcing and MMSE Precoding.   236
           22.2.1   Zero-Forcing Precoder   236
           22.2.2   MMSE Precoder.   238
    22.3   Precoding Based on Singular Value Decomposition   240
           22.3.1   SVD-Based Precoder and Receiver   240

    22.3.2 Comparison of Zero-Forcing and SVD-Based
        Precoding 243
  References 244

23 Principles of Space-Time Coding 245
 23.1 Space-Time Block Coding........................... 245
 23.2 Spatial Multiplexing 249
 23.3 Orthogonal, Linear Space-Time Block Coding 250
    23.3.1 The Alamouti Encoder for MISO System
        with Two Transmit Antennas 251
    23.3.2 The Alamouti Space-Time Encoder for a 2×2
        MIMO System.............................. 255
    23.3.3 Orthogonal Space-Time Block Codes for More
        Than Two Transmit Antennas.................. 257
 23.4 Principle of Space-Time Trellis Coding 260
 23.5 Layered Space-Time Architecture 261
    23.5.1 Vertical Layered Space-Time Coding 262
    23.5.2 Horizontal Layered Space-Time Coding........... 264
    23.5.3 Diagonal Layered Space-Time Coding............ 265
    23.5.4 Iterative Receivers for Layered Space-Time
        Systems................................... 265
  References 266

24 Principles of Multi-user MIMO Transmission 269
 24.1 Introduction 269
 24.2 Precoding for Multi-user MIMO Downlink Transmission 270
    24.2.1 Precoding by "Channel Inversion"................ 270
    24.2.2 Precoding with Block Diagonalization 274
    24.2.3 Alternative Multi-user MIMO Precoding 280
 24.3 Beamforming for Multi-user Downlink................... 281
 24.4 Principles of Multi-user MIMO Uplink Transmission 286
    24.4.1 System Model of the Uplink.................... 286
    24.4.2 Receive Signal at the Base Station 287
    24.4.3 Zero-Forcing Receiver for Multi-user Uplink
        Interference Reduction 287
 24.5 Outlook: Massive MIMO for Multi-user Applications 289
 References 290

Correction to: Introduction to Digital Communications E1

**Appendix A: Some Fundamentals of Random Variables
     and Stochastic Processes**........................ 293

Appendix B: Some Fundamentals of Linear Algebra................. 313

Index 327

About the Author

Joachim Speidel earned his Dipl.-Ing. (M.Sc.) in Electrical Engineering with a major in telecommunications theory from the University of Stuttgart in 1974 followed by a Dr.-Ing. degree (Ph.D.) in 1980 in the field of signal processing and system theory. From 1980 to 1992, he worked in various R&D positions for Philips in the broad field of video and data transmission. Among others, he and his team contributed significantly to algorithms and standards for video data compression and transmission. In 1992, he became a Professor at the Faculty of Computer Science, Electrical Engineering and Information Technology at the University of Stuttgart and was appointed Director of the Institute of Telecommunications. His research area includes telecommunications in wireless, fixed, electrical, and optical networks with a focus on encoding, modulation, detection, and MIMO systems. Since 2017, he is a Professor Emeritus. Through his numerous publications and patents, Prof. Speidel has made extensive contributions in the advancement of the field of telecommunications, the success of its products, and international standards. He is a member of various national and international organizations and advisory and review boards.

Part I
Digital Communications Over Single Input Single Output Channels

Chapter 1
Transmission System with Quadrature Amplitude Modulation

1.1 Introduction

This chapter presents an overview on the principles of digital communications. We focus on a system with one transmitter and one receiver, i.e., for a channel with a single input and a single output (SISO). This will also provide the necessary basics for multiple input multiple output (MIMO) systems investigated in Part III. Depending on the characteristics of the transmission medium, we have to differentiate between a wire-line and a wireless connection. Both channel types exhibit different properties and therefore will be treated separately. We start with the wire-line transmission link and in Chap. 5 the wireless system will be discussed in detail. Also, the transfer functions of the transmission media differ in general. They can have a lowpass or a bandpass characteristic. An electrical line, e.g., a twisted pair or a coaxial cable exhibits a lowpass magnitude response, because the DC current can travel from the input to the output. In contrast, a wireless channel is characterized by a bandpass transfer function, because only high-frequency spectral components can be emitted and received by the antennas. The optical transmission on a glass or a plastic fiber is similar but the transmission spectrum lies in the multi-THz frequency region. However, it should be noted that a connection over an electrical line may contain transformers at the transmitter or at the receiver side for galvanic isolation between the transmission line and the electronic equipment. Then, the overall transfer function becomes a bandpass characteristic. The same is true, if decoupling capacitors are connected in series to the electrical transmission line. As the source signal at the transmitter normally has a lowpass spectrum, a frequency shift into the passband of the channel by dedicated pulse shaping or modulation is required. In the following, we focus on a bandpass channel with a passband around the mid frequency f_0 and employ a modulator with a carrier signal

$$e^{j2\pi f_0 t} = \cos(2\pi f_0 t) + j\sin(2\pi f_0 t) \qquad (1.1)$$

© Springer Nature Switzerland AG 2019
J. Speidel, *Introduction to Digital Communications*, Signals and Communication
Technology, https://doi.org/10.1007/978-3-030-00548-1_1

at the transmitter, where the carrier frequency is f_0. This complex carrier contains an "in-phase" component with $\cos(2\pi f_0 t)$ and a "quadrature" component with $\sin(2\pi f_0 t)$, which are orthogonal. Therefore, the scheme is called quadrature amplitude modulation (QAM) or I-Q modulation. We will introduce an equivalent baseband (or lowpass) system model providing the overall characteristic of a lowpass transfer function. With this approach, lowpass and bandpass transmission schemes can be treated elegantly in a uniform way and the transmission scheme, which does not require modulation, is included for $f_0 = 0$. Then, only real-valued transmit symbols are applicable also called (multi-level) pulse amplitude modulated (PAM) symbols. We will also see that the principle methods for wire-line and wireless transmission have quite a lot in common and we focus on the next chapters on the basic principles of both.

1.2   The Transmitter

Figure 1.1 shows the principle block diagram of a transmission system. The output of the data source at the transmitter is a sequence of bits $b_S(l')$. The forward error correction (FEC) encoder allocates redundant bits to the input, which are thereafter

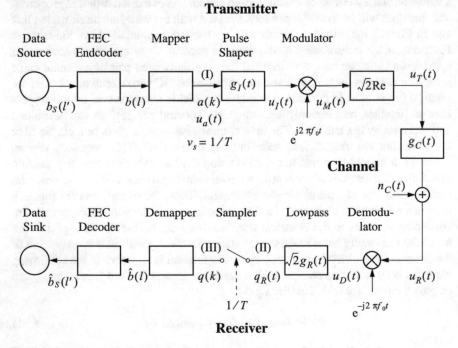

Fig. 1.1 Principle block diagram for digital transmission over a bandpass channel

temporally interleaved to prepare for burst errors. The resulting output bit sequence is denoted as $b(l)$. l and l' are discrete-time variables. The amount of redundancy can be defined by the temporal code rate $r_t \leq 1$, which is the ratio between the number of input and output bits in the same time frame. Then, the output bit sequence contains $(1 - r_t) \cdot 100\%$ redundancy with respect to the input. The large number of methods and codes for forward error correction are not considered here. The reader is referred to the dedicated literature, e.g., [1–9]. The mapper periodically allocates tuples of κ successive bits to a symbol $a(k)$. The mapper outputs the complex symbol sequence

$$a(k) = \text{Re}[a(k)] + \text{jIm}[a(k)] \tag{1.2}$$

at periodic intervals of duration T with the symbol rate $v_S = \frac{1}{T}$. The Fourier spectrum of the discrete-time signal $a(k)$ can be calculated with the help of the z-transform of $a(k)$, which is

$$A(z) = \sum_{k=-\infty}^{\infty} a(k) z^{-k} \tag{1.3}$$

and for $z = e^{j2\pi fT}$ the Fourier spectrum

$$A\left(e^{j2\pi fT}\right) = \sum_{k=-\infty}^{\infty} a(k) e^{-j2\pi fT} \tag{1.4}$$

is obtained. Obviously, $A\left(e^{j2\pi fT}\right)$ is a periodic function of the natural frequency f with period $\frac{1}{T}$, because the argument $e^{j2\pi fT}$ is periodic with $f = \frac{1}{T}$. Consequently, the spectrum of the symbol sequence $a(k)$ exhibits an infinite bandwidth, which has to be limited by a transmit lowpass filter with cut-off frequency f_I. This filter shall be linear and time-invariant with the impulse response $g_I(t)$ and can be used to shape the output impulses $u_I(t)$, for which reason the filter is called impulse shaper or pulse shaper. As is well known, the transfer function of a linear, time-invariant system is defined by the Fourier transform of its impulse response and, in general, the spectrum of a continuous-time signal can be obtained by the Fourier transform, [10]. In the following, we denote the transform by the symbol $\rightarrowtail$ and assume that the reader is familiar with the Fourier transform calculus.

Thus, we obtain the spectrum of $g_I(t)$ as

$$g_I(t) \rightarrowtail G_I(f) = \int_{-\infty}^{\infty} g_I(t) e^{-j2\pi ft} dt \tag{1.5}$$

and the inverse transform is given by

$$G_I(f) \leftarrowtail g_I(t) = \int_{-\infty}^{\infty} G_I(f) e^{j2\pi ft} df \tag{1.6}$$

To check the existence of the integral (1.5), one of the sufficient Dirichlet conditions can be applied

$$\int_{-\infty}^{\infty} |g_I(t)|\, dt \leq M_1 < \infty \quad \text{or} \quad \int_{-\infty}^{\infty} |g_I(t)|^2\, dt \leq M_2 < \infty \qquad (1.7)$$

where M_1 and M_2 are finite real numbers. The second condition confirms that all signals with finite energy are equipped with a Fourier spectrum. $G_I(f)$ is also called the transfer function of the pulse shaper and shall have the lowpass property

$$G_I(f) \quad \begin{cases} \neq 0 \; ; \; |f| \leq f_I \\ = 0 \; ; \quad else \end{cases} \qquad (1.8)$$

where f_I is the cut-off frequency. The continuous-time signal at the input of the pulse shaper is described as

$$u_a(t) = \sum_{k=-\infty}^{\infty} a(k)\delta(t - kT) \qquad (1.9)$$

where $\delta(t)$ is the Dirac impulse. Then, it follows the signal at the pulse shaper output

$$u_I(t) = u_a(t) * g_I(t) = \sum_{k=-\infty}^{\infty} a(k)g_I(t - kT) \qquad (1.10)$$

where $*$ denotes the convolution

$$u_a(t) * g_I(t) = \int_{-\infty}^{\infty} u_a(\tau)g_I(t - \tau)d\tau \qquad (1.11)$$

We obtain the output signal of the modulator with (1.10)

$$u_M(t) = u_I(t)e^{j2\pi f_0 t} = \left(\sum_{k=-\infty}^{\infty} a(k)g_I(t - kT) \right) e^{j2\pi f_0 t} \qquad (1.12)$$

Any physical channel exhibits a real-valued impulse response denoted as $g_C(t)$ in Fig. 1.1. If not otherwise stated, the channel shall be characterized as a linear, time-invariant system. In Chap. 5, we will focus on the wireless channel, which is characterized as linear and time-variant.[1]

If the complex signal $u_M(t)$ is directly input to the channel, a separation of the real and imaginary part at the receiver is not possible. Thus, only the real or the

---

[1]In case of a baseband transmission system with a lowpass channel, no modulation is required, thus we set $f_0 = 0$. Figure 1.1 still holds, wherein the modulator, the real-part operator Re[...], the demodulator, and the gain factors $\sqrt{2}$ are dropped. All signals including the symbols $a(k)$ take on real values only.

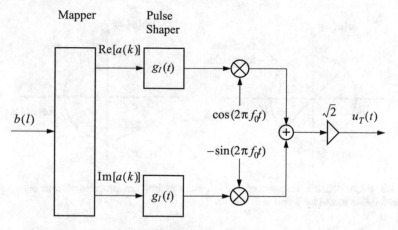

Fig. 1.2 Implementation of a QAM transmitter with exclusively real signals

imaginary part of $u_M(t)$ is feasible as the channel input signal. In Fig. 1.1 the real part is selected and $\sqrt{2}$ is just an amplification factor to achieve an overall amplification of one between the transmitter and the receiver. In the following, we apply the general property for complex numbers

$$\mathrm{Re}[u_M] = \frac{1}{2}\left(u_M + u_M^*\right) \tag{1.13}$$

where the superscript denotes the conjugate complex operation. With (1.12), (1.13), and assuming a real impulse response $g_I(t)$ a straightforward calculation yields the transmit signal

$$u_T(t) = \sqrt{2}\left(\sum_{k=-\infty}^{\infty} \mathrm{Re}[a(k)]\, g_I(t-kT)\right)\cos(2\pi f_0 t)$$
$$-\sqrt{2}\left(\sum_{k=-\infty}^{\infty} \mathrm{Im}[a(k)]\, g_I(t-kT)\right)\sin(2\pi f_0 t) \tag{1.14}$$

From (1.14), we recognize that the transmit signal can carry two independent symbol sequences $\mathrm{Re}[a(k)]$ and $\mathrm{Im}[a(k)]$. Furthermore, as depicted in Fig. 1.2, we can implement an alternative QAM transmitter, which contains only real-valued signals and provides the same output signal $u_T(t)$.

The reader can convince oneself easily that the spectrum $U_T(f)$ of $u_T(t)$ in (1.14) satisfies

$$U_T(f) \begin{cases} \neq 0\,; & f_0 - f_I \leq |f| \leq f_0 + f_I \\ = 0\,; & else \end{cases} \tag{1.15}$$

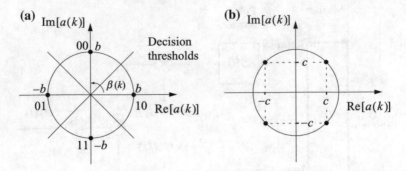

Fig. 1.3 a Constellation diagram of 4-PSK. Pairs of bits are allocated to four signal points **b** Constellation turned by $\frac{\pi}{4}$ and with $c = \frac{b}{\sqrt{2}}$

1.3 Signal Constellation Diagrams

The mapper in Figs. 1.1 and 1.2 assigns to each tuple of κ incoming bits $b(l), b(l-1), \ldots, b(l-\kappa+1)$ a complex symbol $a(k)$. The allocation of these symbols in the complex plane constitutes the signal constellation diagram. The $a(k)$ is also denoted as signal point and encoded by a codeword with κ bits. Hence, the number of different symbols is

$$L = 2^\kappa \tag{1.16}$$

and therefore this scheme is referred to as L-ary quadrature amplitude modulation or L-QAM. There is an infinite number of possible distributions of L signal points in the complex plane. The example in Fig. 1.3 shows the 4-level phase shift keying (4-PSK), where the signal points are distributed on a circle. In the following, we present some more examples of important constellation diagrams used for digital transmission. The relation between the bitrate v_B and the symbol rate $v_S = \frac{1}{T}$ obviously is

$$v_S = \frac{v_B}{\kappa} = \frac{v_B}{\log_2 L} \tag{1.17}$$

and is measured in "symbols per second", which is also referred to as "Baud" according to the French engineer Baudot. Consequently, with L-QAM the clock rate is by factor $\frac{1}{\log_2 L}$ lower than the bitrate, which also holds for the required transmission bandwidth.

4-QAM and 4-PSK

Figure 1.3 shows the constellation diagram of 4-PSK as a special case of 4-QAM. The four signal points are located on a circle and thus they differ only in their angles $\beta(k) \in \left\{0, \frac{\pi}{2}, \pi, \frac{3\pi}{2}\right\}$, which justifies the name phase shift keying. Another special case is 2-PSK, where only two signal points are used. If they are allocated on the real axis of the complex plane, the scheme is very simple to implement. The symbol alphabet

Fig. 1.4 Constellation diagram of 16-QAM

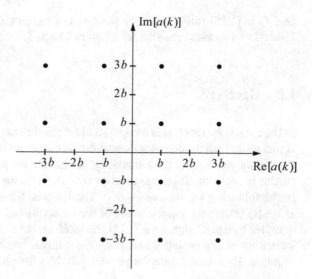

then is $\mathcal{B} = \{-b, b\}$ and the angles are $\beta(k)\epsilon\{0, \pi, \}$. Consequently, for the transmit signal in (1.14) follows

$$u_T(t) = \sqrt{2} \sum_{k=-\infty}^{\infty} \text{Re}[a(k)] \, g_I(t - kT) \, \cos(2\pi f_0 t) \; ; \; \text{with Re}\,[a(k)] \in \{-b, b\}$$

$$(1.18)$$

16-QAM

Figure 1.4 shows another constellation diagram, the 16-QAM, where 4-tuples of bits $b(l), b(l - 1), b(l - 2), b(l - 3)$ are allocated to the 16 different signal points $a(k)$. The scheme is frequently used in many applications for wire-line and wireless digital transmission.

1.4 Transmission Channel

The transmission channel shall be characterized by its time-invariant and real-valued impulse response $g_C(t)$. The bandpass-shaped transfer function $G_C(f)$ is obtained by the Fourier transform of $g_C(t)$ and characterized as

$$g_C(t) \rightarrowtail G_C(f) \begin{cases} \neq 0 \; ; \; f_0 - f_I \leq |f| \leq f_0 + f_I \\ = 0 \; ; \qquad\qquad else \end{cases} \qquad (1.19)$$

The channel passband is located around the center frequency f_0 and the bandwidth is $2f_I$. Thus, the carrier frequency f_0 of the modulator and the cut-off frequency f_I of the pulse shaper have to be determined in such a way that the transmit spectrum

$U_T(f)$ in (1.15) fully covers the passband of the channel transfer function $G_C(f)$. Models of a wireless channel are given in Chap. 5.

1.5 Receiver

In Figure 1.1, the receiver is composed of a demodulator, which multiplies the receive signal $u_R(t)$ with the complex demodulation carrier $e^{-j(2\pi f_0 t + \varphi_0)}$. The frequency f_0 is exactly the same as for the modulating carrier and the phase φ_0 of the demodulation carrier is constant. Thus, the receiver operates synchronously with the transmitter. In the following we assume $\varphi_0 = 0$. The lowpass filter with real impulse response $\sqrt{2}g_R(t)$ selects the baseband out of the demodulated signal $u_D(t)$ resulting in the complex baseband signal $q_R(t)$. At the receiver, there is no interest in the complete waveform of this analog signal. Only the samples taken with the symbol rate $\frac{1}{T}$ are required. Therefore, a sampling device provides the sequence

$$q(k) = q_R(t_0 + kT) \tag{1.20}$$

where t_0 is the signal delay between the transmitter and receiver, which has to be estimated at the receiver. The sampling clock with frequency $\frac{1}{T}$ has to be synchronous with the symbol rate v_S at the transmitter. It is extracted by a special clock recovery circuit from the receive signal to guarantee synchronism and to control deviations. Synchronization methods [11] as well as sampling rates higher than $\frac{1}{T}$, which can meet the sampling theorem for advanced digital signal processing at the receiver, are not considered here.

The receiver input signal is corrupted by real-valued additive noise $n_C(t)$ coming from the channel and the electronic equipment of the receiver. We will just call it channel noise. Using the superposition principle, we obtain the total receive signal

$$u_R(t) = u_T(t) * g_C(t) + n_C(t) \tag{1.21}$$

We will show that the output signal of the receive lowpass filter with the impulse response $\sqrt{2}g_R(t)$ and the transfer function

$$\sqrt{2}G_R(f) \begin{cases} \neq 0 \ ; \ |f| \leq f_R \\ = 0 \ ; \quad else \end{cases} \tag{1.22}$$

is given by

$$q_R(t) = u_a(t) * h_e(t) + n_R(t) \tag{1.23}$$

where

$$h_e(t) = g_I(t) * \left(g_C(t)e^{-j2\pi f_0 t}\right) * g_R(t) \tag{1.24}$$

and

$$n_R(t) = \sqrt{2}\left(n_C(t)e^{-j2\pi f_0 t}\right) * g_R(t) \tag{1.25}$$

$n_R(t)$ represents the demodulated and lowpass filtered noise. Equation (1.23) can be recognized as the input–output relation of the signals between the node (I) and the node (II) in Fig. 1.1. Consequently, $h_e(t)$ is the equivalent impulse response between (I) and (II) in case of no noise, $n_C(t) = 0$. Although this section is time-variant on the first glance due to the synchronous modulation and demodulation, the convolution operation $u_a(t) * h_e(t)$ still holds, as will be proven in the following. This kind of convolution operation will change, if the phase of the demodulation carrier varies with time due to some phase noise or in case of a time-varying wireless channel, as outlined in Chap. 5. Please note that $u_I(t)$, $u_M(t)$, $h_e(t)$, and $n_R(t)$ are complex-valued signals and noise in general, whereas $g_I(t)$, $g_C(t)$, $g_R(t)$, and $n_C(t)$ are real-valued, respectively.

We also show that the spectrum $Q_R(f)$ of the signal $q_R(t)$ at the output of the receiver lowpass is given in the case of no noise as

$$Q_R(f) = U_I(f)G_C(f + f_0)G_R(f) \tag{1.26}$$

where $u_I(t) \rightarrowtail U_I(f)$ is the lowpass spectrum of the output signal $u_I(t)$ of the pulse shaper

$$U_I(f) = G_I(f)U_a(f) \tag{1.27}$$

where $u_a(t) \rightarrowtail U_a(f)$ holds. Obviously, $Q_R(f)$ is a lowpass spectrum. From (1.26) follows with (1.27)

$$H_e(f) = \frac{Q_R(f)}{U_a(f)} = G_I(f)G_C(f + f_0)G_R(f) \tag{1.28}$$

which is the overall transfer function between the nodes (I) and (II). We recognize that $H_e(f)$ represents a lowpass and is called the transfer function of the *equivalent baseband (or lowpass) system*.[2]

Proof of (1.24) **and** (1.26)

We start with the output $u_I(t) \rightarrowtail U_I(f)$ of the pulse shaper, which has a lowpass spectrum with the cut-off frequency f_I given by (1.8). Then, we obtain the modulator output signal

$$u_M(t) = u_I(t)e^{j2\pi f_0 t} \tag{1.29}$$

and with (1.13) the transmit signal

$$u_T(t) = \frac{\sqrt{2}}{2}\left[u_I(t)e^{j2\pi f_0 t} + u_I^*(t)e^{-j2\pi f_0 t}\right] \tag{1.30}$$

---

[2] sometimes referred to as "equivalent baseband (or lowpass) channel".

We proceed in the frequency domain and get with

$$u_I^*(t) \rightarrowtail U_I^*(-f) \tag{1.31}$$

and with the frequency shifting property of the Fourier transform the transmit signal

$$u_T(t) \rightarrowtail U_T(f) = \frac{\sqrt{2}}{2} \left[U_I(f - f_0) + U_I^*(-f - f_0) \right] \tag{1.32}$$

As expected, the spectrum $U_T(f)$ is located in the passband of the channel transfer function around the mid frequencies $-f_0$ and f_0.

The spectrum of the receive signal in case of no noise at the output of the channel can be found as

$$u_R(t) \rightarrowtail U_R(f) = \frac{\sqrt{2}}{2} \left[U_I(f - f_0) + U_I^*(-f - f_0) \right] G_C(f) \tag{1.33}$$

After demodulation, the spectrum at the input of the receiver lowpass filter in the case of no noise is $U_D(f) = U_R(f + f_0)$ and finally, the spectrum of $q_R(t)$ follows

$$Q_R(f) = \sqrt{2} U_R(f + f_0) G_R(f) \tag{1.34}$$

Plugging in (1.33) yields

$$Q_R(f) = \left[U_I(f) G_C(f + f_0) + U_I^*(-f - 2f_0) G_C(f + f_0) \right] G_R(f) \tag{1.35}$$

The second term $U_I^*(-f - 2f_0) G_C(f + f_0)$ is a spectrum with passband around the center frequency $-2f_0$. Multiplied with the lowpass transfer function $G_R(f)$, the product is zero and we obtain from (1.35) $Q_R(f) = U_I(f) G_C(f + f_0) G_R(f)$, which finalizes the proof of (1.26).

We recognize that the overall amplification factor for $Q_R(f)$, excluding the noise, is $\sqrt{2}\sqrt{2}\frac{1}{2} = 1$, which justifies the introduction of the amplification factor $\sqrt{2}$ both at transmitter and receiver.

The inverse Fourier transform of $H_e(f)$ in (1.28) results in
$h_e(t) = g_I(t) * \left(g_C(t) e^{-j2\pi f_0 t} \right) * g_R(t)$ and the proof of (1.24) is finished.

Proof of (1.23) **and** (1.25)

We directly conclude from the left-hand side of (1.28) $Q_R(f) = H_e(f) U_a(f)$ and the inverse Fourier transform yields $q_R(t) = u_a(t) * h_e(t)$. This finalizes the proof of (1.23) in case of no noise.

The proof of (1.25) is straightforward. If the transmitter sends no signal, i.e., $u_T(t) = 0$, then the output signal of the receiver lowpass is the filtered noise $q_R(t) = n_R(t) = \sqrt{2} \left(n_C(t) e^{-j2\pi f_0 t} \right) * g_R(t)$. If $n_C(t) = 0$, then $q_R(t) = u_a(t) * h_e(t)$ holds. In the general case that the transmit signal and the noise are present, the superposition principle can be applied to this linear system yielding $q_R(t) = u_a(t) * h_e(t) + n_R(t)$, which completes the proof of (1.23).

Fig. 1.5 **a** Continuous-time
equivalent baseband system
model between nodes (I) and
(II) with reference to Fig. 1.1
b Discrete-time equivalent
baseband system model
between nodes (I) and
(III) with reference to
Fig. 1.1

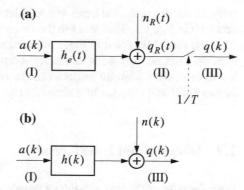

1.6 Equivalent Baseband System Model

The block diagram in Fig. 1.1 is rather detailed. We take advantage of (1.23) and find
the much simpler structure in Fig. 1.5a. Details of modulation, demodulation, and
filtering does not show up anymore. The input–output relation of the "black box"
between nodes (I) and (II) are given by (1.23)–(1.25). $h_e(t)$ is denoted as the impulse
response of the continuous-time equivalent baseband system model. Finally, we are
only interested in the discrete-time relation between the input sequence $a(k)$ and the
sampled output sequence $q(k)$ defined already in (1.20).

To this end, we insert (1.9) into (1.23) and obtain

$$q_R(t) = \sum_{m=-\infty}^{\infty} a(m)h_e(t - mT) + n_R(t) \tag{1.36}$$

Now we take samples at $t = t_0 + kT$ and get

$$q_R(t_0 + kT) = \sum_{m=-\infty}^{\infty} a(m)h_e(t_0 + (k - m)T) + n_R(t_0 + kT) \tag{1.37}$$

With

$$h(k) = h_e(t_0 + kT) = \left\{ g_I(t) * \left(g_C(t)e^{-j2\pi f_0 t} \right) * g_R(t) \right\}_{t=t_0+kT} \tag{1.38}$$

and

$$n(k) = n_R(t_0 + kT) \tag{1.39}$$

follows from (1.37) with (1.20)

$$q(k) = \sum_{m=-\infty}^{\infty} a(m)h(k - m) + n(k) \tag{1.40}$$

which is the discrete-time input–output relation of the "black box" between nodes (I) and (III) in Fig. 1.1. This leads to the discrete-time equivalent baseband system (or channel) model depicted in Fig. 1.5b, which is very helpful, because it focuses our consideration on only a few characteristic parameters of the system. $h(k)$ is called the discrete-time impulse response of the equivalent baseband system model. The noises $n_R(t)$ and $n(k)$ are investigated in quite some detail in Chap. 3.

1.7 Intersymbol Interference

With the help of the discrete-time equivalent baseband system model, we can now get insight into the two major impairments a signal incurs from the transmitter to the receiver, namely intersymbol interference and noise. For that purpose, we separate the term for $m = k$ from the sum in (1.40) and obtain

$$q(k) = a(k)h(0) + \sum_{\substack{m = -\infty \\ m \neq k}}^{\infty} a(m)h\,(k - m) + n(k) \tag{1.41}$$

We see that the receive sample $q(k)$ is composed of the transmit symbol $a(k)$ multiplied by $h(0)$ of the discrete-time impulse response $h(k)$, the distortion term

$$I(k) = \sum_{\substack{m = -\infty \\ m \neq k}}^{\infty} a(m)h\,(k - m) \tag{1.42}$$

and the noise $n(k)$. $I(k)$ is called intersymbol interference, because it consists of the previous and in case of a non-causality of $h(k)$ the future $a(m)$ weighted by the samples $h(k - m)$ of the equivalent baseband impulse response.

References

1. C. Shannon, A mathematical theory of communication. Bell Syst. Tech. J. **27** (1948)
2. R. Gallager, Low-density parity-check codes. IRE Trans. Inf. Theory (1962)
3. J. Hagenauer, E. Offer, L. Papke, Iterative decoding of binary block and convolutional codes. IEEE Trans. Inf. Theory **42** (1996)
4. M. Bossert, *Channel Coding for Telecommunications* (Wiley, New York, 1999)
5. S. ten Brink, Design of concatenated coding schemes based on iterative decoding convergence, Ph.D. dissertation, University of Stuttgart, Institute of Telecommunications, Shaker Publications. ISBN 3-8322-0684-1 (2001)
6. D. Costello, G. Forney, Channel coding: the road to channel capacity. Proc. IEEE (2007)

7. T. Richardson, R. Urbanke, *Modern Coding Theory* (Cambridge University Press, Cambridge, 2008)
8. W. Ryan, S. Lin, *Channel Codes - Classical and Modern* (Cambridge University Press, Cambridge, 2009)
9. D. Declercq, M. Fossorier, E. Biglieri, *Channel Coding: Theory, Algorithms, and Applications* (Academic Press, New York, 2014)
10. A. Papoulis, *The Fourier Integral and Its Applications* (McGraw Hill, New York, 1976)
11. M. Simon, S. Hinedi, W. Lindsey, *Digital Communications: Synchronization* (Prentice Hall, New Jersey, 2000)

Chapter 2
Removal of Intersymbol Interference

2.1 Nyquist's First Criterion in the Time Domain

To remove the intersymbol interference $I(k)$ defined in (1.42) we may not impose any constraint on the symbol sequence $a(k)$, because the system design should hold for any sequence given by the user at the transmitter. Therefore we can only touch upon the impulse response $h(k)$. Looking at (1.38), the system is prepared already with two degrees of freedom, $g_I(t)$ and $g_R(t)$. Hence, for a given impulse response $g_C(t)$ of the physical channel, we can design the overall impulse response in such a way that

$$h(k-m) = h_e(t_0 + (k-m)T) = \begin{cases} 0 & ; \; m \in \mathbb{Z} \; ; \; m \neq k \\ h(0) = h_e(t_0) \neq 0 \; ; & m = k \end{cases}$$
(2.1)

(2.1) is called Nyquist's first criterion in the time domain [1] and the corresponding impulse is referred to as Nyquist impulse. An example of a real-valued impulse response satisfying (2.1) is depicted in Fig. 2.1. Obviously, $h_e(t)$ owns equidistant zeros except at $t = t_0$.

Inserting the Nyquist condition (2.1) into (1.42) yields

$$I(k) = 0 \; ; \; \forall k \in \mathbb{Z}$$
(2.2)

and we obtain from (1.41)

$$q(k) = a(k)h(0) + n(k)$$
(2.3)

As expected, the signal $q(k)$ at the receiver output suffers not anymore from intersymbol interference and the symbol sequence $a(k)h(0)$ is only corrupted by additive noise $n(k)$.

© Springer Nature Switzerland AG 2019
J. Speidel, *Introduction to Digital Communications*, Signals and Communication Technology, https://doi.org/10.1007/978-3-030-00548-1_2

Fig. 2.1 Example of a
real-valued impulse $h_e(t)$
satisfying Nyquist's first
criterion (2.1)

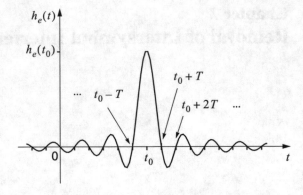

2.2 Nyquist's First Criterion in the Frequency Domain

An interesting question is: "How does the spectrum $H_e(f)$ of a Nyquist impulse look like?" We give the solution for the case of a real-valued spectrum $H_e(f)$ with the consequence that the corresponding impulse is real and an even function, $h_e(-t) = h_e(t)$. Thus, in the following $t_0 = 0$ is assumed. Given these prerequisites, the necessary and sufficient condition for the spectrum $H_e(f)$ is

$$\sum_{m=-\infty}^{\infty} H_e\left(f - m\frac{1}{T}\right) = h_e(0)T \; ; \; \forall f \; ; \; h_e(0)T = H_e(0) \qquad (2.4)$$

which is called Nyquist's first criterion in the frequency domain and will be proven in the following. Equation (2.4) requires that the sum of all periodic repetitions of $H_e(f)$ is a constant, $H_e(0) = h_e(0)T$. A lowpass satisfying the condition (2.4) is also called a Nyquist lowpass. From (2.4) directly follows the solution for a lowpass spectrum with cut-off frequency

$$f_c = f_N + \Delta f \qquad (2.5)$$

$$H_e(f) \begin{cases} = A & ; \; |f| \le f_N - \Delta f \\ H_e(|f_N| - x) + H_e(|f_N| + x) = A ; & 0 < x \le \Delta f \\ = 0 & ; \; |f| > f_N + \Delta f \end{cases} \qquad (2.6)$$

where we have set $A = H_e(0)$. Obviously, $H_e(f)$ is an even function, $H_e(-f) = H_e(f)$. The principle spectrum of $H_e(f)$ is depicted in Fig. 2.2

The relation between the Nyquist frequency f_N and the symbol interval T are given by

$$f_N = \frac{1}{2T} \qquad (2.7)$$

Fig. 2.2 Real-valued
transfer function $H_e(f)$ of a
Nyquist lowpass. Roll-offs
exhibit odd symmetry with
respect to points
P_1 $(f_N, 0.5)$ and
P_2 $(-f_N, 0.5)$

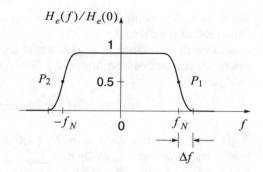

The second condition in (2.6) requires a roll-off function from the passband to the
stopband, which exhibits odd symmetry with respect to the points P_1 and P_2. Δf
defines half of the roll-off bandwidth

$$0 \leq \Delta f \leq f_N \tag{2.8}$$

For $\Delta f = 0$, the spectrum $H_e(f)$ is an ideal lowpass filter and shows the minimal
cut-off frequency $f_c = f_N$. On the contrary, the maximum cut-off frequency is $2 f_N$
and is obtained for $\Delta f = f_N$. In that case, the roll-off starts already at $f = 0$ and
covers the whole passband. For a given symbol rate $v_S = \frac{1}{T}$ the Nyquist frequency
(2.7) results in

$$f_N = \frac{1}{2} v_S \tag{2.9}$$

For example, if we strive for a symbol rate of 1 Gsymbol/s a minimal cut-off frequency
of 500 MHz is required to satisfy Nyquist's criterion.

For the filter design the following definition of a roll-off factor is helpful

$$\alpha = \frac{\Delta f}{f_N} \tag{2.10}$$

From (2.8) we see that $0 \leq \alpha \leq 1$ holds and the cut-off frequency becomes
$f_c = (1 + \alpha) f_N$.

If we drop the previous prerequisite and consider an impulse with $t_0 > 0$, as
depicted in Fig. 2.1, then $H_e(f)$ turns into $H_e(f)e^{-j2\pi f t_0}$ according to the shifting
property of the Fourier transform. Then the spectrum is equipped with a linear phase
term $e^{-j2\pi f t_0}$.

As a conclusion, we can achieve the transmission of the symbol sequence $a(k)$
without intersymbol interference, if the overall transfer function $H_e(f)$ of the system
is a Nyquist lowpass defined in (2.6). An adequate design is

$$f_I = f_R = (1 + \alpha) f_N \tag{2.11}$$

where f_R is the cut-off frequency of the receive lowpass filter $G_R(f)$ and $2f_I$ the bandwidth of the channel.

It is worth mentioning that also transfer functions with other than lowpass characteristics can be deduced from (2.4). For illustration the following example of a bandpass

$$B(f) = \frac{1}{2}H_e(f + 2f_N) + \frac{1}{2}H_e(f - 2f_N) \tag{2.12}$$

is given, which is composed of the Nyquist lowpass $H_e(f)$ according to (2.6). The reader can assure oneself easily that $\sum_{m=-\infty}^{\infty} B(f - m2f_N) = H_e(0)$; $\forall f$ holds and thus $B(f)$ fulfills the first Nyquist criterion (2.4).

Proof of (2.4)

In the following, we assume $t_0 = 0$ and first prove that (2.1) is a sufficient condition for (2.4), i.e., that (2.4) follows from (2.1). Ideal sampling of $h_e(t)$ yields $h_e(t)\sum_{k=-\infty}^{\infty}\delta(t - kT)$ and with (2.1) follows $h_e(t)\sum_{k=-\infty}^{\infty}\delta(t - kT) = h_e(0)\delta(t)$. Applying the Fourier transform on both sides results in $H_e(f) * \frac{1}{T}\sum_{m=-\infty}^{\infty}\delta(f - m\frac{1}{T}) = h_e(0)$, where we have used the transform pairs $\sum_{k=-\infty}^{\infty}\delta(t - kT) \longmapsto \frac{1}{T}\sum_{m=-\infty}^{\infty}\delta(f - m\frac{1}{T})$ and $\delta(t) \longmapsto 1$. With the convolution integral follows $H_e(f) * \frac{1}{T}\sum_{m=-\infty}^{\infty}\delta(f - m\frac{1}{T}) = \frac{1}{T}\sum_{m=-\infty}^{\infty}\int_{-\infty}^{\infty}H_e(\tau)\delta(f - m\frac{1}{T} - \tau)d\tau = \frac{1}{T}\sum_{m=-\infty}^{\infty}H_e(f - m\frac{1}{T})$. Hence, we end up with $\sum_{m=-\infty}^{\infty}H_e(f - m\frac{1}{T}) = h_e(0)T$, which validates (2.4).

Next we prove that (2.1) is a necessary condition for (2.4), i.e., we have to show that from (2.4) follows (2.1). This is easily done by starting from (2.4) and executing the steps done before in reverse direction using the inverse Fourier transform, which finally results in (2.1). This finalizes the proof that (2.1) and (2.4) are necessary and sufficient conditions.

2.3 Raised Cosine Nyquist Lowpass Filter

Now we consider a special Nyquist filter $H_e(f)$, which is frequently used as a design goal for digital communication systems

$$\frac{H_e(f)}{H_e(0)} = \begin{cases} 1 & ; & |f| \leq f_N(1 - \alpha) \\ \frac{1}{2}\left[1 + \cos\left(\frac{\pi}{2}\frac{|f| - f_N(1-\alpha)}{\alpha f_N}\right)\right] & ; & f_N(1 - \alpha) < |f| < f_N(1 + \alpha) \\ 0 & ; & |f| \geq f_N(1 + \alpha) \end{cases} \tag{2.13}$$

and portrait in Fig. 2.3. For an overall gain of one we have to determine $H_e(0) = 1$. The function $1 + \cos(\ldots)$ is called raised cosine. Significant are the roll-offs, which possess an odd symmetry with respect to the points $(\pm f_N, \frac{1}{2})$. The corresponding Nyquist impulses are obtained by inverse Fourier transform

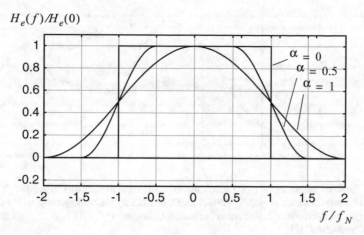

$H_e(f)/H_e(0)$

Fig. 2.3 Raised cosine transfer function $H_e(f)$ of a Nyquist lowpass filter with various roll-off factors α

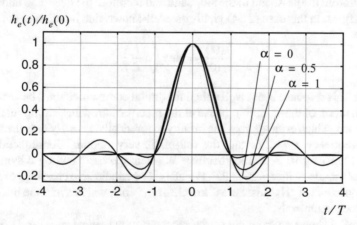

$h_e(t)/h_e(0)$

Fig. 2.4 Impulse response $h_e(t)$ of raised cosine Nyquist lowpass filter with various roll-off factors α

$$\frac{h_e(t)}{h_e(0)} = \mathrm{sinc}\left(\pi\frac{t}{T}\right)\frac{\cos\left(\pi\alpha\frac{t}{T}\right)}{1-\left(2\alpha\frac{t}{T}\right)^2} \tag{2.14}$$

where $h_e(0) = \frac{1}{T}H_e(0)$ and

$$\mathrm{sinc}(x) = \frac{\sin(x)}{x} \tag{2.15}$$

In Fig. 2.3 $H_e(f)$ is shown for various roll-off factors α and the corresponding impulse responses $h_e(t)$ are depicted in Fig. 2.4.

As expected from the properties of the Fourier transform, the smoother the roll-off of $H_e(f)$ the smaller the magnitudes of the over- and undershoots of the impulse responses are. The ideal lowpass filter with $\alpha = 0$ exhibits a step function at the

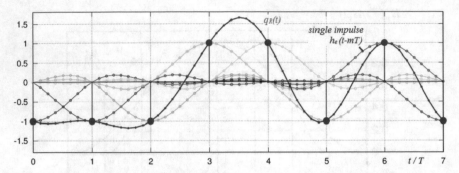

Fig. 2.5 Receiver output signal $q_R(t)$ according to (1.36) with raised cosine Nyquist impulses $h_e(t)$ from (2.14), roll-off factor $\alpha = 0.3$, without noise, and binary transmit symbols $a(k) \in \{-1, +1\}$. No intersymbol interference is present at the sampling instants $t/T = 0, 1, 2, \ldots$. Source: Online platform "webdemo" [2]

transition from the passband to the stopband and therefore the over- and undershoots are the largest. In this case (2.14) yields the well-known sinc function as the impulse response

$$\frac{h_e(t)}{h_e(0)} = \mathrm{sinc}\left(\pi \frac{t}{T}\right) \tag{2.16}$$

However, this response is not well suited for digital communications, because even a small deviation of the sampling phase at the receiver introduces strong intersymbol interference, which even approaches infinity theoretically. $\alpha = 1$ yields the maximal cut-off frequency $f_c = 2 f_N$ and the roll-off is very smooth. Consequently, $h_e(t)$ is almost zero for $|t| > T$ and therefore is often approximated by a symmetrical triangular impulse with duration $2T$. For all α the periodic zero crossings of $h_e(t)$ for $|t| \geq T$, also called "Nyquist zeros" are clearly visible, which avoid the intersymbol interference completely.

Figure 2.5 shows an example of the receiver output signal $q_R(t)$ in (1.36) without noise, simulated with the online platform "webdemo" [2], which also enables to adjust various system parameters online. The impulse response $h_e(t)$ of the equivalent baseband system is chosen as a raised cosine Nyquist impulse (2.14) with roll-off factor $\alpha = 0.3$. The transmit symbol sequence $a(m)$ is binary with the alphabet $\{-1, +1\}$. Apparently, the Nyquist impulse guarantees that no intersymbol interference is present in $q_R(t)$ at the sampling instants $t/T = 0, 1, 2, \ldots$. Hence, the transmit symbol sequence can be recovered without distortion in case of no noise.

References

1. H. Nyquist, Certain topics in telegraph transmission theory (reprint from Transactions of the A.I.E.E., Feb 1928), *Proceedings of the IEEE* (2002)
2. S. ten Brink, Pulse shaping, webdemo, Technical report, Institute of Telecommunications, University of Stuttgart, Germany (2018), http://webdemo.inue.uni-stuttgart.de

Chapter 3
Characterization of the Noise
at the Receiver

Before we are going to discuss various detection algorithms for the receiver output signal $q(k)$, we have to characterize the noise $n_C(t)$ in Fig. 1.1. The main sources of noise are the resistors and the electronic components such as transistors or the photodiode in an optical receiver. In a wireless system, the receive antenna collects noise coming from the channel. In most cases, the first stage of a receiver is composed of an amplifier associated with a bandpass filter to limit the noise spectrum to the passband of the transmit signal, which is given by the channel transfer function in (1.19). The resulting real-valued noise $n_C(t)$ is demodulated and lowpass filtered with $\sqrt{2}G_R(f)$ yielding the complex-valued noise $n_R(t)$ with a lowpass spectrum. The noise sequence $n(k)$ results after sampling and is depicted in the discrete-time equivalent baseband system in Fig. 1.5b. In the following, we analyze the noise at the receiver step by step. For the basics on stochastic processes we refer to the Appendix A.

3.1 Channel Noise $n_C(t)$

A very likely model for the noise $n_C(t)$ at the receiver is the zero-mean additive Gaussian noise with a bandpass shaped power spectral density. $n_C(t)$ is regarded as a sample function of a stationary stochastic bandpass process. In the Appendix A, we show that this noise can be characterized by the following properties[1]:

In general the real-valued stationary bandpass noise is given in the time-domain by

$$n_C(t) = x_1(t)\cos(2\pi f_0 t) - x_2(t)\sin(2\pi f_0 t) \tag{3.1}$$

---

[1]Different from Appendix A we denote the stochastic processes and their sample functions with the same lowercase letters to simplify the notation.

© Springer Nature Switzerland AG 2019
J. Speidel, *Introduction to Digital Communications*, Signals and Communication
Technology, https://doi.org/10.1007/978-3-030-00548-1_3

where $x_1(t)$ and $x_2(t)$ are real-valued Gaussian lowpass noises with the same probability density function

$$p_{x_i}(x_i) = \frac{1}{\sqrt{2\pi}\sigma_x}e^{-\frac{x_i^2}{2\sigma_x^2}} \; ; \; i = 1, 2 \tag{3.2}$$

It goes without saying that the probability density function holds for every fixed time instant t of the stationary stochastic process. $n_C(t)$, $x_1(t)$, and $x_2(t)$ possess zero mean

$$\mathbf{E}[n_C(t)] = \mathbf{E}[x_1(t)] = \mathbf{E}[x_2(t)] = 0 \tag{3.3}$$

and the same mean power

$$\sigma_x^2 = \mathbf{E}\left[n_C^2(t)\right] = \mathbf{E}\left[x_1^2(t)\right] = \mathbf{E}\left[x_2^2(t)\right] \tag{3.4}$$

Moreover we assume $x_1(t)$ and $x_2(t)$ as uncorrelated, i.e., the cross-correlation function is

$$R_{x_1x_2}(\tau) = \mathbf{E}[x_1(t)x_2(t+\tau)] = \mathbf{E}[x_1(t)]\mathbf{E}[x_2(t)] \tag{3.5}$$

and because of (3.3) $R_{x_1x_2}(\tau) = 0$ holds. As shown in the Appendix A, Gaussian processes with this property are even statistically independent.

The power spectral densities of $x_1(t)$ and $x_2(t)$ are identical. They are given by

$$S_{x_ix_i}(f) = \begin{cases} N_0 \; ; \; |f| \le f_I \\ 0 \; ; \quad else \end{cases} \; ; \; i = 1, 2 \tag{3.6}$$

where $2f_I$ is the bandwidth of the channel transfer function $G_C(f)$ in (1.19). The noise spectrum is step-wise constant. Although misleading, it is therefore sometimes called "band-limited white" Gaussian noise. The relation between $S_{x_ix_i}(f)$ and the mean noise power is

$$\sigma_x^2 = \int_{-\infty}^{\infty} S_{x_ix_i}(f)df = 2f_I N_0 \; ; \; i = 1, 2 \tag{3.7}$$

As shown in the Appendix A, the power spectral density of $n_C(t)$ can be determined with $S_{x_ix_i}(f)$ as

$$S_{n_C n_C}(f) = \frac{1}{2}\left[S_{x_ix_i}(f-f_0) + S_{x_ix_i}(f+f_0)\right] = \begin{cases} \frac{N_0}{2} \; ; \; f_0 - f_I \le |f| \le f_0 + f_I \\ 0 \; ; \qquad\qquad else \end{cases}$$
$$\tag{3.8}$$

Thus, we have a strictly band-limited bandpass shaped noise spectrum with the passband in the region of the transmission channel (1.19). This noise can be simulated by a frequency flat broadband noise at the input of an ideal bandpass filter with cut-off frequencies given in (1.19). Sometimes $n_c(t)$ is also called "band-limited white" noise, although misleading.

3.2 Noise After Demodulation and Lowpass Filtering

According to Fig. 1.1 after demodulation and lowpass filtering with $\sqrt{2}g_R(t)$, the real-valued channel noise $n_C(t)$ turns into a complex noise $n_R(t)$ given by (1.25). Rewriting (3.1) using $\cos(2\pi f_0 t) = \frac{1}{2}\left[e^{j2\pi f_0 t} + e^{-j2\pi f_0 t}\right]$ and $\sin(2\pi f_0 t) = \frac{1}{2j}\left[e^{j2\pi f_0 t} - e^{-j2\pi f_0 t}\right]$ we obtain from (1.25)

$$
\begin{aligned}
n_R(t) = \quad & \frac{\sqrt{2}}{2}\left[x_1(t) + jx_2(t)\right] * g_R(t) + \\
& + \frac{\sqrt{2}}{2}\left\{\left[x_1(t) - jx_2(t)\right]e^{-j4\pi f_0 t}\right\} * g_R(t)
\end{aligned}
\tag{3.9}
$$

The term in the second line of (3.9) represents the convolution of a bandpass noise in the frequency range around $f = 2f_0$ with a lowpass impulse response $g_R(t)$ with a cut-off frequency $f_R \ll 2f_0$. Thus, the result is zero. Consequently we obtain

$$
n_R(t) = \frac{\sqrt{2}}{2}\left[x_1(t) + jx_2(t)\right] * g_R(t)
\tag{3.10}
$$

Now we assume that $g_R(t)$ is the impulse response of an ideal lowpass filter with cut-off frequency f_R. From the viewpoint of noise reduction we wish to make f_R small. However, please note that in this case we limit the bandwidth of the signal, which is f_I, at the same time. As this is not adequate, we have to accept

$$
f_R = f_I
\tag{3.11}
$$

according to (2.11), where $2f_I$ is the bandwidth of the bandpass channel transfer function. Let

$$
G_R(f) = \begin{cases} 1 & ; \ |f| \le f_I \\ = 0 & ; \ else \end{cases}
\tag{3.12}
$$

be the ideal lowpass receive filter. Then its output is the complex-valued lowpass noise

$$
n_R(t) = \frac{\sqrt{2}}{2}\left[x_1(t) + jx_2(t)\right]
\tag{3.13}
$$

$n_R(t)$ and thus $x_1(t)$ as well as $x_2(t)$ remain Gaussian, because a complex Gaussian process passing through a linear system remains Gaussian, only the variance may change. Noting that the real and the imaginary part of $n_R(t)$ are statistically independent $n_R(t)$ has the mean power

$$\mathbf{E}\left[|n_R|^2\right] = \sigma_x^2 \tag{3.14}$$

To simplify the notation we introduce new components of the complex noise incorporating the factor $\frac{\sqrt{2}}{2}$

$$\tilde{n}_i(t) = \frac{\sqrt{2}}{2} x_i(t) \; ; \; i = 1, 2 \tag{3.15}$$

Then the output noise of the receive lowpass can be written as

$$n_R(t) = \tilde{n}_1(t) + j\tilde{n}_2(t) \tag{3.16}$$

Obviously, $\tilde{n}_1(t)$ and $\tilde{n}_2(t)$ have identical Gaussian probability density functions

$$p_{\tilde{n}_i}(\tilde{n}_i) = \frac{1}{\sqrt{2\pi}\sigma_n} e^{-\frac{\tilde{n}_i^2}{2\sigma_n^2}} \; ; \; i = 1, 2 \tag{3.17}$$

and the variance σ_n^2 of $\tilde{n}_i(t)$ is obtained from (3.15) with (3.4)

$$\sigma_n^2 = \mathbf{E}\left[\tilde{n}_i^2\right] = \frac{1}{2}\mathbf{E}\left[x_i^2\right] = \frac{1}{2}\sigma_x^2 \; ; \; i = 1, 2 \tag{3.18}$$

The power spectral density of the noise at the output of the receive lowpass then follows from (3.6)

$$S_{\tilde{n}_i\tilde{n}_i}(f) = \begin{cases} \frac{N_0}{2} \; ; \; |f| \le f_I \\ 0 \; ; \quad else \end{cases} \; ; \; i = 1, 2 \tag{3.19}$$

As shown in the Appendix A, the autocorrelation function follows from the power spectral density by applying the inverse Fourier transform. Thus, we obtain the autocorrelation function $R_{\tilde{n}_i\tilde{n}_i}(\tau)$ of $\tilde{n}_i(t)$

$$S_{\tilde{n}_i\tilde{n}_i}(f) \hookleftarrow R_{\tilde{n}_i\tilde{n}_i}(\tau) = N_0 f_I \text{sinc}\left(2\pi f_I \tau\right) \; ; \; i = 1, 2 \tag{3.20}$$

3.3 Noise After Sampling

For the detection of the QAM symbols, the statistics of the real and imaginary part of the noise after sampling is important to know. At the output of the sampling device in Fig. 1.1, we obtain the noise sequence from (3.16)

$$n_R(t_0 + kT) = \tilde{n}_1(t_0 + kT) + j\tilde{n}_2(t_0 + kT) \tag{3.21}$$

To simplify the notation we introduce

$$n(k) = n_R(t_0 + kT) \tag{3.22}$$

and

$$n_i(k) = \tilde{n}_i(t_0 + kT) \; ; \; i = 1, 2 \tag{3.23}$$

Then (3.21) turns into

$$n(k) = n_1(k) + jn_2(k) \tag{3.24}$$

As the Gaussian noise is stationary, the probability density function is independent of any time instant. Consequently, the samples $n_i(k)$ possess the same probability density functions as $\tilde{n}_i(t)$; $i = 1, 2$, which are given in (3.17) and (3.18)

$$p_{n_i}(n_i) = \frac{1}{\sqrt{2\pi}\sigma_n} e^{-\frac{n_i^2}{2\sigma_n^2}} \; ; \; \mathbf{E}\left[n_i^2\right] = \sigma_n^2 = \frac{1}{2}\sigma_x^2 \; ; \; i = 1, 2 \tag{3.25}$$

Furthermore, both noise components have zero mean

$$\mathbf{E}[n_i] = 0 \; ; \; i = 1, 2 \tag{3.26}$$

The probability density function of $n(k)$ can be obtained with the following consideration. As $x_1(t)$ and $x_2(t)$ are statistically independent, this also holds for $n_1(k)$ and $n_2(k)$. Consequently, we obtain the density function of the complex noise $n(k)$ as

$$p_n(n) = p_{n_1}(n_1)p_{n_2}(n_2) = \frac{1}{2\pi\sigma_n^2} e^{-\frac{n_1^2 + n_2^2}{2\sigma_n^2}} = \frac{1}{2\pi\sigma_n^2} e^{-\frac{|n|^2}{2\sigma_n^2}} \tag{3.27}$$

As shown in the Appendix A, sampling of a stochastic process with the sampling rate $\frac{1}{T}$ results in a discrete-time autocorrelation function with samples at $\tau = mT$; $m \in \mathbb{Z}$.[2] Thus, the autocorrelation function of $n_i(k)$ is $R_{\tilde{n}_i\tilde{n}_i}(mT) = R_{n_i n_i}(m)$ and with (3.20) we obtain

$$R_{n_i n_i}(m) = N_0 f_I \operatorname{sinc}(2\pi f_I T m) \; ; \; i = 1, 2 \; ; \; m \in \mathbb{Z} \tag{3.28}$$

If we choose the cut-off frequency $f_I = (1 + \alpha) f_N = \frac{1+\alpha}{2T}$ as discussed in (2.11) we get

$$R_{n_i n_i}(m) = N_0 (1 + \alpha) f_N \operatorname{sinc}(\pi(1 + \alpha)m) \; ; \; i = 1, 2 \; ; \; m \in \mathbb{Z} \tag{3.29}$$

The reader can convince oneself easily that for $\alpha = 0$ and $\alpha = 1$ the autocorrelation function is zero for $m = \pm1, \pm2, \ldots$, thus,

---

[2]As mentioned in the footnote of Appendix A, the sampling function is defined as $T \sum_{k=-\infty}^{\infty} \delta(t - kT)$.

$$R_{n_i n_i}(m) = \begin{cases} N_0 (1 + \alpha) f_N ; & m = 0 \\ 0 & ; m = \pm 1, \pm 2, \ldots \end{cases} \tag{3.30}$$

We conclude from (3.30) that two samples $n_i(k)$ and $n_i(k + m)$, which can be considered as random variables, are uncorrelated for $m \neq 0$. Hence, the sampling operation with the sampling frequency $\frac{1}{T}$ acts as a decorrelation. Because the uncorrelated random variables $n_i(k)$ and $n_i(k + m)$ are also Gaussian, they are even statistically independent. However, $\alpha = 0$ is not feasible for a practical implementation, as discussed earlier.

According to Appendix A, the corresponding power spectral density of a discrete-time stochastic process is periodic and given by

$$S_{n_i n_i}(f) = \sum_{\nu=-\infty}^{\infty} S_{\tilde{n}_i \tilde{n}_i}\left(f - \nu \frac{1}{T}\right) \tag{3.31}$$

Example 1

One expects that the power spectral density of an uncorrelated stochastic process is white. Consequently, $S_{n_i n_i}(f) = const.$ must hold for $\alpha = 0$ and $\alpha = 1$.

Solution: For $\alpha = 0$ the cut-off frequency of $S_{\tilde{n}_i \tilde{n}_i}(f)$ in (3.19) is $f_I = \frac{1+\alpha}{2T} = \frac{1}{2T}$. The periodic repetition of $S_{\tilde{n}_i \tilde{n}_i}(f)$ in (3.31) yields $S_{n_i n_i}(f) = \frac{N_0}{2} = const.$ In case of $\alpha = 1$ follows $f_I = \frac{1+\alpha}{2T} = \frac{1}{T}$ and the spectral parts in (3.31) overlap resulting in $S_{n_i n_i}(f) = N_0 = const.$

3.4 Summary

We summarize the results to complete the noise properties for the equivalent baseband system models given in Sect. 1.6. The original block diagram and that of the equivalent system are shown in Figs. 1.1 and 1.5, respectively.

Continuous-Time Equivalent Baseband System Model

The bandpass noise $n_C(t)$ at the input of the receiver is characterized by (3.1). The lowpass in-phase $x_1(t)$ and quadrature component $x_2(t)$ are assumed to be statistically independent Gaussian noises with the same probability density function defined in (3.2), zero mean (3.3), and identical mean power (3.4). After demodulation and lowpass filtering $n_C(t)$ turns into a complex-valued Gaussian noise $n_R(t) = \tilde{n}_1(t) + j\tilde{n}_2(t)$ according to (3.16). Due to the amplification factor $\sqrt{2}$ of the receive lowpass the noise components possess the mean power $\mathbf{E}\left[\tilde{n}_i^2\right] = \frac{1}{2}\mathbf{E}\left[x_i^2\right]$ in (3.18) and the power spectral density given in (3.19) with lowpass characteristics. $\tilde{n}_1(t)$ and $\tilde{n}_2(t)$ are statistically independent with Gaussian probability density function (3.17).

Discrete-Time Equivalent Baseband System Model

The complex-valued, discrete-time noise $n(k) = n_1(k) + jn_2(k)$ is given by (3.24), in which $n_i(k) = \tilde{n}_i(t_0 + kT)$; $i = 1, 2$. The $n_1(k)$ and $n_2(k)$ exhibit the same zero-mean Gaussian probability density function (3.25) and they are statistically independent. If the cut-off frequency of an ideal lowpass filter (3.12) at the receiver is chosen as $f_I = \frac{(1+\alpha)}{2T}$ with $\alpha = 0$ or $\alpha = 1$ and the noise input spectrum is flat, then the output samples $n_i(k)$ and $n_i(k + m)$ are statistically independent ($m \neq 0$) and the power spectral density $S_{n_i n_i}(f)$ in (3.31) is white ($i = 1, 2$).

Chapter 4
Detection Methods

4.1 Receive Signal Under Detection

In the following, a survey on the most important detection methods is presented. We differentiate in principle between the symbol-by-symbol and the sequence or sequential detection. With the first method, the receive signal $q(k)$ in Figs. 1.1 and 1.5 is decided at every time instant k. The sequential detection scheme takes decisions periodically after the observation of K past samples, e.g., after $q(0), q(1), \ldots, q(K-1)$. In this section, we illustrate the key detection methods and consider 4-PSK depicted in Fig. 4.1 as an example. Assume that the intersymbol interference is completely removed and that the signal at the input of the detector is $q(k)$ given by (2.3). For simplification let $h(0) = 1$. Then, we obtain from (2.3) the signal under decision as

$$q(k) = a(k) + n(k) \tag{4.1}$$

which is composed of the symbol sequence $a(k)$ sent by the transmitter and the additive Gaussian noise $n(k) = n_1(k) + jn_2(k)$ with zero mean and probability density function $p_{n_i}(n_i)$; $i = 1, 2$ in (3.25). Each $a(k)$ can represent one signal point out of the symbol alphabet $\mathcal{B} = \{a_1, a_2, a_3, a_4\}$. Figure 4.1 illustrates (4.1) in the complex plane for the case that the symbol $a(k) = a_1$ was sent.

4.2 Maximum Likelihood Symbol-by-Symbol Detection

4.2.1 Maximum Likelihood Detection

To show the principle, let us assume that the symbol $a(k) = a_\nu$ was sent at time instant k. In the following, we drop k to simplify notation. For maximum likelihood detection, a likelihood function is defined, which is the conditional probability density function $p_L(q \mid a_\nu)$ or a monotonic function of it, e.g., the logarithmic function. $p_L(q \mid a_\nu)$

© Springer Nature Switzerland AG 2019
J. Speidel, *Introduction to Digital Communications*, Signals and Communication
Technology, https://doi.org/10.1007/978-3-030-00548-1_4

Fig. 4.1 QAM signal points
$a_1, a_2, \ldots, a_4$, receive signal
$q(k)$, and noise $n(k)$ for
4-PSK

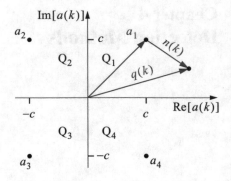

relates the density function of q under the condition that the symbol a_ν was sent.
With $p_n(n)$ in (3.27) and with (4.1) follows

$$p_L(q \mid a_\nu) = p_n(q - a_\nu) = \frac{1}{2\pi\sigma_n^2} e^{-\frac{|q - a_\nu|^2}{2\sigma_n^2}} \ ; \quad \nu = 1, 2, \ldots, 4 \qquad (4.2)$$

Please note that (4.2) describes for the signal constellation in Fig. 4.1 a set of four
density functions, each one shifted by a_ν. The maximum likelihood detection (or
decision) rule is as follows.

Select that symbol a_ν, which is associated with the largest $p_L(q \mid a_\nu)$. Thus, the
output of the detector is

$$\hat{a} = \arg \left\{ \max_{\nu=1,\ldots4} [p_L(q \mid a_\nu)] \right\} \qquad (4.3)$$

Looking at (4.2) the maximal density function results for minimal $|q - a_\nu|^2$, as the
exponent of the exponential function is negative. Consequently, (4.3) is equivalent
to

$$\hat{a} = \arg \left\{ \min_{\nu=1,\ldots4} [|q - a_\nu|^2] \right\} \qquad (4.4)$$

and the probabilistic approach turns into the calculation of the squared Euclidian
distance between two complex numbers, namely, q and a_ν for $\nu = 1, 2, \ldots, 4$. It is
also worth noting that the solution (4.4) does not depend on the mean power σ_n^2 of
the noise n.

4.2.2 Threshold Detection

From the regular signal constellation given in Fig. 4.1, we see that the distance
between the various a_ν and a given q is minimal in the first quadrant Q_1 of the
complex plane. Consequently, we can formulate a detection criterion alternative to
(4.4) by introducing decision regions in the complex plane. In our example, these are

the four quadrants $Q_1, Q_2, \ldots, Q_4$, if the symbols have equal a-priori probabilities. The decision regions are separated by decision thresholds, which are the real and the imaginary axes in the example of Fig. 4.1. The decision rule for a correct decision then is

$$\hat{a} = a_\nu \text{ if } q \in Q_\nu \; ; \;\; \nu = 1, 2, \ldots, 4 \tag{4.5}$$

otherwise, the decision is in error. If q is located on the decision threshold, no unique result is achieved. For low magnitudes of n, we recognize a more reliable decision when the q is located closer to a symbol. Therefore, advanced detection techniques take this fact into consideration and introduce "soft decision" as opposed to the described "hard decision" in particular together with forward error correction encoding [1–3].

4.2.3 Symbol Error Probability for Threshold Detection

The Gaussian Q-Function

The symbol error probability can be formulated with the Gaussian Q-function, which is introduced in the following. Let X be a random variable with probability density function $p_x(x)$ given in (3.2), where we set $x_i = x$. As shown in Appendix A, the probability that the random variable X is larger than a given real number α is

$$P[X > \alpha] = \int_\alpha^\infty p_x(x)dx = \frac{1}{\sqrt{2\pi}\sigma_x} \int_\alpha^\infty e^{-\frac{x^2}{2\sigma_x^2}} dx \tag{4.6}$$

where $P[\ldots]$ denotes the probability operator. With the substitution $u = \frac{x}{\sigma_x}$ follows

$$P[X > \alpha] = \frac{1}{\sqrt{2\pi}} \int_{\frac{\alpha}{\sigma_x}}^\infty e^{-\frac{u^2}{2}} du = Q\left(\frac{\alpha}{\sigma_x}\right) \tag{4.7}$$

with the Q-function

$$Q(\alpha) = \frac{1}{\sqrt{2\pi}} \int_\alpha^\infty e^{-\frac{u^2}{2}} du \tag{4.8}$$

It is straightforward to show the following properties of the Q-function

$$Q(-\infty) = 1; \;\; Q(0) = \frac{1}{2}; \;\; Q(\infty) = 0; \;\; Q(-\alpha) = 1 - Q(\alpha) \tag{4.9}$$

and the relation with the error function $\text{erf}(x) = \frac{2}{\sqrt{\pi}} \int_0^x e^{-u^2} du$ is

$$Q(\alpha) = \frac{1}{2}\left[1 - \text{erf}(\frac{\alpha}{\sqrt{2}})\right] = \frac{1}{2}\text{erfc}(\frac{\alpha}{\sqrt{2}}) \tag{4.10}$$

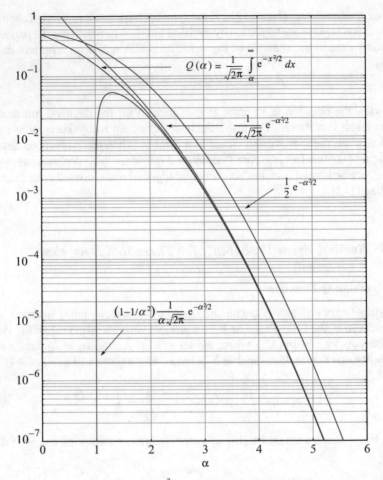

$$Q(\alpha) = \frac{1}{\sqrt{2\pi}} \int\limits_{\alpha}^{\infty} e^{-x^2/2} \, dx$$

$$\frac{1}{\alpha\sqrt{2\pi}} e^{-\alpha^2/2}$$

$$\frac{1}{2} e^{-\alpha^2/2}$$

$$\left(1 - 1/\alpha^2\right) \frac{1}{\alpha\sqrt{2\pi}} e^{-\alpha^2/2}$$

Fig. 4.2 Q-function $Q(\alpha) = \frac{1}{\sqrt{2\pi}} \int_{\alpha}^{\infty} e^{-\frac{x^2}{2}} \, dx$ together with approximations and upper bounds

where $\mathrm{erfc}(x) = 1 - \mathrm{erf}(x)$ is defined. There is no closed-form solution for the integral $Q(\alpha)$ for arbitrary α. Hence, we have to rely on numerical calculations. The result is depicted in Fig. 4.2 together with some helpful approximations and upper bounds. Obviously, the Q-function is declining strongly with increasing α.

Example 2: Symbol Error Probability for 4-PSK

We consider Fig. 4.1, which depicts the signal points $a_1, \ldots, a_4$ for 4-PSK in the constellation diagram. Obviously, these points are located on a circle with radius $b = \sqrt{2}c$ around the origin. Figure 4.3 shows the surface diagram of the superposition of the conditional probability density functions $p_L(q \mid a_\nu)$ in (4.2) of the receiver output signal q for the 4-PSK symbols a_ν with equal a-priori probabilities ($\nu = 1, 2, 3, 4$). In this diagram $10 \log \left(\frac{1}{2\sigma_n^2} \right) = 5$ dB is assumed. It is generated with the

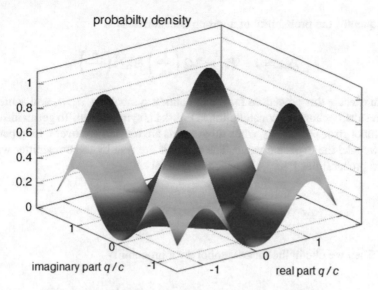

Fig. 4.3 Superposition of the Gaussian conditional probability density functions $p_L(q \mid a_\nu)$ in (4.2) of the receiver output signal q for the 4-PSK symbols a_ν with equal a-priori probabilities ($\nu = 1, 2, 3, 4$) and $10 \log \left(\frac{1}{2\sigma_n^2} \right) = 5\,\mathrm{dB}$. Source: Online platform "webdemo" [4]

online platform "webdemo" [4], which also enables to adjust different parameters. Apparently, the individual probability density functions centered around each PSK symbol overlap and hence we expect quite some decision errors.

Assume for the moment that only the symbol $a_1 = c + jc$ is transmitted, which shall be known at the receiver. Then, the signal (4.1) under decision will be with (3.24)

$$q(k) = a_1 + n(k) = c + n_1(k) + j(c + n_2(k)) \tag{4.11}$$

at time instant k.

According to the decision rule (4.5), the decision for a_1 is correct, if $q(k)$ is located in the first quadrant Q_1 of the complex plane. The following equivalent relations are true (k is dropped to simplify notation)

$$q \in Q_1 \Leftrightarrow \mathrm{Re}\,[q] > 0 \;;\; \mathrm{Im}\,[q] > 0 \Leftrightarrow n_1 > -c \;;\; n_2 > -c \tag{4.12}$$

Then follows for the probability P_{a_1} that a_1 is decided correctly

$$P_{a_1} = \mathrm{P}\,[n_1 > -c,\, n_2 > -c] \tag{4.13}$$

Using the assumption that n_1 and n_2 are statistically independent follows with (3.25) and the Q-function (4.7), (4.9)

$$P_{a_1} = \mathrm{P}\,[n_1 > -c]\,\mathrm{P}\,[n_2 > -c] = \left[1 - Q\left(\frac{c}{\sigma_n} \right) \right]^2 \tag{4.14}$$

Consequently, the probability of a wrong decision of a_1 is

$$P_{S,1} = 1 - P_{a_1} = 2Q\left(\frac{c}{\sigma_n}\right) - Q^2\left(\frac{c}{\sigma_n}\right) \tag{4.15}$$

We can execute this procedure for the remaining symbols a_2, a_3, a_4 and achieve the same results, because the constellation in Fig. 4.1 is symmetrical. To get an idea about the symbol error probability when all symbols are transmitted over a long period of time, we take the expected value. The symbols are sent by the transmitter with the a-priori probabilities $P[a_\nu]$, for which in general

$$\sum_{\nu=1}^{4} P[a_\nu] = 1 \tag{4.16}$$

holds. Then we obtain the mean symbol error probability

$$P_S = \sum_{\nu=1}^{4} P[a_\nu] P_{S,1} = P_{S,1} = 2Q\left(\frac{c}{\sigma_n}\right) - Q^2\left(\frac{c}{\sigma_n}\right) \tag{4.17}$$

Obviously, the minimal distance c from the decision threshold of each PSK symbol in Fig. 4.1 and σ_n determine the mean symbol error probability. It can be shown that this also holds in principle for higher order PSK or QAM constellations.

For high signal-to-noise ratios, $\frac{c}{\sigma_n} \gg 1$, we conclude $Q^2\left(\frac{c}{\sigma_n}\right) \ll 2Q\left(\frac{c}{\sigma_n}\right)$ and the mean symbol error probability approximately is

$$P_S = 2Q\left(\frac{c}{\sigma_n}\right) \tag{4.18}$$

For statistical transmit data each symbol a_ν is a discrete random variable with probability of occurrence $P[a_\nu]$. The mean power of the symbol alphabet $\mathcal{B} = \{a_1, \ldots, a_4\}$ is $\sum_{\nu=1}^{4} |a_\nu|^2 P[a_\nu] = \sum_{\nu=1}^{4} \left(\sqrt{2}c\right)^2 P[a_\nu] = 2c^2$, in which we have used (4.16). Hence, the mean symbol error probability can also be expressed as $P_S = 2Q\left(\sqrt{\frac{\gamma}{2}}\right) - Q^2\left(\sqrt{\frac{\gamma}{2}}\right)$ where $\gamma = \frac{2c^2}{\sigma_n^2}$ is the signal-to-noise power ratio.

4.3 Maximum A-Posterior Symbol-by-Symbol Detection

The signal under decision at time instant k is q given in (4.1) and depicted in Fig. 4.1 for the example of 4-PSK. We define the a-posterior probability

$$P_{APP}(a_\nu \mid q) = P[a_\nu \mid q] \; ; \; \nu = 1, 2, \ldots, 4 \tag{4.19}$$

as the conditional probability of the symbol a_ν to be detected under the condition that the signal q is observed. Equation (4.19) defines a set of probabilities. The maximum a-posterior probability (MAP) detection rule is as follows.

Select that symbol a_ν, which is associated with the largest $P_{APP}(a_\nu \mid q)$. Then the output of the MAP detector is

$$\hat{a} = \arg\left\{\max_{\nu=1,\ldots4}[P_{APP}(a_\nu \mid q)]\right\} \tag{4.20}$$

According to the decision theory, this method provides the best symbol-by-symbol detection of unknown events $a_1, a_2, \ldots$ in the stochastic signal q, because the detector deduces the cause a_ν from the effect q, which is expressed by the term a-posterior or "a posteriori". Obviously, the likelihood decision strategy in (4.3) argues with $p_L(q \mid a_\nu)$ the other way round.

Using a special form of the Bayes rule given in the Appendix A, we can rewrite (4.19) as

$$P_{APP}(a_\nu \mid q) = \frac{p_L(q \mid a_\nu)\, P[a_\nu]}{p_q(q)} \; ; \; \nu = 1, 2, \ldots, 4 \tag{4.21}$$

where $p_q(q)$ is the probability density function of q. For the maximization of $P_{APP}(a_\nu \mid q)$, we have to find the maximal numerator in (4.21), because $p_q(q)$ does not depend on the decision of the detector. Consequently, we can rewrite the decision rule (4.20) as

$$\hat{a} = \arg\left\{\max_{\nu=1,\ldots4}[p_L(q \mid a_\nu)\, P[a_\nu]]\right\} \tag{4.22}$$

Unlike the maximum likelihood detection rule in (4.3), we have to look for the maximal product composed of the likelihood probability density function $p_L(q \mid a_\nu)$ and the a-priori probabilities $P[a_\nu]$. Consequently, the detector at the receiver has to know all $P[a_\nu]$, which are normally only available at the transmitter, if at all. This is the main hurdle for the application of a maximum posterior probability detector. However, in many cases, the transmit symbols have equal a-priori probabilities $P[a_\nu] = \frac{1}{L}$ for an L-ary QAM. Then the maximum posterior turns into the maximum likelihood criterion (4.3), because from (4.22) follows

$$\hat{a} = \arg\left\{\frac{1}{L}\max_{\nu=1,\ldots4}[p_L(q \mid a_\nu)]\right\} \tag{4.23}$$

where the factor $\frac{1}{L}$ can be dropped, as it does not depend on ν. Another critical point is that the maximum posterior detection requires the knowledge of the mean power σ_n^2 of the noise n, which has to be estimated at the receiver. This will be evident with an example using the Gaussian noise. To this end, we replace p_L in (4.22) by (4.2) yielding the rather complex maximum posterior decision rule

$$\hat{a} = \arg\left\{ \max_{\nu=1,\dots 4}\left[\frac{1}{2\pi\sigma_n^2}e^{-\frac{|q-a_\nu|^2}{2\sigma_n^2}}P\left[a_\nu\right]\right]\right\} \tag{4.24}$$

To execute (4.24), we can use the fact that the maximization is equivalent to the minimization of the reciprocal. In addition, we take a monotonic function, such as the natural logarithm ln(...) and obtain

$$\hat{a} = \arg\left\{ \min_{\nu=1,\dots 4}\left[\frac{|q-a_\nu|^2}{2\sigma_n^2} - \ln\left(P\left[a_\nu\right]\right)\right]\right\} \tag{4.25}$$

where we have dropped the term $\ln\left(2\pi\sigma_n^2\right)$, because it does not affect the result. (4.25) clearly reveals that the result $\hat{a}$ of the minimization procedure depends on the a-priori probabilities $P\left[a_i\right]$ and on the mean noise power σ_n^2. The MAP detector is optimal in the sense that it can minimize the symbol error probability [5].

4.4 Maximum Likelihood Sequence Detection

A sequence detector takes a decision after considering K receive samples $q(0), q(1), \dots, q(K-1)$. The algorithm for maximum likelihood sequence detection will be best explained with the help of an example as follows.

4.4.1 System Model

Example 3

Given the simplified discrete-time transmission system depicted in Fig. 4.4 with the real-valued symbol sequence $a(k)$ and the symbol alphabet $\mathcal{B} = \{-1, 1\}$. Hence, the modulation scheme is 2-PSK, which allocates symbols only on the real axis of the complex plain. The equivalent baseband system is modeled as a finite impulse response (FIR) filter with real-valued impulse response

$$h(k) = \begin{cases} 1 & ; \quad k=0 \\ -\frac{1}{2} & ; \quad k=1 \\ 0 & ; k=-1, \pm 2, \pm 3, \dots \end{cases} \tag{4.26}$$

The input signal to the sequential detector is

$$q(k) = w(k) + n(k) \tag{4.27}$$

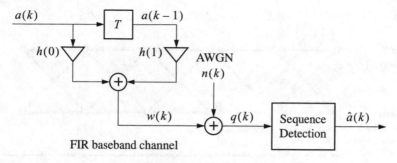

Fig. 4.4 Discrete-time equivalent baseband model of a digital transmission system with sequence detection (Viterbi equalizer)

where $n(k)$ is real-valued Gaussian noise with zero mean and probability density function

$$p_n(n) = \frac{1}{\sqrt{2\pi}\sigma_n} e^{-\frac{n^2}{2\sigma_n^2}} \tag{4.28}$$

The noise samples are statistically independent. The output of the decoder is referred to as $\hat{a}(k)$. The FIR system outputs

$$w(k) = a(k) * h(k) = a(k) - \frac{1}{2}a(k-1) \tag{4.29}$$

which contains strong intersymbol interference owing to the term $-\frac{1}{2}a(k-1)$.

4.4.2 State Space Trellis Diagram

Now we are going to describe the input–output relation in the state space. For that purpose, a trellis diagram depicted in Fig. 4.5 is used, which is a state transition diagram annotated by the discrete time k. The key component is the state variable, which is always associated with the output of the system memories. If each state variable can take on L different values and if M is the number of independent memory elements of the system model, then the number of states is $N_S = L^M$. In our example, the state variable is $a(k-1)$, as there is only a single memory. Consequently, $M = 1$ and with $L = 2$ we have $N_S = 2$ states, which we indicate freely as S_ν, $\nu = 1, 2$.

In general, the trellis diagram consists of nodes and transitions. The nodes represent the states and the transitions, which are labeled by $(a(k) \mid w(k))$, indicate the change of the output signal $w(k)$ depending on the input signal $a(k)$. As already defined, $w(k)$ is the output of the equivalent channel model without noise. The detector has to know the channel impulse response $h(k)$ in (4.26). To illustrate the trellis diagram in Fig. 4.5 assume that the system is in state S_1, thus $a(k-1) = 1$. If the input $a(k)$ keeps to be 1 for the next time instances $k = 1, 2, \ldots$, then the system

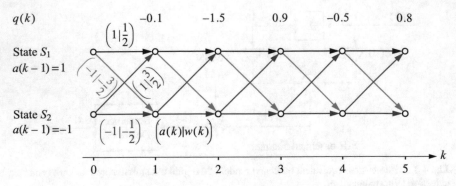

Fig. 4.5 Trellis diagram for the input–output relation of the equivalent baseband system in Fig. 4.4

remains in state S_1 indicated by the horizontal arrows and $w(k)$ remains $\frac{1}{2}$. In case the input $a(k)$ changes to -1, the system moves from state S_1 to state S_2 and provides the output $w(k) = -\frac{3}{2}$. The system remains in state S_2, when the input keeps going to be -1 until the input changes to 1 resulting in a state transition from S_2 to S_1 and the output will be $w(k) = \frac{3}{2}$. We recognize that the trellis diagram in Fig. 4.5 is periodic with time period 1 and the same annotation repeats for $k > 1$, which is not shown.

4.4.3 Maximum Likelihood Sequence Detection

The detector estimates the transmit symbol sequence $a(0), a(1), \ldots, a(K-1)$ from the receive sequence $q(0), q(1), \ldots, q(K-1)$ in the maximum likelihood sense. Hence, the detection method is also called maximum likelihood sequence estimation (MLSE). As the receiver knows the channel coefficients $h(k)$, it can calculate the sequence $w(0), w(1), \ldots, w(K-1)$ for all possible symbols $a(k)$ of the symbol alphabet $\mathcal{B}$. This is indicated in the trellis diagram in Fig. 4.5, where $w(k)$ can take on four different values $\frac{1}{2}, -\frac{3}{2}, \frac{3}{2}, -\frac{1}{2}$. For a given symbol sequence $a(0), a(1), \ldots, a(K-1)$, the trellis illustrates the sequence $w(0), w(1), \ldots, w(K-1)$ as a path through the diagram from $k = 0$ to $K-1$ and the trellis shows the set of all such sequences. In the example, we have chosen $K - 1 = 5$.

As for the symbol-by-symbol detection described previously, we also define a likelihood probability density function

$$p_K\left(q(0), q(1), \ldots, q(K-1) \mid w(0), w(1), \ldots, w(K-1)\right) \qquad (4.30)$$

which is the multivariate density function of the observation $q(0), q(1), \ldots, q(K-1)$ conditioned on $w(0), w(1), \ldots, w(K-1)$. There is a combinatorial multitude of sequences in the argument of (4.30). For the detection algorithm, the set of density functions is relevant, which reflects all possible paths through the trellis diagram. Then the sequence detector is looking for that sequence $w(0), w(1), \ldots, w(K-1)$, which is maximal likely to the observation $q(0), q(1), \ldots, q(K-1)$ and from that dedicated path in the trellis diagram the sequence of optimal estimates $\hat{a}(0), \hat{a}(1), \ldots, \hat{a}(K-1)$ is derived. Assuming $q(0), q(1), \ldots$ as statistically independent and also independent of $w(0), w(1), \ldots$ yields the density function

$$p_K\left(q(0), q(1), \ldots, q(K-1) \mid w(0), w(1), \ldots, w(K-1)\right) = \prod_{k=0}^{K-1} p_n\left(q(k) - w(k)\right)$$

(4.31)

with p_n given in (4.28). Then follows

$$p_K\left(\ldots \mid \ldots\right) = \left(\frac{1}{\sqrt{2\pi}\sigma_n}\right)^K e^{-\frac{1}{2\sigma_n^2}\sum_{k=0}^{K-1}[q(k)-w(k)]^2}$$

(4.32)

Now we introduce the branch metric

$$d_k = [q(k) - w(k)]^2$$

(4.33)

and the path metric

$$D_K = \sum_{k=0}^{K-1} d_k$$

(4.34)

and obtain

$$p_K\left(\ldots \mid \ldots\right) = \left(\frac{1}{\sqrt{2\pi}\sigma_n}\right)^K e^{-\frac{1}{2\sigma_n^2}D_K}$$

(4.35)

The calculated numbers for the branch metrics d_k are indicated in the trellis diagram in Fig. 4.6. Consequently, the path metrics D_K for all possible paths through the trellis can be found with (4.34) and also the set of conditional probability density functions in (4.35). Please note that D_K is positive.

The criterion for maximum likelihood detection is to select the maximal density function p_K out of the set. Equivalently, we can search for that path in the trellis, which exhibits minimal path metric D_K owing to the negative exponent in (4.35). Thus, the most likely symbol sequence $\hat{a}(0), \hat{a}(1), \ldots, \hat{a}(K-1)$ is deduced from the sequence of transitions in the trellis in Fig. 4.5

$$\{(a(0) \mid w(0)), (a(1) \mid w(1)), \ldots, (a(K-1) \mid w(K-1))\}_{\min[D_K]}$$

(4.36)

where the procedure is executed over all paths starting at $k = 0$ and ending at $k = K - 1$.

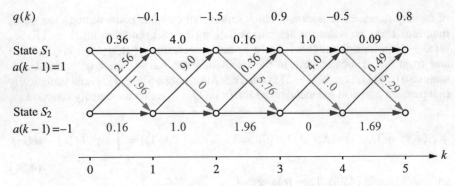

Fig. 4.6 Trellis diagram with branch metrics $d_k = [q(k) - w(k)]^2$; $k = 1, 2, \ldots, 5$

4.4.4 Solution Using the Viterbi Algorithm

A straightforward way to find the solution (4.36) is the calculation of all branch metrics d_k and all path metrics D_K of the trellis. Then we take the path with the smallest metric and deduce the estimated symbol sequence $\hat{a}(0), \hat{a}(1), \ldots, \hat{a}(K-1)$ using the trellis in Fig. 4.5. However, the computational amount with this "brute search" can be very large and this procedure is not effective, because we calculate all path metrics although we just require one, the minimal one. Consequently, more effective procedures have been the topic of research. A task similar to sequential detection is known from operations research as the "traveling salesman problem," where a salesman has to visit customers at various places spending minimal total traveling cost. The solution for the salesman problem was first given by Bellman in 1957. He formulated an optimality principle saying that "an optimal policy has the property that whatever the initial state and initial decision are, the remaining decisions must constitute an optimal policy with regard to the state resulting from the first decision" [6, 7]. The procedure to get the solution is called "dynamic programming." Similar tasks are known from computer science and solved with the algorithm proposed by Dijkstra [8].

For digital communications, A. Viterbi has found a similar algorithm for the decoding of data, which are encoded with convolutional codes [9]. The Viterbi algorithm can be applied for many optimization problems based on a trellis diagram to find the most probable sequence of a hidden Markov model. For example, it is applied for pattern recognition in bioinformatics and artificial intelligence. Here, we focus on the maximum likelihood sequence detection. There are several formulations of the mathematical algorithm and as a program. Let us indicate the nodes in the trellis by its state S_ν and the time instant k, i.e., by (S_ν, k); $\nu = 0, 1, \ldots, N_S$, in which N_S is the number of states, in our example $N_S = 2$. All paths, which start at the beginning of the trellis and enter that node, are called partial paths. The principle strategy of the Viterbi algorithm is not to calculate all paths through the trellis up to the end $k = K - 1$, but to dynamically dismiss all those partial paths entering the node (S_ν, k), which exhibit larger partial path metrics compared to the partial path with minimal metric at that node. If there are more than one partial paths with the same minimum, the final decoding result will be ambiguous. The discarded "dead paths" are skipped from the trellis diagram. This procedure has to be executed for all nodes (S_ν, k); $\nu = 1, 2 \ldots, N_S$ and $k = 1, 2, \ldots, K - 1$ step by step. For the initialization at $k = 0$, the partial path metrics are set to zero at all nodes $(S_\nu, 0)$. Of course, the partial path metrics can be calculated recursively by adding the new branch metric in each step. At the end, all branches are traced back from the end of the trellis and the "surviving" branches provide the optimal path (or paths), from which the sequence (4.36) is deduced.

Programs are best explained with an example. Therefore, we go back to the trellis in Fig. 4.6. The various steps of the algorithm are explained in Fig. 4.7. In step 1, the branch metrics at nodes $(S_1, 1)$ and $(S_2, 1)$ are calculated. We notice that the partial path with metric 2.56 can be discarded at node $(S_1, 1)$, because $2.56 > 0.36$. With the similar argument, we discard at node $(S_2, 1)$ the partial path with metric 1.96. In step 2, the partial path metrics are updated by adding the respective branch metrics of the second segment of the trellis. At node $(S_1, 2)$, the partial path with metric 9.16 can be dismissed as dead path. Similarly at node $(S_2, 2)$, we drop the partial path with metric 1.16. This procedure is continued until step 5, in which we cancel two partial paths with metrics 2.21 and 7.01, respectively. The algorithm finally reveals the result in step 6 from the path with minimal path metric 1.81 terminating in node $(S_1, 5)$. From the trellis diagram in Fig. 4.5, we allocate the input symbols $a(k)$ to the nodes crossed by the final path from $k = 1, \ldots, 5$ resulting in the estimated symbol sequence $\hat{a}(k)$

$$1, -1, 1, 1, 1 \tag{4.37}$$

For comparison, a symbol-by-symbol threshold detector applied on $q(k)$ with threshold level $q = 0$ and decision rule

$$\hat{a}(k) = \begin{cases} 1 & ; q(k) \geq 0 \\ -1 & ; q(k) < 0 \end{cases} \tag{4.38}$$

shall be employed, which yields the sequence $\hat{a}(k)$ after the decision

$$-1, -1, 1, -1, 1 \tag{4.39}$$

Both results differ in the first and next to last symbol indicating that the symbol-by-symbol and the sequence detection can provide different results.

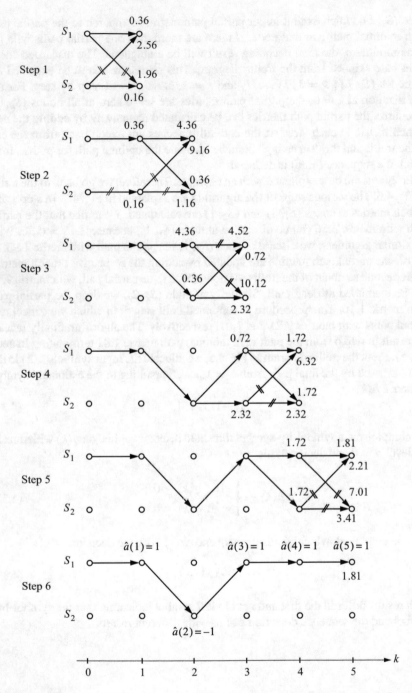

Fig. 4.7 Steps of the Viterbi algorithm with partial paths and associated metrics. Step 6 shows path with minimal metric and the detected sequence $\hat{a}(k)$

4.4.5   Viterbi Equalizer

In several practical applications, the impulse response $g_R(t)$ of the receive lowpass in Fig. 1.1 is designed in such a way that the overall impulse response of the equivalent baseband system satisfies the first Nyquist criterion to eliminate the intersymbol interference. In this case, the receive lowpass is called an *equalizer*. In the example of Fig. 4.4, the overall impulse response $h(k)$ shows a strong post cursor $h(1) = -\frac{1}{2}$, which causes severe intersymbol interference. No separate equalizer is present. However, the sequence detector takes this intersymbol interference together with the noise $n(k)$ into account and provides an optimal decision. Consequently, the tasks of equalization and detection under the impact of noise and intersymbol interference have been merged with this approach and the presented sequence detector is therefore often denoted as a *Viterbi equalizer*.

References

1. J. Hagenauer, E. Offer, L. Papke, Iterative decoding of binary block and convolutional codes. IEEE Trans. Inf. Theory **42** (1996)
2. J. Hagenauer, P. Hoeher, A Viterbi algorithm with soft-decision outputs and its applications, in *IEEE Internal Conference on Global Communications (GLOBECOM)* (1989)
3. J. Hagenauer, The turbo principle: tutorial introduction and state of the art, in *Proceedings of 1st International Symposium on Turbo codes* (1997)
4. M. Bernhard, QAM, webdemo, Technical report, Institute of Telecommunications, University of Stuttgart, Germany (2018), http://webdemo.inue.uni-stuttgart.de
5. J.R. Barry, E.A. Lee, D.G. Messerschmitt, *Digital Communication* (Springer, Berlin, 2012)
6. R.E. Bellman, *Dynamic Programming* (Princeton University Press, Princeton, 1957)
7. R.E. Bellman, *Dynamic Programming* (Princeton University Press, Princeton, 2010)
8. E.W. Dijkstra, A note on two problems in connexion with graphs. Numerische Mathematik **1**, 269 (1959)
9. A. Viterbi, Error bounds for convolutional codes and an asymptotically optimum decoding algorithm. IEEE Trans. Inf. Theory **13**, 260 (1967)

Chapter 5
Digital Transmission over Wireless, Time-Variant Channels

5.1 Transmission System with Time-Variant Channel

Digital signal transmission over a wireless channel has become an important field due to the high flexibility and comfort of wireless connections for many users with the motto "telecommunications anytime and anywhere." Therefore, in the following chapters, we describe the principles of such systems and their design in some detail. A significant part is devoted to the wireless channel. The parameters of electrical cables or optical fibers are approximately constant over time. Consequently, they have been characterized as time-invariant and described by an impulse response $g_c(t)$ in Fig. 1.1 of Sect. 1.2. As we will see in detail, a wireless channel is significantly different and multifaceted. The transmit signal travels on a multitude of different paths from the transmitter to the receiver and undergoes reflections and scattering at objects, such as buildings. Moreover, if the transmitter or the receiver are moving, the signal suffers by the Doppler effect. There is quite a lot of knowledge on the basis of wave propagation and electromagnetic field theory. We will build upon these findings and emphasize a system-theoretical approach, which allows together with adequate channel models the effective design of algorithms for signal transmission and reception. We focus on single input single output (SISO) channels in this Part I and prepare the ground for the wireless multiple input multiple output (MIMO) systems discussed in Part III well knowing that the MIMO channel is composed of many channels of the SISO type.

We start with the block diagram for the wireless transmission scheme and emphasize the mathematical description of the input–output relation. The wireless channel is described as a time-variant system with the two-dimensional impulse response $w(t, s)$ or the delay spread function $g_C(t, \tau)$, also called modified impulse response. We will use and resume the principle results from Part II on the theory of time-variant systems. Part II is self-contained and readers not familiar with time-variant systems are encouraged to switch to that part beforehand or on demand during the study of the following chapters, because we shall merely quote those results here.

© Springer Nature Switzerland AG 2019
J. Speidel, *Introduction to Digital Communications*, Signals and Communication Technology, https://doi.org/10.1007/978-3-030-00548-1_5

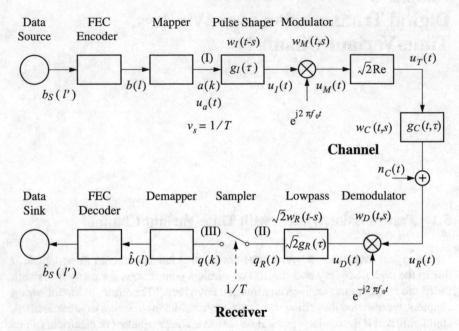

Fig. 5.1 Block diagram of a transmission system with a time-variant channel with impulse response $w_C(t, s)$ and delay spread function $g_C(t, \tau)$

Figure 5.1 shows the block diagram of a system for wireless transmission. We recognize the strong similarity in Fig. 1.1 in Sect. 1.2. However, the differences are the time-variant channel and the different notations for the impulse responses. As outlined in Part II, linear and time-variant systems are completely characterized by their response $w(t, s)$ at observation time t to an input Dirac impulse $\delta(t - s)$, which is active at the time instant $s \leq t$. One denotes $w(t, s)$ as time-variant impulse response. s and t are independent variables. It can be observed that w as a function of t exhibits different shapes depending on the initial time instant s of the Dirac impulse. This is quite in contrast to a linear time-invariant system, where the shape does not change and the response is just shifted by s on the t-axis. Thus, a time-invariant system is characterized by the impulse response $w(t, s) = w(t - s)$.

As outlined in Part II, the Fourier transform of $w(t, s)$ does not provide meaningful system functions in the frequency domain, such as a transfer function. Therefore, the transformation of variables $s = t - \tau$ was proposed, which yields the delay spread function also called modified impulse response

$$w(t, t - \tau) = g(t, \tau) \tag{5.1}$$

This function can be regarded as the response of the time-variant system at observation time t to an input Dirac impulse effective at $t - \tau \leq t$.

For time-invariant systems $w(t - s) = w(\tau) = g(\tau)$ holds. $g(\tau)$ and not $g(t)$ are formally required, if a time-invariant system is described in the framework of time-variant systems, e.g., in case of a cascade of such systems. If the system is composed solely of time-invariant building blocks, τ can be replaced by t. Of course, $g(t)$ and $g(\tau)$ exhibit the same shape, because mathematically we can use any variable to describe a function. In Part II, various input–output relations and Fourier spectra are derived in quite some detail, which we will use also in the following.

5.2 Overall Time-Variant Impulse Response

As in Sect. 1.6 for the time-invariant channel, we are now going to derive the impulse response between the nodes (I) and (II) in Fig. 5.1. For that purpose, we excite the system at the input of the pulse shaper by $\delta(t - s)$ and allocate to all blocks impulse responses. The time-variant systems get impulse responses with the argument t, s and the time-invariant systems with $t - s$. Thus, the impulse responses of the time-invariant pulse shaper and receive lowpass filter are $w_I(t, s) = w_I(t - s)$ and $w_R(t, s) = w_R(t - s)$, respectively. The reader assures oneself easily of the time-variant impulse response of the modulator

$$w_M(t, s) = \delta(t - s)e^{j2\pi f_0 t} \tag{5.2}$$

and the demodulator

$$w_D(t, s) = \delta(t - s)e^{-j2\pi f_0 t} \tag{5.3}$$

We merely quote some more properties from Part II of time-variant systems before going into the details in Fig. 5.1. Let $x(t)$ be the input signal of a time-variant system with impulse response $w(t, s)$. Then the output signal $y(t)$ is given by the "generalized convolution integral" or "time-variant convolution"

$$y(t) = x(t) \circledast w(t, s) = \int_{-\infty}^{\infty} x(s)w(t, s)ds \tag{5.4}$$

Equation (5.4) actually incorporates the "conventional" or "time-invariant" convolution, if we set $w(t, s) = w(t - s)$ resulting in

$$y(t) = x(t) \circledast w(t - s) = \int_{-\infty}^{\infty} x(s)w(t - s)ds \tag{5.5}$$

The overall time-variant impulse response of a cascade of systems with $w_1(t, s)$ and $w_2(t, s)$ is given by

$$w_1(t, s) \circledast w_2(t, s) = \int_{-\infty}^{\infty} w_1(\zeta, s)w_2(t, \zeta)d\zeta \tag{5.6}$$

If one system is time-invariant, e.g., $w_1(t, s)$, then we just write $w_1(t, s) = w_1(t - s)$ and replace the comma by the minus sign. In Part II, we also show that the time-variant convolution is noncommutative in general. Hence, we have to respect the sequential arrangement of time-variant systems. This is in contrast to the time-invariant convolution.

Furnished with these basics, we can focus on the system in Fig. 5.1. The cascade of the pulse shaper and the modulator with (5.2) owns the time-variant impulse response

$$w_I(t - s) \circledast w_M(t, s) = \int_{-\infty}^{\infty} w_I(\zeta - s) w_M(t, \zeta) d\zeta = w_I(t - s) e^{j2\pi f_0 t} \quad (5.7)$$

which we could have directly concluded from Fig. 5.1. Then follows the time-variant impulse response at the output of the real part operator $\sqrt{2}\text{Re}[...]$

$$w_1(t, s) = \frac{\sqrt{2}}{2} \left[w_I(t - s) e^{j2\pi f_0 t} + w_I^*(t - s) e^{-j2\pi f_0 t} \right] \quad (5.8)$$

Next, we determine the impulse response of the cascade of the channel $w_C(t, s)$ and the demodulator $w_D(t, s)$ yielding with (5.3) and (5.6)

$$w_2(t, s) = w_C(t, s) \circledast w_D(t, s) = w_C(t, s) e^{-j2\pi f_0 t} \quad (5.9)$$

Finally, the overall time-variant impulse response between the nodes (I) and (II) in Fig. 5.1 is

$$w_e(t, s) = w_1(t, s) \circledast w_2(t, s) \circledast w_R(t - s)\sqrt{2} \quad (5.10)$$

where we have used the associative property of the time-variant convolution. Plugging in (5.8), we get

$$w_e(t, s) = w_{e1}(t, s) + w_{e2}(t, s) \quad (5.11)$$

with

$$\begin{aligned} w_{e1}(t, s) &= \left(w_I(t - s) e^{j2\pi f_0 t} \right) \circledast \left(w_C(t, s) e^{-j2\pi f_0 t} \right) \circledast w_R(t - s) \\ w_{e2}(t, s) &= \left(w_I^*(t - s) e^{-j2\pi f_0 t} \right) \circledast \left(w_C(t, s) e^{-j2\pi f_0 t} \right) \circledast w_R(t - s) \end{aligned} \quad (5.12)$$

In Sect. 5.4.2, we consider the spectrum $G_{e2}(f_t, f_\tau)$ of the delay spread function $g_{e2}(t, \tau)$ corresponding to $w_{e2}(t, s)$ and show that $G_{e2}(f_t, f_\tau)$ is zero for usual system parameters. As a consequence, one can assume $g_{e2}(t, \tau) = 0$ and thus $w_{e2}(t, s) = 0$ in the following. Then we obtain the overall time-variant impulse response between the nodes (I) and (II) in Fig. 5.1 using (5.11) and (5.12)

$$w_e(t, s) = \left(w_I(t - s) e^{j2\pi f_0 t} \right) \circledast \left(w_C(t, s) e^{-j2\pi f_0 t} \right) \circledast w_R(t - s) \quad (5.13)$$

5.3 Overall Delay Spread Function

Now, we are interested in the overall delay spread function between the nodes (I) and (II) in Fig. 5.1. As pointed out in Part II, the time-variant impulse response has no meaningful Fourier spectrum for signal processing. This can be overcome with the transformation of variables

$$s = t - \tau \tag{5.14}$$

which turns the time-variant impulse responses into delay spread functions as follows:

$$w_e(t, s) = g_e(t, \tau), \quad w_I(t - s) = g_I(\tau), \quad w_c(t, s) = g_C(t, \tau), \quad w_R(t - s) = g_R(\tau) \tag{5.15}$$

We show at the end of this section that from (5.11) and (5.12) the overall delay spread function follows as

$$g_e(t, \tau) = g_{e1}(t, \tau) + g_{e2}(t, \tau) \tag{5.16}$$

with

$$\begin{aligned} g_{e1}(t, \tau) &= g_I(\tau) \circledast \left(g_C(t, \tau) e^{-j2\pi f_0 \tau}\right) \circledast g_R(\tau) \\ g_{e2}(t, \tau) &= g_I^*(\tau) \circledast \left(g_C(t, \tau) e^{-j4\pi f_0 t} e^{j2\pi f_0 \tau}\right) \circledast g_R(\tau) \end{aligned} \tag{5.17}$$

In Sect. 5.4, the associated Fourier spectra are discussed. There we show that a usual choice of the cut-off frequencies results in

$$g_{e2}(t, \tau) = 0 \tag{5.18}$$

yielding from (5.16)

$$g_e(t, \tau) = g_I(\tau) \circledast \left(g_C(t, \tau) e^{-j2\pi f_0 \tau}\right) \circledast g_R(\tau) \tag{5.19}$$

Please note, we use the same symbol $\circledast$ for the time-variant convolution of delay spread functions and for the time-variant impulse responses. However, their integral representations are different.

As shown in Part II, the time-variant convolution of two delay spread functions $g_1(\tau)$ and $g_2(t, \tau)$ is given by

$$g_{12}(t, \tau) = g_1(\tau) \circledast g_2(t, \tau) = \int_{-\infty}^{\infty} g_1(\tau - \zeta) g_2(t, \zeta) d\zeta = \int_{-\infty}^{\infty} g_1(\eta) g_2(t, \tau - \eta) d\eta \tag{5.20}$$

and

$$g_{21}(t, \tau) = g_2(t, \tau) \circledast g_1(\tau) = \int_{-\infty}^{\infty} g_2(t - \zeta, \tau - \zeta) g_1(\zeta) d\zeta \tag{5.21}$$

Both relations apply for (5.17) and apparently are noncommutative. Thus, the sequential arrangement may not be interchanged. For the input–output relation, we obtain from (5.4) with $s = t - \tau$ and $w(t, t - \tau) = g(t, \tau)$

$$y(t) = x(t) \circledast g(t, \tau) = \int_{-\infty}^{\infty} x(t - \tau)g(t, \tau)d\tau \qquad (5.22)$$

Proof of (5.16) **and** (5.17)

We use the definition of the time-variant convolution (5.6) and obtain for the upper line in (5.12) $w_{e1}(t, s) = \left(\int_{-\infty}^{\infty} w_I(\zeta - s)w_C(t, \zeta)e^{-j2\pi f_0(t-\zeta)}d\zeta\right) \circledast w_R(t - s)$. Introducing the substitution $\zeta = t - u$; $d\zeta = -du$ yields $w_{e1}(t, s) = \left(\int_{-\infty}^{\infty} w_I(t - u - s)w_C(t, t - u)e^{-j2\pi f_0 u}du\right) \circledast w_R(t - s)$. With $s = t - \tau$ and (5.15) follows $w_{e1}(t, t - \tau) = g_{e1}(t, \tau) = \left(\int_{-\infty}^{\infty} g_I(\tau - u)g_C(t, u)e^{-j2\pi f_0 u}du\right) \circledast g_R(\tau)$. Using the definition (5.20), we obtain $g_{e1}(t, \tau) = g_I(\tau) \circledast \left(g_C(t, \tau)e^{-j2\pi f_0 \tau}\right) \circledast g_R(\tau)$ and the first equation in (5.17) is proven.

In a similar way, we prove $g_{e2}(t, \tau)$. We start with the lower line in (5.12) $w_{e2}(t, s) = \left(\int_{-\infty}^{\infty} w_I^*(\zeta - s)e^{-j2\pi f_0 \zeta}w_C(t, \zeta)e^{-j2\pi f_0 t}d\zeta\right) \circledast w_R(t - s)$ and the substitution $\zeta = t - u$; $d\zeta = -du$ yields $w_{e2}(t, s) = \left(\int_{-\infty}^{\infty} w_I^*(t - u - s)w_C(t, t - u)e^{-j4\pi f_0 t}e^{j2\pi f_0 u}du\right) \circledast w_R(t - s)$. With $s = t - \tau$, $w_{e2}(t, t - \tau) = g_{e2}(t, \tau)$, $w_C(t, t - u) = g_C(t, u)$, $g_C'(t, u) = g_C(t, u)e^{-j4\pi f_0 t}e^{j2\pi f_0 u}$, and $w_R(t - s) = g_R(\tau)$ follows $g_{e2}(t, \tau) = \left(\int_{-\infty}^{\infty} g_I^*(\tau - u)g_C'(t, u)du\right) \circledast g_R(\tau)$ from which we obtain with (5.20) the symbolic notation $g_{e2}(t, \tau) = g_I^*(\tau) \circledast g_C'(t, \tau) \circledast g_R(\tau) = g_I^*(\tau) \circledast \left(g_C(t, \tau)e^{-j4\pi f_0 t}e^{j2\pi f_0 \tau}\right) \circledast g_R(\tau)$ and the proof is finished.

5.4   Overall Doppler Spread Function

5.4.1   Fourier Transform of the Overall Delay Spread Function

To get inside into (5.17) and to define alternative system functions, we consider the frequency domain. To this end, we will apply the Fourier transform with respect to the variables t and τ. Therefore, we define the corresponding variables as

$$t \leftrightarrow f_t \; ; \; \tau \leftrightarrow f_\tau$$

and use the symbol $\rightarrowtail$ for the transform. In wireless communications, f_t is called Doppler frequency, as it illustrates the time variance of the channel. f_τ is the "natural" frequency, also used for the ordinary frequency response of a time-invariant system or the spectrum of a signal. In Part II, the delay spread functions and their Fourier transforms are discussed in detail, where the table in Sect. 12.6 presents a summary. We shall merely quote the results here without proof. We define with capital letters the transfer function of the pulse shaper

$$g_I(\tau) \overset{\tau}{\rightarrowtail} G_I(f_\tau) \tag{5.23}$$

the transfer function of the receive filter

$$g_R(\tau) \overset{\tau}{\rightarrowtail} G_R(f_\tau) \tag{5.24}$$

the Doppler spread function of the channel

$$g_C(t, \tau) \overset{t,\tau}{\rightarrowtail} G_C(f_t, f_\tau) \tag{5.25}$$

and the overall Doppler spread function

$$g_e(t, \tau) \overset{t,\tau}{\rightarrowtail} G_e(f_t, f_\tau) \tag{5.26}$$

On top of the arrows, we indicate the direction of the Fourier transform with respect to the t- and the τ-coordinate.

We quote from Sect. 12.5 of Part II the following transform pairs:
Let $g_1(\tau) \overset{\tau}{\rightarrowtail} G_1(f_\tau)$ and $g_2(t, \tau) \overset{t,\tau}{\rightarrowtail} G_2(f_t, f_\tau)$, then

$$g_1(\tau) \circledast g_2(t, \tau) \overset{t,\tau}{\rightarrowtail} G_1(f_\tau)G_2(f_t, f_\tau) \tag{5.27}$$

$$g_2(t, \tau) \circledast g_1(\tau) \overset{t,\tau}{\rightarrowtail} G_2(f_t, f_\tau)G_1(f_t + f_\tau) \tag{5.28}$$

With these findings the following Fourier transforms of (5.16) and (5.17) are proven at the end of this Section,

$$g_e(t, \tau) \overset{t,\tau}{\rightarrowtail} G_e(f_t, f_\tau) = G_{e1}(f_t, f_\tau) + G_{e2}(f_t, f_\tau) \tag{5.29}$$

with

$$\begin{aligned}
g_{e1}(t, \tau) &\overset{t,\tau}{\rightarrowtail} G_{e1}(f_t, f_\tau) = \quad G_I(f_\tau)G_C(f_t, f_\tau + f_0)G_R(f_t + f_\tau) \\
g_{e2}(t, \tau) &\overset{t,\tau}{\rightarrowtail} G_{e2}(f_t, f_\tau) = G_I^*(-f_\tau)G_C(f_t + 2f_0, f_\tau - f_0)G_R(f_t + f_\tau)
\end{aligned} \tag{5.30}$$

In the following, it will be shown that the cut-off frequencies of the pulse shaper and the receive filter can be determined in such a way that $G_{e2}(f_t, f_\tau) \approx 0$, which results in the overall Doppler spread function

$$G_e(f_t, f_\tau) = G_I(f_\tau)G_C(f_t, f_\tau + f_0)G_R(f_t + f_\tau) \tag{5.31}$$

Apparently, $G_e(f_t, f_\tau)$ is composed of the frequency-shifted Doppler spread function of the time-variant channel, $G_C(f_t, f_\tau + f_0)$, filtered by $G_I(f_\tau)$ and $G_R(f_t + f_\tau)$.

5.4.2   Principal System Model Parameters

Cut-Off Frequencies

To bring the relevance of (5.29) and (5.30) to light, we impose the following model parameters. The impulse shaper and the receive lowpass shall possess transfer functions with ideal stopbands

$$G_I(f_\tau) \begin{cases} \neq 0 \; ; \; |f_\tau| \leq f_I \\ = 0 \; ; \quad else \end{cases} \; ; \quad G_R(f_\tau) \begin{cases} \neq 0 \; ; \; |f_\tau| \leq f_R \\ = 0 \; ; \quad else \end{cases} \tag{5.32}$$

f_I and f_R are the cut-off frequencies of the pulse shaper and the receive filter, respectively. The Doppler spread function of the channel shall have a bandpass shape with a passband around the carrier frequency $f_\tau = \pm f_0$ and shall be strictly band-limited as follows:

$$G_C(f_t, f_\tau) \begin{cases} \neq 0 \; ; \; f_0 - f_{\tau,C} \leq |f_\tau| \leq f_0 + f_{\tau,C} \\ \neq 0 \; ; \quad -f_{t,C} \leq |f_t| \leq f_{t,C} \\ 0 \; ; \quad\quad\quad\quad else \end{cases} \tag{5.33}$$

$2f_{\tau,C}$ defines the transmission bandwidth and $f_{t,C}$ the maximal Doppler frequency of the time-variant channel.

Overall Doppler Spread Function Without Receive Filter

For a better understanding, we first define an intermediate delay spread function and a Doppler spread function from the input of the pulse shaper (node I) to the receive filter input in Fig. 5.1 as $g_{e,D}(t, \tau)$ and $G_{e,D}(f_t, f_\tau)$. With (5.16) and (5.17), we obtain

$$g_{e,D}(t, \tau) = g_I(\tau) \circledast \left(g_C(t, \tau) e^{-j2\pi f_0 \tau} \right) + g_I^*(\tau) \circledast \left(g_C(t, \tau) e^{-j4\pi f_0 t} e^{j2\pi f_0 \tau} \right) \tag{5.34}$$

With (5.29) and (5.30) follows

$$G_{e,D}(f_t, f_\tau) = G_I(f_\tau) G_C(f_t, f_\tau + f_0) + G_I^*(-f_\tau) G_C(f_t + 2f_0, f_\tau - f_0) \tag{5.35}$$

where we do not take account of the gain factor $\sqrt{2}$ of the receive filter at the moment. In Fig. 5.2, the top view of the spectral components of $G_{e,D}(f_t, f_\tau)$ in (5.35) is illustrated. This busy figure needs some comments. To show the principle, we assume real-valued spectra. As defined in (5.33), the passbands of the channel Doppler spread function $G_C(f_t, f_\tau)$ are squares located around $f_\tau = \pm f_0$ and exhibit a bandwidth of $2f_{\tau,C}$ and $2f_{t,C}$. The passband of $G_I(f_\tau)$ is shown as a horizontal stripe. Please note, if a one-dimensional function $G_I(f_\tau)$ is plotted in a two-dimensional diagram, we have to keep $G_I(f_\tau)$ constant along the f_t-axis. The inside of the stripe represents the passband and the outside the stopband of this filter. $G_I^*(-f_\tau)$ in principle has the same shape.

The frequency-shifted Doppler spread function $G_C(f_t + 2f_0, f_\tau - f_0)$ is illustrated in the upper left part of Fig. 5.2. For the defined cut-off frequencies, we can substantiate that all spectral parts for $|f_\tau| > f_I$ are canceled by $G_I(f_\tau)$ and $G_I^*(-f_\tau)$. The important baseband $G_I(f_\tau)G_C(f_t, f_\tau + f_0)$ lies as a rectangular around the origin in the frequency range $|f_\tau| \leq f_I$ and $|f_t| \leq f_{t,C}$. Apparently, also a main part of the annoying two-dimensional spectrum $G_I^*(-f_\tau)G_C(f_t + 2f_0, f_\tau - f_0)$ around $(-2f_0, 0)$ contributes to $G_{e,D}(f_t, f_\tau)$. It does not overlap with the baseband, because normally the carrier frequency is much larger than the maximal Doppler shift of the channel, $2f_0 \gg f_{t,C}$.

Impact of the Receive Filter

Now, we introduce the receive filter $g_R(\tau)$ and consider the overall delay spread function, which is

$$g_e(t, \tau) = g_{e,D}(t, \tau) \circledast g_R(\tau) \qquad (5.36)$$

as well as the overall Doppler spread function

$$G_e(f_t, f_\tau) = G_{e,D}(f_t, f_\tau)G_R(f_t + f_\tau) \qquad (5.37)$$

Figure 5.3 illustrates the remaining spectral components from Fig. 5.2. The transfer function of the receive lowpass is defined in (5.32) with the cut-off frequency f_R. In (5.37), it is represented as $G_R(f_t + f_\tau)$ and thus as a diagonal stripe in the top view of Fig. 5.3. The inside of the stripe indicates the passband and the outside is the stopband with $G_R(f_t + f_\tau) = 0$. Apparently, the receive filter cancels the remaining part of $G_I^*(-f_\tau)G_C(f_t + 2f_0, f_\tau - f_0)$ located around $(-2f_0, 0)$ and cuts even some corners of the channel baseband. As a result,

$$G_I^*(-f_\tau)G_C(f_t + 2f_0, f_\tau - f_0)G_R(f_t + f_\tau) = G_{e2}(f_t, f_\tau) = 0 \qquad (5.38)$$

and $G_e(f_t, f_\tau)$ in (5.31) is verified. From $G_{e2}(f_t, f_\tau) = 0$ follows $g_{e2}(t, \tau) = 0$ and (5.18) is proven as well.

The top view of $G_e(f_t, f_\tau)$ in Fig. 5.3 results as a baseband spectrum with a rectangle or a hexagon shape in our example depending on the cut-off frequencies and indicated by dashed lines. The interesting question is, whether or not the filter $G_R(f_t + f_\tau)$ can limit the maximal Doppler shift $f_{t,C}$ of the fading channel. We see from Fig. 5.3 that the border lines of $G_R(f_t + f_\tau)$ cut the f_t-axis at $\pm f_R$ as well as the f_τ-axis. A reduction of f_R is able to limit the Doppler frequency of the channel on the f_t-axis, however, simultaneously also the transmission bandwidth along the f_τ- axis. Therefore, a time-invariant receive lowpass is not effective in reducing the impact of the fading other than the spectral parts of $G_I^*(-f_\tau)G_C(f_t + 2f_0, f_\tau - f_0)$.

Finally, please note that f_I in Fig. 5.3 cannot be considered as the final bandwidth of the receiver output signal $q_R(t)$. The reason is the Doppler shift $f_{t,C}$ of the fading channel, which increases the bandwidth of $q_R(t)$ beyond f_I, as will be discussed in the next Section.

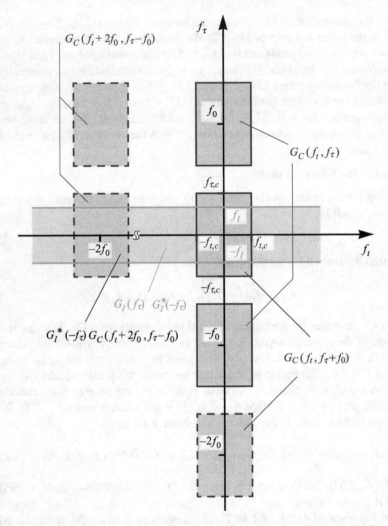

Fig. 5.2 Top view of spectral parts of the Doppler spread function $G_{e,D}(f_t, f_\tau)$ in (5.35) between node I and the input of the receive lowpass in Fig. 5.1. All spectra shall be real-valued. $G_C(f_t, f_\tau)$, Doppler spread function of time-variant channel; $G_C(f_t, f_\tau + f_0)$, version of $G_C(f_t, f_\tau)$ shifted on f_τ-axis; $G_C(f_t + 2f_0, f_\tau - f_0)$, version of $G_C(f_t, f_\tau)$ shifted on f_t- and f_τ-axis; $G_I(f_\tau)$, transfer function of the pulse shaper

Proof of (5.29) **and** (5.30)

We start with the first line in (5.17) and obtain with the frequency shifting property of the Fourier transform $g_C(t, \tau)e^{-j2\pi f_0\tau} \overset{t,\tau}{\rightharpoonup} G_C(f_t, f_\tau + f_0)$. Using (5.27) yields $g_I(\tau) \circledast (g_C(t, \tau)e^{-j2\pi f_0\tau}) \overset{t,\tau}{\rightharpoonup} G_I(f_\tau)G_C(f_t, f_\tau + f_0)$ and with the help of (5.28) we obtain $[g_I(\tau) \circledast (g_C(t, \tau)e^{-j2\pi f_0\tau})] \circledast g_R(\tau) \overset{t,\tau}{\rightharpoonup} [G_I(f_\tau)G_C(f_t, f_\tau + f_0)]$ $G_R(f_t + f_\tau)$ which proves the first line of (5.30).

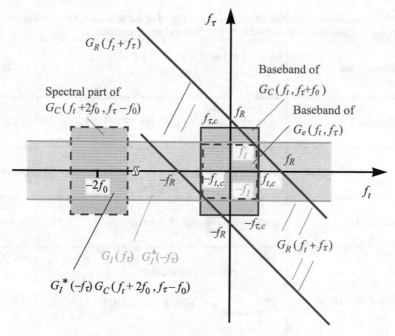

Fig. 5.3 Top view of spectral parts of the overall Doppler spread function $G_e(f_t, f_\tau)$ in (5.29) with (5.30). $G_R(f_t + f_\tau)$, Doppler spread function of receive filter; $G_I(f_\tau)$, transfer function of the pulse shaper. All spectra shall be real-valued

For the second line, we go back to (5.17) and start with $g_C(t, \tau) \mathrm{e}^{-\mathrm{j}4\pi f_0 t} \mathrm{e}^{\mathrm{j}2\pi f_0 \tau} \overset{t,\tau}{\longmapsto} G_C(f_t + 2f_0, f_\tau - f_0)$. With $g_I^*(\tau) \overset{\tau}{\longmapsto} G_I^*(-f_\tau)$ and using (5.27) results in $g_I^*(\tau) \circledast g_C(t, \tau) \mathrm{e}^{-\mathrm{j}4\pi f_0 t} \mathrm{e}^{\mathrm{j}2\pi f_0 \tau} \overset{t,\tau}{\longmapsto} G_I^*(-f_\tau) G_C(f_t + 2f_0, f_\tau - f_0)$ and with the help of (5.28) we obtain $\left[g_I^*(\tau) \circledast \left(g_C(t, \tau) \mathrm{e}^{-\mathrm{j}4\pi f_0 t} \mathrm{e}^{\mathrm{j}2\pi f_0 \tau} \right) \right] \circledast g_R(\tau) \overset{t,\tau}{\longmapsto} \left[G_I^*(-f_\tau) G_C(f_t + 2f_0, f_\tau - f_0) \right] G_R(f_t + f_\tau)$, which proves the second line of (5.30).

The proof of (5.29) is straightforward.

5.5 Equivalent Time-Variant Baseband System and Receiver Output Signal

5.5.1 Equivalent Time-Variant Baseband System

Equipped with the overall impulse response $w_e(t, s)$, the delay spread function $g_e(t, \tau)$, and the overall Doppler spread function $G_e(f_t, f_\tau)$ summarized in Table 5.1, we can illustrate the previous results with the block diagram in Fig. 5.4. It shows the equivalent time-variant baseband system between nodes (I) and (II) in Fig. 5.1. In the

Table 5.1 Summary—system functions and output signal of the equivalent time-variant baseband system, input signal $x(t)$ with spectrum $X(f_t)$, time-variant convolution $\circledast$

Impulse response	$w_e(t, s) = \left(w_I(t - s)e^{j2\pi f_0 t}\right) \circledast$ $\left(w_C(t, s)e^{-j2\pi f_0 t}\right) \circledast w_R(t - s)$	(5.13)
Delay spread function	$g_e(t, \tau) = g_I(\tau) \circledast \left(g_C(t, \tau)e^{-j2\pi f_0 \tau}\right) \circledast g_R(\tau)$	(5.19)
Doppler spread function	$G_e(f_t, f_\tau) = G_I(f_\tau)G_C(f_t, f_\tau + f_0)G_R(f_t + f_\tau)$	(5.31)
Output signal	$q_R(t) = x(t) \circledast g_e(t, \tau) + n_R(t)$	(5.40)
Sampled output signal	$q(k) = a(k)g(k, 0) +$ $\sum\limits_{\substack{m = -\infty \\ m \neq k}}^{\infty} a(m)g(k, k - m) + n(k)$	(5.44)
Output spectrum	$Q_R(f_t) = \int_{-\infty}^{\infty} X(u)G_e(f_t - u, u)du$	(5.46)
Noise	$n_R(t) = \sqrt{2}\left(n_C(t)e^{-j2\pi f_0 t}\right) * g_R(t)$	(5.41)

last column of Table 5.1, the equation numbers are given as a reference. The noise $n_R(t)$ is investigated in quite some detail in Chap. 3. This equivalent model provides a condensed system description between the input and the output as a "black box" with just one system function, without details of filtering, modulation, and demodulation. We will use this scheme in the next section to determine the output signal $q_R(t)$ and its Fourier spectrum $Q_R(f_t)$.

5.5.2   Receiver Output Signal

Receiver Output Signal in the Time Domain

With the time-variant convolution (5.4) quoted from Part II and using the overall impulse response $w_e(t, s)$ of the equivalent time-variant baseband system model in

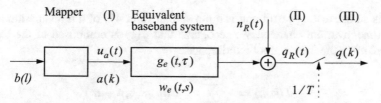

Fig. 5.4 Equivalent time-variant baseband system for wireless transmission with impulse response $w_e(t, s)$ and delay spread function $g_e(t, \tau)$

Fig. 5.4, we obtain

$$q_R(t) = u_a(t) \circledast w_e(t, s) + n_R(t) = \int_{-\infty}^{\infty} u_a(s) w_e(t, s) ds + n_R(t) \qquad (5.39)$$

With the introduction of $s = t - \tau$ and the delay spread function $g_e(t, \tau) = w_e(t, t - \tau)$ follows

$$q_R(t) = u_a(t) \circledast g_e(t, \tau) + n_R(t) = \int_{-\infty}^{\infty} u_a(t - \tau) g_e(t, \tau) d\tau + n_R(t) \qquad (5.40)$$

where $n_R(t)$ is the lowpass noise given earlier in (1.25) as

$$n_R(t) = \sqrt{2} \left(n_C(t) e^{-j2\pi f_0 t} \right) * g_R(t) \qquad (5.41)$$

With the input signal $u_a(t)$ of the pulse shaper (1.9) follows from (5.40)
$q_R(t) = \left(\sum_{m=-\infty}^{\infty} a(m) \delta(t - mT) \right) \circledast g_e(t, \tau) + n_R(t)$ and the time-variant convolution yields $q_R(t) = \sum_{m=-\infty}^{\infty} a(m) \int_{-\infty}^{\infty} \delta(t - \tau - mT) g_e(t, \tau) d\tau + n_R(t)$ resulting in

$$q_R(t) = \sum_{m=-\infty}^{\infty} a(m) g_e(t, t - mT) + n_R(t) \qquad (5.42)$$

As depicted in Fig. 5.4, $q_R(t)$ is sampled at $t = t_0 + kT$. With

$$g_e(t_0 + kT, t_0 + kT - mT) = g(k, k - m) \qquad (5.43)$$

$q_R(t_0 + kT) = q(k)$, and $n_R(t_0 + kT) = n(k)$ the receive signal at observation instant k is obtained as

$$q(k) = a(k) g(0, 0) + \sum_{\substack{m = -\infty \\ m \neq k}}^{\infty} a(m) g(k, k - m) + n(k) \qquad (5.44)$$

This is an interesting result and is quite similar to (1.41) of a transmission system with time-invariant channel. We recognize that $q(k)$ is composed of the transmit symbol $a(k)$, the intersymbol interference

$$I(k, k) = \sum_{\substack{m = -\infty \\ m \neq k}}^{\infty} a(m)g(k, k - m) \tag{5.45}$$

and the additive noise $n(k)$. The receive sample $a(k)$ is weighted by $g(k, 0)$. In contrast to $h(0)$ of the time-invariant system in (1.41), $g(k, 0)$ changes with time due to the fading of the channel. Also, the intersymbol interference $I(k, k,)$ is characterized by fading compared to $I(k)$ in (1.42).

In summary, the receive signal $q(k)$ of a wireless time-variant connection suffers from three impairments, namely, the fading of the gain coefficient $g(k, 0)$, the time-variant intersymbol interference $I(k, k)$, and the additive noise $n(k)$.

Spectrum of the Receiver Output Signal

Now we are interested to see the change of the transmit spectrum when passing through the time-variant channel. To this end, we input the signal $u_a(t) = x(t)$ with spectrum $X(f_t)$ to the pulse shaper and determine the spectrum $Q_R(f_t)$ of the receive signal $q_R(t)$. Of course, we take advantage of the equivalent time-variant system model in Fig. 5.4 with the Doppler spread function $G_e(f_t, f_\tau)$.

Please note, in contrast to time-invariant systems, $Q_R(f_t)$ is not the product of a "transfer function" $G_e(f_t, f_\tau)$ and the input spectrum $X(f_t)$. In Part II, we show in quite some detail that the correct input–output relation is given by an integral as follows:

$$Q_R(f_t) = \int_{-\infty}^{\infty} X(u)G_e(f_t - u, u)du \tag{5.46}$$

This equation has some similarities with a convolution integral (except for the second argument u in G_e) and is the reason why the output spectrum $Q_R(f_t)$ can have a wider bandwidth than the input spectrum $X(f_t)$, as will be demonstrated in Example 4. We insert (5.31) into (5.46)

$$Q_R(f_t) = \int_{-\infty}^{\infty} X(u)G_I(u)G_C(f_t - u, u + f_0)G_R(f_t - u + u)du \tag{5.47}$$

and obtain

$$Q_R(f_t) = G_R(f_t) \int_{-\infty}^{\infty} X(u)G_I(u)G_C(f_t - u, u + f_0)du = G_R(f_t)U_D(f_t) \tag{5.48}$$

where

$$U_D(f_t) = \int_{-\infty}^{\infty} X(u)G_I(u)G_C(f_t - u, u + f_0)du \tag{5.49}$$

is the Fourier spectrum of the output signal $u_D(t)$ of the demodulator in Fig. 5.1. Thus, (5.46) boils down to (5.48), which clearly reveals that the receive lowpass $G_R(f_t)$ filters its input signal $U_D(f_t)$, as expected. Moreover, in contrast to time-invariant systems, we cannot determine a quotient $\frac{Q_R(f_t)}{X(f_t)}$ from (5.48), such as the transfer function of a time-invariant system in (1.28). All findings are briefly summarized in Table 5.1.

Example 4

The top view of the real-valued Doppler spread function $G_e(f_t, f_\tau)$ of an *equivalent time-variant baseband system* is given in Fig. 5.5a. The time variance is revealed by the maximum Doppler frequency (Doppler shift) $f_{t,G}$, which is the cut-off frequency

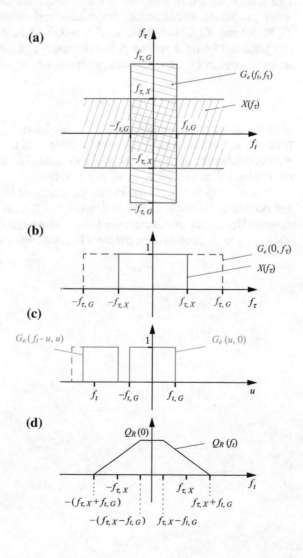

Fig. 5.5   a Top view of the Doppler spread function $G_e(f_t, f_\tau)$ of the equivalent time-variant baseband system together with the spectrum $X(f_\tau)$ of its input signal **b** $G_e(0, f_\tau)$ and $X(f_\tau)$ as a function of f_τ **c** Steps to determine the integrand in (5.46) **d** Spectrum $Q_R(f_t)$ of the output signal of the equivalent time-variant baseband system

on the f_t- axis. The higher the speed of the transmitter or receiver the larger $f_{t,G}$ will be. If the physical channel would be time-invariant and thus refrains from temporal fading, then $f_{t,G} = 0$. The cut-off frequency of the Doppler spread function in the f_τ-direction shall be $f_{\tau,G}$ and indicates the transmission bandwidth. We have assumed an input signal $x(t)$ of the pulse shaper with real-valued spectrum $X(f_\tau)$ and the cut-off frequency $f_{\tau,X} < f_{\tau,G}$. In principle, for any one-dimensional spectrum X, we can allocate either frequency variable, f_t or f_τ, because the functions are mathematically the same. However, in a two-dimensional diagram, the argument indicates in what frequency direction the filtering is effective. When we plot $X(f_\tau)$ as a two-dimensional function, we have to consider $X(f_\tau)$ as constant with respect to f_t. Consequently, $X(f_\tau)$ is a horizontal stripe in Fig. 5.5a. The magnitudes of the two spectra are unequal to zero in the shaded areas and outside they are zero with sharp transitions. For simplicity, we assume real-valued spectra. Figure 5.5b shows $G_e(0, f_\tau)$ and $X(f_\tau)$ as a function of f_τ and Fig. 5.5c illustrates $G_e(f_t - u, u)$ in the integrand of (5.46) for various f_t to calculate $Q_R(f_t)$. Fig. 5.4d depicts the resulting output spectrum $Q_R(f_t)$. Apparently, its cut-off frequency

$$f_{t,Q} = f_{\tau,X} + f_{t,G} \tag{5.50}$$

is by the quantity $f_{t,G}$ larger than $f_{\tau,X}$ of the input $X(f_\tau)$. The reason clearly is the time variance of the channel, because for the case of time invariance, $f_{t,G} = 0$ and no excess bandwidth can be observed. As a conclusion, the stronger the time variance of the channel, the larger $f_{t,G}$ and $f_{t,Q}$ will be.

According to (5.48), the bandwidth f_R of the receive filter must cover $f_{\tau,X} + f_{t,G}$ and not just the bandwidth of the transmit signal $f_{\tau,X}$, as in the case of a time-invariant system. Hence, the receive lowpass must be designed with the cut-off frequency $f_R \geq f_{\tau,X} + f_{t,G}$; otherwise, the filter limits its input spectrum.

Chapter 6
Basic Parameters of Wireless Channels

In this chapter, we summarize the main facts, which characterize a wireless single input single output channel. Such channels can be partitioned into different segments. The inner part is the wave propagation channel, which is characterized by the free space between the output of the transmit antenna and the input to the receive antenna. The next level includes the characteristics of the transmit and the receive antenna, such as radiation pattern and antenna gains. Finally, the equivalent baseband system incorporates modulation, demodulation, and filtering, as described in Sect. 5.1. In the following, we characterize the main transmission effects of wireless and mobile channels with adequate models. There are several physical details for refinements, which are beyond our scope here, such as specific indoor and outdoor scenarios as well as details of electromagnetic field theory. The interested reader is referred to dedicated material, such as [1–4].

6.1 Path Loss

An electromagnetic wave traveling from the transmit to the receive antenna undergoes free-space path loss, scattering, refraction, and diffraction from surfaces of buildings, hills, vegetation, rain, and various objects. These effects are well understood and investigated theoretically and by measurements using two- and tree-dimensional modeling of the landscape, ray tracing, and Snell´s law [2, 5]. Friis´ law [6] for free-space transmission relates the receive mean power P_r to the mean transmit power P_t as

$$\frac{P_r}{P_t} = G_t G_r \frac{1}{P_L} \tag{6.1}$$

© Springer Nature Switzerland AG 2019
J. Speidel, *Introduction to Digital Communications*, Signals and Communication
Technology, https://doi.org/10.1007/978-3-030-00548-1_6

where G_t and G_r are the effective antenna gains for the transmit and receive antenna, respectively. $P_L = \left(\frac{4\pi d}{\lambda}\right)^2$ denotes the path loss. $\lambda = \frac{c}{f_0}$ is the wavelength of the electromagnetic waves, which are assumed to be planar waves in the far field. c denotes the speed of light in the air and f_0 is the carrier frequency of the modulator. The path loss P_L increases proportional to the square of the distance d between the transmitter and the receiver. However, for more complex propagation environments, highly sophisticated and empirical models are required, such as the COST 231 Hata model or the Walfisch–Ikegami model, which are frequently used for the planning of cellular networks. Often, simple and experimental path loss models will do, in which the proportionality

$$P_L \sim \left(\frac{d}{d_0}\right)^n \tag{6.2}$$

is used approximately, where n is the path loss exponent and d_0 a reference. The range of n is reported to be about 2.....6, and it depends on the environment for indoor and outdoor communications. As can be seen from (6.1) and (6.2), on a logarithmic scale the receive power P_r declines linearly with the distance d. On the basis of this path loss, the necessary power budget between the transmit and the receive antenna can be calculated including antenna gains and also some margin for additional effects, such as small-scale fading. This approach provides the bottom line of the system design and mainly reflects an area denoted as path loss region, which is beyond the multitude of local scattering and reflections close to the antennas. On top of that, statistical fluctuations of the receive power are observed, which is called fading. The main sources of fading are scattering and reflections of the propagating waves at objects on the transmitter and the receiver side as well as due to statistical movement of the transmitter and/or the receiver. This is briefly outlined in the next section.

6.2 Shadowing

An effect, which causes a fluctuation of the receive power, is shadowing also called shadow fading. It can be characterized by a long-term power loss ψ, and the main reason is the movement of the transmitters and/or receiver behind some shielding objects in the propagation path. Also changing surfaces of objects, scatterers, and reflectors contribute to this effect. Shadow fading normally is a stochastic effect, and the resulting path loss $\psi > 0$ is often modeled satisfactorily by a log-normally distributed random variable with probability density function

$$p_0(\psi) = \frac{10}{\ln(10)\sqrt{2\pi}\sigma_0\psi} \mathrm{e}^{-\frac{(10\log_{10}(\psi)-\mu_0)^2}{2\sigma_0^2}} \tag{6.3}$$

in which μ_0 and σ_0^2 are the mean and the variance, respectively, [5]. μ_0 and σ_0 in dB and are the result of measurements or analytical models. Typical values are $\mu_0 = 0\,\mathrm{dB}$

and $\sigma_0 = 8\,\text{dB}$, [3]. Shadow fading loss ψ and path loss P_L can be multiplied to get approximately the overall path loss.

References

1. A.F. Molisch, H. Asplund, R. Heddergott, M. Steinbauer, T. Zwick, The COST259 directional channel model - part i: overview and methodology. IEEE Trans. Wirel. Commun. **5**, 3421–3433 (2006)
2. A.F. Molisch, *Wireless Communications* (Wiley and IEEE press, New York, 2009)
3. Physical layer aspects for evolved universal terrestrial radio access (UTRA), 3GPP TR 25.814 v7.1.0, Technical report, 2006
4. S. Haykin, M. Moher, *Modern Wireless Communications* (Pearson Prentice Hall, New Jersey, 2005)
5. A. Goldsmith, *Wireless Communications* (Cambridge University Press, New York, 2005)
6. H.T. Friis, A note on a simple transmission formula. Proc. IRE **34**, 254–256 (1946)

Chapter 7
Wireless System with Multipath Propagation

In particular, for broadband applications, where the signal bandwidth is a noticeable fraction of the carrier frequency and where sophisticated modulation and detection methods have to be used, the receiver does not only require a sufficient mean power but also a reasonable approximation of the impulse response $w_e(t, s)$ or delay spread function $g_e(t, \tau)$ of the equivalent time-variant baseband system. In the following, we consider the link budget design as given and focus on additional impairments, such as the frequency- and the time-dependent fading. Practical examples are modern cellular networks, where the carrier frequency and the signal bandwidth are around 2 and 0.1 GHz, respectively. Given an omnidirectional antenna pattern, the electromagnetic waves are continuously distributed over the whole space. For a typical wave propagation area in a city with many buildings, there is no line of sight between the base stations and the mobile stations. In this scenario, reflections and scattering of the waves are favorable for sufficient signal reception. Other propagation scenarios, e.g., according to the COST 207 channel model already used for GSM and the COST 259 model differentiate between "rural area," "typical urban" with approximately no significant reflections and scattering, as well as "bad urban." The model "hilly terrain" accounts for significant reflections, [1–3]. Difficult propagation conditions can occur, if there is just a narrow "key hole" between the transmitter and the receiver, [4].

7.1 Multipath Model of Time-Invariant Channel

Let us now consider the propagation scenario depicted in Fig. 7.1, which is a model for finding the propagation channel parameters approximately by measurement with ray tracing. Assume that the transmitter emits an electromagnetic wave as a narrow beam received by the mobile station. Consequently, discrete paths ν from the transmitter

The original version of this chapter was revised: a few typographical errors were corrected. The correction to this chapter can be found at https://doi.org/10.1007/978-3-030-00548-1_25

© Springer Nature Switzerland AG 2019
J. Speidel, *Introduction to Digital Communications*, Signals and Communication Technology, https://doi.org/10.1007/978-3-030-00548-1_7

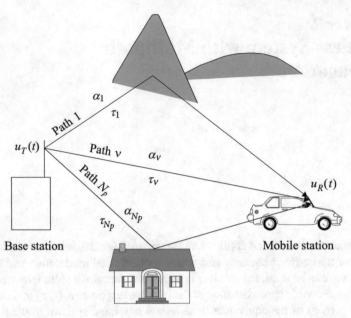

Fig. 7.1 Wireless downlink multipath propagation

to the receiver result, where the waves undergo path losses and all kinds of scattering and reflections, which are investigated using the wave propagation theory. Often, there is also a direct path between transmitter and receiver, called line of sight path. Each ray ν can be characterized approximately by its path loss coefficient α_ν and its delay time τ_ν. The delay $\tau_\nu = \frac{l_\nu}{c}$ is given by the path length l_ν and the speed of light c in the air. The path loss coefficient $0 < \alpha_\nu \leq 1$ depends on l_ν as given by the path loss model in (6.2), thus $\alpha_\nu \sim \left(\frac{1}{\tau_\nu}\right)^n$ holds approximately with n as the path loss exponent. Consequently, a long path exhibits a high loss and thus can often be dropped for approximation. Now assume that the transmit antenna emits the signal $u_T(t)$, which is traveling on ray ν and arrives at the receive antenna as $\alpha_\nu u_T(t - \tau_\nu)$. At the receiver, the signals of all paths superimpose yielding the receive signal

$$u_R(t) = \sum_{\nu=1}^{N_P} \alpha_\nu u_T(t - \tau_\nu) \qquad (7.1)$$

where in (7.1) only the N_P strongest paths are kept.

This model characterizes a linear and time-invariant system. Consequently, it can be fully described by its impulse response. Let $u_T(t) = \delta(t)$, then the impulse response is

$$h(t) = \sum_{\nu=1}^{N_P} \alpha_\nu \delta(t - \tau_\nu) \qquad (7.2)$$

We substantiate that the single Dirac impulse at the input results in a chain of impulses, which superimpose and the output can be considered as a broad impulse with the duration given approximately by the difference between the largest and the smallest delay, also called delay spread. Hence, the multipath channel is broadening the input signal in the time domain.

It should be mentioned that the presented model characterizes a channel with an infinite transmission bandwidth, which is not realistic. However, the pulse shaper at the transmitter and the lowpass filter at the receiver finally limit the bandwidth of the overall system.

7.2 Multipath Model of Time-Variant Channel

7.2.1 Delay Spread Function of the Time-Variant Multipath Channel

If the receiver moves and/or scattering, reflections, and shadowing effects are time-variant, the path loss coefficients and the delays depend on time, $\alpha_\nu = \alpha_\nu(t)$ and $\tau_\nu = \tau_\nu(t)$, respectively. Hence, the change of the receive signal $u_R(t)$ is not only caused by the variations of the transmit signal $u_T(t)$ but also by the temporal variations of the channel parameters. Consequently, the channel model has to be refined and described as a time-variant system. As outlined in Part II, time-variant systems are characterized by the delay spread function $g_C(t, \tau)$. By definition, $g_C(t, \tau)$ is the response at observation time t to a Dirac impulse at the time instant $t - \tau \leq t$. We first consider the input Dirac impulse $u_T(t) = \delta(t - (t_0 - \tau))$ active at the instant $t_0 - \tau$. Then, we proceed with $t_0 \to t$. Hence, on the basis of the model in Fig. 7.1, we obtain the delay spread function

$$g_C(t, \tau) = \sum_{\nu=1}^{N_P} \alpha_\nu(t)\delta(t - (t_0 - \tau) - \tau_\nu(t))\mid_{t_0 \to t} = \sum_{\nu=1}^{N_P} \alpha_\nu(t)\delta(\tau - \tau_\nu(t)) \quad (7.3)$$

which shows some differences compared to (7.2). The multipath model in Fig. 7.1 with (7.3) is widely accepted for wireless communications. As all path loss coefficients $\alpha_\nu(t)$ are real-valued, also the delay spread function $g_C(t, \tau)$ shows this property and is in line with the fact that any physical channel exhibits a real-valued impulse response or delay spread function. It is readily appreciated that the delay spread function in (7.3) of the multipath channel is much simpler than the general form $g_C(t, \tau)$ treated in Sect. 5.1, because (7.3) is almost separated into the product of a function of t and a function of τ. The separation is perfect, if we can assume $\tau_\nu(t) = \tau_\nu = const.$ that is often fulfilled approximately. We follow this special case later and will see that the results are becoming rather simple.

The input–output relation is given by the time-variant convolution defined in (5.22). Consequently, we obtain the channel output signal $u_R(t)$ as a function of the input signal $u_T(t)$ with the help of (5.22)

$$u_R(t) = u_T(t) \circledast g_C(t, \tau) = \int_{-\infty}^{\infty} u_T(t - \tau) g_C(t, \tau) d\tau \tag{7.4}$$

Plugging in (7.3) yields $u_R(t) = \sum_{\nu=1}^{N_P} \int_{-\infty}^{\infty} u_T(t - \tau)\alpha_\nu(t)\delta(\tau - \tau_\nu(t)) d\tau$ and the result is

$$u_R(t) = \sum_{\nu=1}^{N_P} \alpha_\nu(t) u_T(t - \tau_\nu(t)) \tag{7.5}$$

Hence, the channel output signal for path ν is $\alpha_\nu(t) u_T(t - \tau_\nu(t))$ determined by the input signal, which is delayed by the time-variant path delay and attenuated by the time-variant path loss coefficient.

As a special case, if all path delays are constant, $\tau_\nu(t) = \tau_\nu$, the delay spread function of the path ν in (7.3) is

$$g_{C,\nu}(t, \tau) = \alpha_\nu(t)\delta(\tau - \tau_\nu) \; ; \; \nu = 1, 2, \ldots, N_P \tag{7.6}$$

and is separated into the product of solely time and delay depending functions $\alpha_\nu(t)$ and $\delta(\tau - \tau_\nu)$, respectively.

7.2.2   Delay Spread Function of the Equivalent Time-Variant Multipath Baseband System

Now we are going to determine the delay spread function $g_e(t, \tau)$ of the equivalent baseband system between the nodes (I) and (II) in Figs. 5.1 and 5.4 for a wireless channel with time-variant multipath propagation. The channel delay spread function is given by (7.3). We prove at the end of this section that

$$g_e(t, \tau) = \sum_{\nu=1}^{N_P} g_{e,\nu}(t, \tau) \tag{7.7}$$

holds, in which

$$g_{e,\nu}(t, \tau) = \left[\tilde{\alpha}_\nu(t) g_I(\tau - \tau_\nu(t))\right] \circledast g_R(\tau) \; ; \; \nu = 1, 2, \ldots, N_P \tag{7.8}$$

is the equivalent baseband delay spread function of path ν and

$$\tilde{\alpha}_\nu(t) = \alpha_\nu(t) e^{-j2\pi f_0 \tau_\nu(t)} \; ; \; \nu = 1, 2, \ldots, N_P \tag{7.9}$$

defines the complex path loss coefficient. As can be seen, $g_e(t, \tau)$ in (7.7) is the superposition of N_P delay spread functions $g_{e,\nu}(t, \tau)$ of the various paths, which are composed of the delay spread function $g_I(\tau - \tau_\nu(t))$ of the pulse shaper weighted by $\tilde{\alpha}_\nu(t)$ and filtered by the receive lowpass $g_R(\tau)$.

Proof of (7.7) **and** (7.8)

We plug (7.3) into the general form of the equivalent delay spread function (5.19) and obtain $g_e(t, \tau) = g_I(\tau) \circledast \left(\sum_{\nu=1}^{N_P} \alpha_\nu(t) \delta\left(\tau - \tau_\nu(t)\right) e^{-j2\pi f_0 \tau} \right) \circledast g_R(\tau)$. As the time-variant convolution is distributive, we obtain $g_e(t, \tau) = \sum_{\nu=1}^{N_P} g_{e,\nu}(t, \tau)$ with $g_{e,\nu}(t, \tau) = g_I(\tau) \circledast \left[\alpha_\nu(t) \delta\left(\tau - \tau_\nu(t)\right) e^{-j2\pi f_0 \tau} \right] \circledast g_R(\tau)$. To execute the first time-variant convolution, we use (5.20) and obtain

$g_I(\tau) \circledast \left[\alpha_\nu(t) \delta\left(\tau - \tau_\nu(t)\right) e^{-j2\pi f_0 \tau} \right] = \int_{-\infty}^{\infty} g_I(\eta) \alpha_\nu(t) \delta\left(\tau - \eta - \tau_\nu(t)\right)$
$e^{-j2\pi f_0 (\tau - \eta)} d\eta =$
$g_I(\tau - \tau_\nu(t)) \tilde{\alpha}_\nu(t)$, with $\tilde{\alpha}_\nu(t) = \alpha_\nu(t) e^{-j2\pi f_0 \tau_\nu(t)}$. Finally,
$g_{e,\nu}(t, \tau) = \left[\tilde{\alpha}_\nu(t) g_I\left(\tau - \tau_\nu(t)\right) \right] \circledast g_R(\tau)$ follows and the proof is finished.

7.2.3 Doppler Spread Function of the Equivalent Time-Variant Multipath Baseband System

The Doppler spread function $G_e(f_t, f_\tau)$ of the equivalent time-variant multipath baseband system is obtained by the Fourier transform of $g_e(t, \tau)$ in (7.7) with respect to t and τ. To this end, we make the following presuppositions. First, all complex path loss coefficients $\tilde{\alpha}_\nu(t)$ exhibit the Fourier spectrum $\tilde{A}_\nu(f_t)$ with lowpass characteristic and second, all path delays are constant, $\tau_\nu(t) = \tau_\nu = const. (\nu = 1, 2, \ldots, N_P)$. We proof at the end of this section that $G_e(f_t, f_\tau)$ is determined by

$$g_e(t, \tau) \overset{t,\tau}{\longmapsto} G_e(f_t, f_\tau) = \sum_{\nu=1}^{N_P} G_{e,\nu}(f_t, f_\tau) \qquad (7.10)$$

with

$$g_{e,\nu}(t, \tau) \overset{t,\tau}{\longmapsto} G_{e,\nu}(f_t, f_\tau) = G_I(f_\tau) \tilde{A}_\nu(f_t) e^{-j2\pi \tau_\nu f_\tau} G_R(f_t + f_\tau) ; \quad \nu = 1, 2, \ldots, N_P \quad (7.11)$$

where the $G_{e,\nu}(f_t, f_\tau)$ are the Doppler spread functions of the individual paths. The term $e^{-j2\pi \tau_\nu f_\tau}$ reflects the signal delay on each path.

Proof of (7.10) **and** (7.11)

We consider (7.8). Under the given prerequisite $\tau_\nu(t) = \tau_\nu = const.$, the Fourier transform is applicable and we get $\tilde{\alpha}_\nu(t) g_I(\tau - \tau_\nu) \overset{t,\tau}{\longmapsto} \tilde{A}_\nu(f_t) G_I(f_\tau) e^{-j2\pi \tau_\nu f_\tau}$. Next, we apply the transform pair (5.28) on (7.8) and obtain

$g_{e,\nu}(t, \tau) = \left[\tilde{\alpha}_\nu(t) g_I(\tau - \tau_\nu) \right] \circledast g_R(\tau) \overset{t,\tau}{\longmapsto} \tilde{A}_\nu(f_t) G_I(f_\tau) e^{-j2\pi \tau_\nu f_\tau} G_R(f_t + f_\tau) = G_{e,\nu}(f_t, f_\tau)$, which proves (7.11). The summation over all N_P paths, $\sum_{\nu=1}^{N_P} G_{e,\nu}(f_t, f_\tau)$, finalizes the proof.

7.2.4   Receiver Output Signal $q_R(t)$

With the help of $g_e(t, \tau)$ in (7.7), (7.8), and the general input–output relation (5.22) for a time-variant system we obtain the signal at the receiver output

$$q_R(t) = \sum_{\nu=1}^{N_P} q_{R,\nu}(t) + n_R(t) \tag{7.12}$$

with the signal component of path ν

$$q_{R,\nu}(t) = u_a(t) \circledast \left[\tilde{\alpha}_\nu(t) g_I (\tau - \tau_\nu(t)) \right] \circledast g_R(\tau) \; ; \; \nu = 1, 2, \ldots, N_P \tag{7.13}$$

and the transmit signal $u_a(t)$.

7.2.5   Fourier Spectrum $Q_R(f_t)$ of the Receiver Output Signal

The Fourier spectrum $Q_R(f_t)$ of $q_R(t)$ is obtained with the general input–output relation (5.46) and the Doppler spread function (7.10) under the prerequisite $\tau_\nu(t) = \tau_\nu = const$. With the spectrum $X(f_t)$ of the signal $u_a(t) = x(t)$ at the input of the pulse shaper, we get without considering the receiver noise $Q_R(f_t) = \int_{-\infty}^{\infty} X(u) \sum_{\nu=1}^{N_P} G_{e,\nu}(f_t - u, u) du$. Interchanging integration and summation yields

$$Q_R(f_t) = \sum_{\nu=1}^{N_P} Q_{R,\nu}(f_t) \tag{7.14}$$

where

$$Q_{R,\nu}(f_t) = \int_{-\infty}^{\infty} X(u) G_{e,\nu}(f_t - u, u) du \tag{7.15}$$

is the receiver output spectrum allocated to path ν. With (7.11) follows

$$\begin{aligned} Q_{R,\nu}(f_t) &= \int_{-\infty}^{\infty} X(u) G_I(u) G_R(f_t) e^{-j2\pi\tau_\nu u} \tilde{A}_\nu(f_t - u) du \\ &= \left[(X(f_t) G_I(f_t) e^{-j2\pi\tau_\nu f_t}) * \tilde{A}_\nu(f_t) \right] G_R(f_t) \end{aligned} \tag{7.16}$$

We recognize that the spectrum $Q_{R,\nu}(f_t)$ of the νth receive signal is given by the convolution of $X(f_t) G_I(f_t) e^{-j2\pi\tau_\nu f_t}$ and the path loss spectrum $\tilde{A}_\nu(f_t)$ with respect to f_t, subsequently filtered by $G_R(f_t)$. As expected from the convolution, the temporal fading of the multipath channel results in an excess bandwidth of the input signal of the receive filter compared to the transmit signal. The difference is given by the bandwidth of the path loss $\tilde{A}_\nu(f_t)$. For more details, please see Example 5. The reason why the time-variant convolution boils down to the time-invariant convolution in (7.16) is the fact that the Doppler spread function $G_{e,\nu}(f_t, f_\tau)$ in (7.11) excluding G_R is a product of isolated functions of f_t and f_τ, respectively, which holds for the given prerequisite of constant delays τ_ν. Again, we recognize that a quotient $\frac{Q_{R,\nu}(f_t)}{X(f_t)}$ like the transfer function of a time-invariant system cannot be defined owing to the integral in (7.15).

Example 5

(a) Let the cut-off frequencies of $X(f_t)$, $G_I(f_t)$, $\tilde{A}_\nu(f_t)$, and $G_R(f_t)$ be f_X, $f_I = f_X$, $f_{\tilde{A}_\nu}$, and $f_R > f_I + f_{\tilde{A}_\nu}$, respectively. Find the maximal cut-off frequency of $Q_{R,\nu}(f_t)$.

(b) Determine the output spectrum $Q_{R,\nu}(f_t)$ of path ν, if the channel is approximately static, i.e., its parameters do not change with time.

Solution:

(a) In general, the convolution operation in (7.16) yields a maximal cut-off frequency, which is the sum of the cut-off frequencies of the spectra under convolution. Hence, $f_I + f_{\tilde{A}_\nu}$ is the cut-off frequency of $Q_{R,\nu}(f_t)$ as a maximum. If we determine $f_R > f_I + f_{\tilde{A}_\nu}$, the receive lowpass $G_R(f_t)$ is not cutting its input spectrum.

(b) If the channel is showing long periods, in which there is no fading, the path loss is almost constant, say $\tilde{\alpha}_\nu(t) = 1$ and consequently $\tilde{A}_\nu(f_t) = \delta(f_t)$. Then follows from (7.16) $Q_{R,\nu}(f_t) = X(f_t)G_I(f_t)e^{-j2\pi\tau_\nu f_t}G_R(f_t)$, which is the output spectrum of a time-invariant system with channel transfer function $G_C(f_t) = e^{-j2\pi\tau_\nu f_t}$, similar to (1.26) of Sect. 1.5. Consequently, $f_R = f_I$ suffices in this case.

7.3 Multipath Channel and Mobile Receiver

7.3.1 System Model

An important practical case is considered in Fig. 7.2 for a wireless time-variant multipath channel. The fading of the parameters $\alpha_\nu(t)$ and $\tau_\nu(t)$ is caused by the movement of the receiver with the velocity $\mathbf{v}_0$. Only the path ν of the multipath propagation is depicted. The receiver starts at time instant $t_0 = 0$ at the location P_0 and moves with the velocity $\mathbf{v}_0$ to arrive at the location P at the time instant $t > 0$. We assume $|\mathbf{v}_0| = v_0 = constant$. Within a short interval $t - t_0$, the receiver moves approximately on a straight line in the direction indicated by φ_ν.

The delay of the receive signal at time instant t in location P is determined approximately by

$$\tau_\nu(t) = \tau_\nu(t_0) + \frac{v_0 \cos(\varphi_\nu)}{c}t \qquad (7.17)$$

Fig. 7.2 Downlink transmission over a multipath wireless channel to a receiver, which moves with velocity v_0

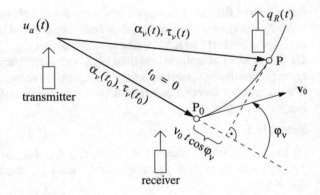

where c is the speed of light in the air.

7.3.2 Doppler Shift

Plugging (7.17) into (7.8) and (7.9) results in the delay spread function of the equivalent baseband system for the path ν

$$g_{e,\nu}(t,\tau) = \left[\alpha_\nu(t)\mathrm{e}^{-\mathrm{j}2\pi f_0\tau_\nu(t_0)}\mathrm{e}^{-\mathrm{j}2\pi f_{D,\nu}t}g_I\left(\tau - \tau_\nu(t_0) - \frac{v_0\cos(\varphi_\nu)}{c}t\right)\right] \circledast g_R(\tau)$$

(7.18)

where

$$f_{D,\nu} = f_0\frac{v_0}{c}\cos(\varphi_\nu)$$

(7.19)

is denoted as the Doppler frequency or the Doppler shift for the path ν. From (7.19), we conclude the following,

- The Doppler shift $f_{D,\nu}$ is proportional to the velocity v_0 of the receiver and to the carrier frequency f_0 of the modulator. Consequently, if the receiver stands still, then $f_{D,\nu} = 0$. The same holds, if no modulation is present, $f_0 = 0$.
- For $-\frac{\pi}{2} < \varphi_\nu < \frac{\pi}{2}$ the receiver moves away from the transmitter and $f_{D,\nu} > 0$. For $\varphi_\nu = 0$ the Doppler shift is maximal,

$$f_{D,max} = f_0\frac{v_0}{c}$$

(7.20)

- For $\frac{\pi}{2} < \varphi_\nu < \frac{3\pi}{2}$, the receiver moves towards the transmitter and $f_{D,\nu} < 0$.
- For $\varphi_\nu = \pm\frac{\pi}{2}$, the receiver does not change the distance to the transmitter and $f_{D,\nu} = 0$.

7.3.3 Delay Spread Function

For a small time interval $t - t_0$ and $\frac{v_0}{c} \ll 1$, we can neglect $\frac{v_0 \cos(\varphi_\nu)}{c} t$ and obtain from (7.18) approximately the delay spread function of the equivalent baseband system for the path ν

$$g_{e,\nu}(t, \tau) = \left[\alpha_\nu(t) e^{-j2\pi f_0 \tau_\nu(t_0)} e^{-j2\pi f_{D,\nu} t} g_I(\tau - \tau_\nu(t_0)) \right] \circledast g_R(\tau) \qquad (7.21)$$

Obviously, the delays $\tau_\nu(t_0) = \tau_\nu$ in (7.21) are constant. Plugging in (7.9) yields

$$g_{e,\nu}(t, \tau) = \left[\tilde{\alpha}_\nu(t) e^{-j2\pi f_{D,\nu} t} g_I(\tau - \tau_\nu) \right] \circledast g_R(\tau) \qquad (7.22)$$

with $\tilde{\alpha}_\nu(t) = \alpha_\nu(t) e^{-j2\pi f_0 \tau_\nu}$.

Regarding (7.22) with (7.19), the movement of the receiver causes a modulation of the path loss coefficient $\tilde{\alpha}_\nu(t)$ with the "carrier" $e^{-j2\pi f_{D,\nu} t}$ and consequently the spectrum $\tilde{A}_\nu(f_t)$ of $\tilde{\alpha}_\nu(t)$ will experience a frequency shift, also called Doppler shift

$$\tilde{\alpha}_\nu(t) e^{-j2\pi f_{D,\nu} t} \overset{t}{\longmapsto} \tilde{A}_\nu(f_t + f_{D,\nu}) \qquad (7.23)$$

As can be seen, the main difference between (7.22) and (7.8) is the Doppler shift $e^{-j2\pi f_{D,\nu} t}$ and the constant delay $\tau_\nu(t) = \tau_\nu(t_0) = \tau_\nu$. Hence, we can use the terms for $g_{e,\nu}(t, \tau)$ and $q_{R,\nu}(t)$ derived earlier, if we just replace $\tilde{\alpha}_\nu(t)$ by $\tilde{\alpha}_\nu(t) e^{-j2\pi f_{D,\nu} t}$ and take $\tau_\nu(t_0) = \tau_\nu$. This will be done in the next section.

7.3.4 Receiver Output Signal $q_R(t)$ with Doppler Shift

For the receiver output signal, $q_R(t)$ (7.12) is valid

$$q_R(t) = \sum_{\nu=1}^{N_P} q_{R,\nu}(t) + n_R(t) \qquad (7.24)$$

and with (7.22) follows

$$q_{R,\nu}(t) = x(t) \circledast \left[\tilde{\alpha}_\nu(t) e^{-j2\pi f_{D,\nu} t} g_I(\tau - \tau_\nu) \right] \circledast g_R(\tau) \; ; \; \nu = 1, 2, \ldots, N_P \qquad (7.25)$$

where $u_a(t) = x(t)$ is the transmit signal at the pulse shaper input. Compared to (7.13), the Doppler shift $e^{-j2\pi f_{D,\nu} t}$ is effective and $\tau_\nu(t) = \tau_\nu(t_0) = \tau_\nu$ holds.

From (7.21) and (7.25), we derive the equivalent time-variant baseband system model depicted in Fig. 7.3. Each path ν is composed of a delay τ_ν, a complex path loss coefficient $\alpha_\nu(t) e^{-j2\pi f_0 \tau_\nu}$ and a modulator with carrier $e^{-j2\pi f_{D,\nu} t}$, which represents the Doppler shift of the spectrum. In this model, the signals at the input and at the

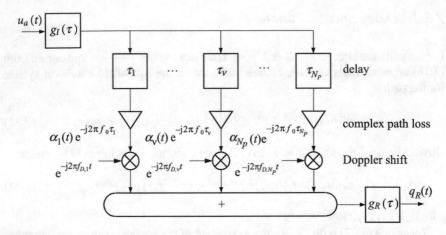

Fig. 7.3 Model of equivalent time-variant baseband system with multipath channel and moving receiver

output are filtered by the pulse shaper $g_I(\tau)$ and the receive lowpass filter $g_R(\tau)$, respectively. The resulting output signal $q_R(t)$ is given by (7.24) and (7.25) without considering the receiver noise.

Example 6

Find the Doppler spread function $G_{e,\nu}(f_t, f_\tau)$ of the equivalent baseband system of path ν for a receiver with speed $\mathbf{v}_0$ and constant path delay in Fig. 7.2. How large is the maximal Doppler shift, if a person is moving with $v_0 = 1$ m/s ($= 3.6$ km/h) and a fast train with 360 km/h?
Determine the receiver output spectrum $Q_R(f_t)$ associated with path ν.

Solution:

We can determine $G_{e,\nu}(f_t, f_\tau)$ from (7.22) by the Fourier transform with respect to t and τ or alternatively from (7.11), if we replace $\tilde{A}_\nu(f_t)$ by $\tilde{A}_\nu(f_t + f_{D,\nu})$. The result is

$$g_{e,\nu}(t, \tau) \overset{t,\tau}{\multimap} G_{e,\nu}(f_t, f_\tau) = G_I(f_\tau)e^{-j2\pi\tau_\nu f_\tau}\tilde{A}_\nu(f_t + f_{D,\nu})G_R(f_t + f_\tau) \quad (7.26)$$

Compared to the baseband of $G_e(f_t, f_\tau)$ in Fig. 5.3, the spectrum $\tilde{A}_\nu(f_t + f_{D,\nu})$ is shifted to the left by $f_{D,\nu}$ on the f_t-axis and subsequently filtered by $G_R(f_t + f_\tau)$. With (7.20) the maximal Doppler shifts are $f_{D,max} \approx 7$ and 700 Hz. Output spectrum: We apply (7.16) and replace $\tilde{A}_\nu(f_t)$ by $\tilde{A}_\nu(f_t + f_{D,\nu})$ yielding

$$Q_{R,\nu}(f_t) = \left[\left(X(f_t)G_I(f_t)e^{-j2\pi\tau_\nu f_t}\right) * \tilde{A}_\nu(f_t + f_{D,\nu})\right]G_R(f_t) = \quad (7.27)$$

Example 7

Consider a harmonic transmit signal $u_a(t) = x(t) = \hat{x}e^{j2\pi f_X t}$ with the frequency $f_X < f_I < f_R$, where f_I and f_R are the cut-off frequencies of the pulse shaper $g_I(t)$ and the receive lowpass $g_R(t)$, respectively. Find the spectrum $Q_{R,\nu}(f_t)$ at the receiver output for path ν, if the receiver moves with the velocity $\mathbf{v}_0$ given in Fig. 7.2.

Solution:

We take advantage of the frequency domain. Knowing that $x(t) \overset{t}{\rightarrowtail} X(f_t) = \hat{x}\delta(f_t - f_X)$, we find with (5.46)
$Q_{R,\nu}(f_t) = \int_{-\infty}^{\infty} X(u)G_{e,\nu}(f_t - u, u)du = \hat{x} \int_{-\infty}^{\infty} \delta(u - f_X)G_{e,\nu}(f_t - u, u)du$.
With the sifting property of the Dirac impulse, we obtain

$$Q_{R,\nu}(f_t) = \hat{x}G_{e,\nu}(f_t - f_X, f_X) \tag{7.28}$$

Plugging in (7.26) yields

$$Q_{R,\nu}(f_t) = \left[\hat{x}G_I(f_X)e^{-j2\pi \tau_\nu f_X}\right] \tilde{A}_\nu(f_t - f_X + f_{D,\nu})G_R(f_t) \tag{7.29}$$

The term in brackets is a constant factor. The remaining part of $Q_{R,\nu}(f_t)$ represents a continuous spectrum rather than just a Dirac impulse as $X(f_t)$ owing to the temporal fading of $\tilde{\alpha}_\nu(t)$.

7.4 Frequency Selective Fading of Multipath Channel

In this section, we consider the time-variant transfer function of the multipath wireless channel and investigate its frequency response for a constant time instant t. We will see that it can change significantly as a function of f_τ. This effect is called frequency selectivity or frequency selective fading. With the transform pair $\delta(\tau - \tau_\nu(t)) \overset{\tau}{\rightarrowtail} e^{-j2\pi\tau_\nu(t)f_\tau}$, the Fourier transform yields from the delay spread function in (7.3) the time-variant transfer function

$$g_C(\tau, t) \overset{\tau}{\rightarrowtail} G_{C,t}(t, f_\tau) = \sum_{\nu=1}^{N_P} \alpha_\nu(t)e^{-j2\pi\tau_\nu(t)f_\tau} \tag{7.30}$$

With the following simple example, we show that this channel transfer function can be strongly frequency selective.

Example 8

Assume a two-path channel model with $N_P = 2$, $\alpha_1(t) = \alpha_2(t) = 1$, and $\tau_1(t) = 0$ for a fixed time instant t. We rename f_τ as f and obtain from (7.30)

$$G_{C,t}(t, f) = 2\cos(\pi f \tau_2(t)) e^{-j2\pi\tau_2(t)f} \tag{7.31}$$

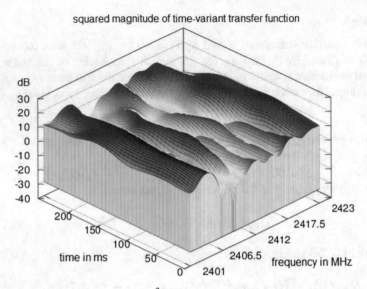

squared magnitude of time-variant transfer function

Fig. 7.4 Squared magnitude $\left|G_{C,t}(t, f_\tau)\right|^2$ of the time-variant transfer function of a typical indoor channel for the WLAN IEEE 802.11 standard in the office environment. Source: Online platform "webdemo" [5]

We recognize deep notches $\left|G_{C,t}(t, f_m)\right| = 0$ for $f = f_m = \frac{1+2m}{2\tau_2(t)}$; $m = 0, \pm 1, \ldots$. This is attributed as a fading effect and denoted as frequency selective fading or multipath fading. Hence, the term fading is not exclusively used for temporal fading.

Example 9: Time and Frequency Selective Fading

We consider a channel incorporating time and frequency selective fading and characterize this channel by its time-variant transfer function $G_{C,t}(t, f_\tau)$. Figure 7.4 shows the squared magnitude $\left|G_{C,t}(t, f_\tau)\right|^2$ of the time-variant transfer function of a typical channel for the wireless local area network standard WLAN IEEE 802.11 in the indoor office environment. The frequency range is at around 2400 MHz. The graph is generated by the open online simulation platform "webdemo" [5]. Apparently, for a fixed time instant, the magnitude response is changing with respect to the frequency and thus is frequency selective. On the other hand, for a given frequency $\left|G_{C,t}(t, f_\tau)\right|^2$ varies with time. Hence, the channel is time and frequency selective often denoted as double selective.

References

1. A.F. Molisch, H. Asplund, R. Heddergott, M. Steinbauer, T. Zwick, The COST259 directional channel model - part i: overview and methodology. IEEE Trans. Wirel. Commun. **5** (2006)
2. Physical layer aspects for evolved universal terrestrial radio access (UTRA), Technical report, 3GPP TR 25.814 v7.1.0 (2006)

3. Physical channels and modulation, Technical Specifications, Technical report, TS 36.211, V11.5.0, 3GPP (2012)
4. A. Mueller, J. Speidel, Performance limits of multiple-input multiple-output keyhole channels with antenna selection. Eur. Trans. Telecommun. **19** (2008)
5. M. Ziegler, 802.11 WLAN spectrum analysis, webdemo, Technical report, Institute of Telecommunications, University of Stuttgart, Germany (2018), http://webdemo.inue.uni-stuttgart.de

Chapter 8
Statistical Description of Wireless Multipath Channel

In the following, we are going to describe the time variance of a multipath channel by statistical parameters. We assume a multipath channel with delay spread function $g_c(t, \tau)$ given in (7.3) and neglect the receiver noise in the following. For the receive signal (7.12), then holds with (7.13) and $u_a(t) = x(t)$

$$q_R(t) = \sum_{\nu=1}^{N_P} x(t) \circledast \left[\tilde{\alpha}_\nu(t) g_I \left(\tau - \tau_\nu(t) \right) \right] \circledast g_R(\tau) \tag{8.1}$$

As can be seen, even for a constant transmit signal $x(t) = const.$, the receiver output signal changes with time, due to the temporal fading of the path loss coefficients $\tilde{\alpha}_\nu(t)$ and the path delays $\tau_\nu(t)$. Because there are many impacts, which are causing this fading, such as the polymorphic environment in the first place, a statistical description is adequate. Nevertheless, the deterministic approach outlined in the previous chapters is helpful to understand the interaction of the various building blocks of the wireless transmission system. For an introduction on stochastic processes, the reader is referred to the Appendix A.

8.1 Complex Gaussian Multipath Model

To derive the complex Gaussian multipath model, the following prerequisites shall apply:

- $\alpha_\nu(t)$ and $\tau_\nu(t)$ change independently at random,
- The N_p paths are independent,
- The number of independent paths grows to infinity, $N_P \to \infty$.

The original version of this chapter was revised: a few typographical errors were corrected. The correction to this chapter can be found at https://doi.org/10.1007/978-3-030-00548-1_25

© Springer Nature Switzerland AG 2019
J. Speidel, *Introduction to Digital Communications*, Signals and Communication Technology, https://doi.org/10.1007/978-3-030-00548-1_8

Then the conditions of the central limit theorem [1] are fulfilled resulting in the following properties:

- $q_R(t)$ in (8.1) becomes a complex-valued Gaussian process,
- $\text{Re}\,[q_R(t)]$ and $\text{Im}\,[q_R(t)]$ are statistically independent, real-valued Gaussian processes each with variance σ^2.

8.2   Channel Model with Rayleigh Fading

The multipath channel model with Rayleigh fading is determined under the prerequisites in Sect. 8.1.

If there is no line of sight or if there are no fixed scatterers or reflectors between the transmit and the receive antenna, then the Gaussian process $q_R(t)$ has zero mean, $\mathbf{E}\,[q_R] = 0$.

Let $z = |q_R(t)|$ and $\phi = \arg\,[q_R(t)]$, then the probability density function $p_z(z)$ is a Rayleigh density given by

$$p_z(z) = \begin{cases} \frac{z}{\sigma^2} e^{-\frac{z^2}{2\sigma^2}} & ;\ z \geq 0 \\ 0 & ;\ z < 0 \end{cases} \tag{8.2}$$

with

$p_z(z) = 0\ \forall\, z < 0$, as z is the absolute value of $q_R(t)$.

$\mathbf{E}\,[z] = \sigma\sqrt{\frac{\pi}{2}}$ the expected value

$\mathbf{E}\,[z^2] = 2\sigma^2$ the mean power

$var\,[z] = \mathbf{E}\,[(z - \mathbf{E}\,[z])^2] = \sigma^2\left(z - \frac{\pi}{2}\right)$ the variance of z

The probability density function $p_\phi(\phi)$ of ϕ is uniform with

$$p_\phi(\phi) = \begin{cases} \frac{1}{2\pi} & ;\ -\pi \leq \phi \leq \pi \\ 0 & ;\quad else \end{cases} \tag{8.3}$$

It should be pointed out that the Rayleigh fading model and also the models described in the following are applicable to each branch of the tapped delay-line model in Fig. 7.3 except for the Doppler shift. This is true, because in reality each branch can be regarded as a composition of an infinite number of independent sub-paths, for which the central limit theorem also applies. For the Doppler shift, a separate statistical model is described later in Sect. 8.5.

8.3 Channel Model with Rician Fading

If there are line of sight or dominant fixed scatterers or reflectors present between the transmit and the receive antenna, the receive signal $q_R(t)$ exhibits a mean value unequal to zero. The probability density function is given by

$$p_z(z) = \begin{cases} \frac{z}{\sigma^2} e^{-\frac{z^2+s^2}{2\sigma^2}} I_0\left(\frac{zs}{\sigma^2}\right) & ; z \geq 0 \\ 0 & ; z < 0 \end{cases} \tag{8.4}$$

where $s \in \mathbb{R}$ is called the non-centrality parameter. The term $I_0(x)$ defines the modified Bessel function of the first kind and zero order. $I_0(x)$ is monotonically increasing with x and not oscillating as the ordinary Bessel functions $J_m(x)$. Please note, $I_0(0) = 1$ and the approximation $I_0(x) \approx \frac{1}{\sqrt{2\pi x}} e^x$ holds for $x \gg 1$, [2] . It is straightforward to show that

$$\mathbf{E}[z] = \sigma\sqrt{\frac{\pi}{2}} + s \text{ expected value}$$
$$\mathbf{E}[z^2] = 2\sigma^2 + s^2 \text{ mean power}$$

We also realize that the Rician density in (8.4) boils down to the Rayleigh density for $s = 0$.

8.4 Channel Model with Nakagami-m Fading

An interesting and versatile statistical model for $z = |q_R(t)|$ is given by the Nakagami-m probability density function [3]

$$p_z(z) = \begin{cases} \frac{2m^m}{\Gamma(m)\Omega^m} z^{2m-1} e^{-\frac{z^2 m}{\Omega}} & ; z \geq 0 \\ 0 & ; z < 0 \end{cases} \tag{8.5}$$

where $\Gamma(m)$ is the gamma function [2] and $\Omega = \mathbf{E}[z^2]$. m is called fading parameter. The Nakagami-m fading model is a generalization of the Rayleigh fading. The parameter m can change the impact of the fading, e.g., for $m = 1$ the probability density function (8.5) turns into the density of the Rayleigh fading given by (8.2), because $\Gamma(1) = 1$ and $\Omega = 2\sigma^2$. For very large m, $m \to \infty$, it can be shown that the Nakagami-m fading model resembles a time-invariant channel [4]. As outlined in [5, 6] for many applications, the Nakagami-m fading also provides a good approximation of the Rician probability density function.

8.5   Channel Model of Clarke and Jakes

Clarke considers in his approach [7] Gaussian distributed I- and Q-signals at the output of the multipath channel. The result is a Rayleigh distributed magnitude of the envelope. It is one of the first implementations of a Rayleigh fading channel models.

With the model of Jakes [8], the receiver is located in the middle of a circle and receives N_P signals from reflectors, which are uniformly distributed on the circle with an angular spacing of $\frac{2\pi}{N_P}$. Hence, the angles of arrival, equivalent to $\pi - \varphi_\nu$ in Fig. 7.2, are equally distributed. Each receive signal consists of the I- and Q-component of the complex envelope, and all signals superimpose at the receiver. The I- and Q-component signals are equipped with constant amplitudes and their phases are assumed to be equally distributed in the interval from 0 to 2π. There is a variety of enhancements of this channel model. Some are using additional filters for a higher decorrelation of the signal components [9].

In the following, we focus again on the scenario in Fig. 7.2 with a receiver moving at the speed $\mathbf{v}_0$. For small noise, the receiver output signal $q_R(t)$ in (7.24) is

$$q_R(t) = \sum_{\nu=1}^{N_P} q_{R,\nu}(t) \tag{8.6}$$

All delays τ_ν are assumed to be constant. From (7.25) follows with $x(t) \circledast g_I (\tau - \tau_\nu)$ $= x(t) * g_I (t - \tau_\nu) = u_I(t - \tau_\nu)$ the receiver output signal for path ν

$$q_{R,\nu}(t) = \left[\tilde{\alpha}_\nu(t) e^{-j2\pi f_{D,\nu} t} u_I(t - \tau_\nu) \right] \circledast g_R(\tau) \; ; \;\; \nu = 1, 2, ..., N_P \tag{8.7}$$

In the following, $g_R(\tau)$ shall represent the delay spread function of an ideal lowpass receive filter with a cut-off frequency large enough to pass its input signal without any distortions to the output. Therefore, we will drop $g_R(\tau)$ and get from (8.7) the receive signal for path ν

$$q_{R,\nu}(t) = \alpha_\nu(t) e^{j\Theta_\nu} e^{-j2\pi f_{D,\nu} t} u_I (t - \tau_\nu) \tag{8.8}$$

where we have substituted $\tilde{\alpha}_\nu(t) = \alpha_\nu(t) e^{j\Theta_\nu}$ with $\Theta_\nu = -2\pi f_0 \tau_\nu$. The Doppler shift $f_{D,\nu}$ is defined by (7.19). The following prerequisites shall apply:

1. All signals are described as wide-sense stationary (WSS) stochastic processes. As outlined in Appendix A, a wide-sense stationary process owns the properties that the (joint) probability density functions and the mean value are independent of the time t. Furthermore, the auto- and cross-correlation functions R depend on a time difference $\zeta = t_2 - t_1$ rather than on the distinct time instants t_1 and t_2, thus $R(t_1, t_2) = R(\zeta)$. The statistical description of the signals holds for any time instant t, which is skipped in some cases to simplify notation.

2. As with the Jakes model, a multitude of signals shall be present at the receiver in Fig. 7.2 with angles of arrival $\pi - \varphi_\nu$. All φ_ν, $\nu = 1, 2, \ldots, N_P$, are uniformly distributed with the probability density function

$$p_\varphi(\varphi) = \begin{cases} \frac{1}{2\pi} \; ; \; |\varphi| \leq \pi \\ 0 \; ; \quad else \end{cases} \tag{8.9}$$

where φ stands for φ_ν.

3. We assume uncorrelated scattering (US), which means according to Appendix A that the multipath receive signals $q_{R,\nu}(t)$ and $q_{R,\mu}(t)$ are uncorrelated $\forall \nu \neq \mu$. Furthermore, each Θ_ν shall be uniformly distributed with the density function (8.9), where φ stands for Θ_ν.

4. α_ν, $e^{j\Theta_\nu}$, $e^{-j2\pi f_{D,\nu}t}$, and $u_I(t - \tau_\nu)$ are uncorrelated $\forall \nu$. Approximately, all path loss coefficients shall be constant.

5. $e^{j\Theta_\nu}$ and $e^{j\Theta_\mu}$ are uncorrelated $\forall \nu \neq \mu$.

6. The transmit signal at the output of the pulse shaper shall be $u_I(t)$ with autocorrelation function

$$R_{u_I u_I}(\zeta) = \mathbf{E}\left[u_I^*(t)u_I(t + \zeta)\right] \tag{8.10}$$

With these prerequisites, we can determine the autocorrelation function of $q_R(t)$ and obtain the final result

$$R_{q_R q_R}(\zeta) = \mathbf{E}\left[q_R^*(t)q_R(t + \zeta)\right] = P_\alpha R_{u_I u_I}(\zeta) J_0\left(2\pi f_{D,max}\zeta\right) \tag{8.11}$$

where

$$P_\alpha = \sum_{\nu=1}^{N_P} \mathbf{E}\left[\alpha_\nu^2\right] \tag{8.12}$$

is the total mean power of the path loss coefficients. $J_0(x)$ denotes the Bessel function of the first kind and zero order defined as

$$J_0(x) = \frac{1}{2\pi} \int_{-\pi}^{\pi} e^{jx \cos(\beta)} d\beta \tag{8.13}$$

with $J_0(0) = 1$ and $J_0(x) \approx \sqrt{\frac{2}{\pi x}} \cos\left(x - \frac{\pi}{4}\right)$; $x \gg \frac{1}{4}$, [2]. The proof of (8.11) is given at the end of this section.

$R_{q_R q_R}(\zeta)$ in (8.11) is determined by the total mean power P_α of the path losses, the autocorrelation function $R_{u_I u_I}(\zeta)$ of the transmit signal at the output of the pulse shaper, and the Bessel function $J_0\left(2\pi f_{D,max}\zeta\right)$.

As outlined in Appendix A, the Fourier transform of the autocorrelation function provides the power spectral density. Thus, for the power spectral density, $S_{u_I u_I}(f_t)$ of $u_I(t)$ holds

$$R_{u_I u_I}(\zeta) \overset{\zeta}{\rightarrowtail} S_{u_I u_I}(f_t) \tag{8.14}$$

and with the convolution operation in the frequency domain follows from (8.11) the power spectral density $S_{q_R q_R}(f_t)$ of $q_R(t)$

$$R_{q_R q_R}(\zeta) \overset{\zeta}{\rightarrowtail} S_{q_R q_R}(f_t) = P_\alpha S_{u_I u_I}(f_t) * \frac{\text{rect}\left(\frac{f_t}{f_{D,max}}\right)}{\sqrt{1 - \left(\frac{f_t}{f_{D,max}}\right)^2}} \tag{8.15}$$

where we have used the Fourier transform pair

$$J_0\left(2\pi a t\right) \overset{t}{\rightarrowtail} \frac{\text{rect}\left(\frac{f_t}{a}\right)}{\sqrt{1 - \left(\frac{f_t}{a}\right)^2}} \; ; \; a \neq 0 \tag{8.16}$$

The rectangular function is defined as

$$\text{rect}\left(\frac{x}{a}\right) = \begin{cases} 1 \; ; \; |x| \leq \frac{a}{2} \\ 0 \; ; \quad else \end{cases} \tag{8.17}$$

The second term of the convolution operation in (8.15) exhibits poles at $f_t = \pm f_{D,max}$. However, the power spectral density $S_{u_I u_I}(f_t)$ of realistic transmit signals $u_I(t)$ owns a lowpass characteristic and thus the result $S_{q_R q_R}(f_t)$ of the convolution shows a smooth transition at the band edges and no poles.

To get a rough idea about the correlation property of the receiver output signal $q_R(t)$, we can discuss $R_{q_R q_R}(\zeta)$ for a simple input signal, which is constant, $u_I(t) = 1$. Its autocorrelation function is $R_{u_I u_I}(\zeta) = 1$. From (8.11) follows

$$R_{q_R q_R}(\zeta) = P_\alpha J_0\left(2\pi f_{D,max}\zeta\right) \tag{8.18}$$

As the envelope of $J_0(x)$ declines approximately with $\sqrt{\frac{2}{\pi x}}$ for increasing x, we conclude that the sample $q_R(t)$ at time instants t and the sample $q_R(t + \zeta)$ at $t + \zeta$ are more and more uncorrelated, the larger the ζ is. Hence, the previous assumption of uncorrelated scattering for the samples of $q_R(t)$ is valid approximately.

Example 10

We consider the special transmit signal $u_I(t) = 1$ with constant amplitude and look for the power spectral densities $S_{u_I u_I}(f_t)$ and $S_{q_R q_R}(f_t)$.

We obtain the autocorrelation function of $u_I(t)$ from (8.10) as $R_{u_I u_I}(\zeta) = \mathbf{E}\left[u_I^*(t) u_I(t+\zeta)\right] = 1$, from which the power spectral density $R_{u_I u_I}(\zeta) \overset{\zeta}{\longrightarrow} S_{u_I u_I}(f_t) = \delta(f_t)$ follows. Then we get from (8.15)

$$S_{q_R q_R}(f_t) = P_\alpha \frac{\operatorname{rect}\left(\frac{f_t}{f_{D,max}}\right)}{\sqrt{1 - \left(\frac{f_t}{f_{D,max}}\right)^2}} \qquad (8.19)$$

Apparently, $S_{q_R q_R}(f_t) \approx P_\alpha$ for $f_t \ll f_{D,max}$ and $S_{q_R q_R}(f_t) \to \infty$ for $f_t \to \pm f_{D,max}$. Measured power spectral densities verify the shape of $S_{q_R q_R}(f_t)$ approximately, however, with maxima rather than poles at $\pm f_{D,max}$, [10]. Anyhow, in reality, transmit signals $u_I(t)$ are not constant and their power spectral density $S_{u_I u_I}(f_t)$ shows a lowpass characteristic which makes $S_{q_R q_R}(f)$ more smooth at the edges of the frequency band.

Proof of (8.11)

For $R_{q_R q_R}(\zeta)$ follows with (8.6)

$$R_{q_R q_R}(\zeta) = \mathbf{E}\left[\sum_{\nu=1}^{N_P} \sum_{\mu=1}^{N_P} q_{R,\nu}^*(t) q_{R,\mu}(t+\zeta)\right] \qquad (8.20)$$

As outlined in Appendix A, the cross-correlation function of two wide-sense stationary and uncorrelated stochastic processes $X(t)$ and $Y(t)$ is $R_{XY}(\zeta) = \mathbf{E}[X^*(t)Y(t+\zeta)] = \mathbf{E}[X^*(t)]\mathbf{E}[Y(t)]$. As the expectation operator is linear, we obtain with (8.8)

$$R_{q_R q_R}(\zeta) = \sum_{\nu=1}^{N_P} \sum_{\mu=1}^{N_P} \left\{\mathbf{E}\left[\alpha_\nu \alpha_\mu\right] \mathbf{E}\left[e^{-j\Theta_\nu(t)} e^{j\Theta_\mu(t+\zeta)}\right]\right\} \cdot$$
$$\left\{\mathbf{E}\left[e^{j2\pi f_{D,\nu} t} e^{-j2\pi f_{D,\mu}(t+\zeta)}\right] \mathbf{E}\left[u_I^*\left(t-\tau_\nu\right) u_I\left(t-\tau_\mu+\zeta\right)\right]\right\} \qquad (8.21)$$

in which we have accounted for the precondition 4 with the consequence that the expected value of the product is equal to the product of the expected values. Next, we make use of the prerequisite 5 and obtain

$$\mathbf{E}\left[e^{-j\Theta_\nu(t)} e^{j\Theta_\mu(t+\zeta)}\right] = \begin{cases} \mathbf{E}\left[e^{-j\Theta_\nu(t)}\right] \mathbf{E}\left[e^{j\Theta_\mu(t)}\right] & ; \nu \neq \mu \\ 1 & ; \nu = \mu \end{cases} \qquad (8.22)$$

Please note, as defined in (8.8), $\Theta_\nu(t)$ is independent of t. However, we maintain the argument. Before we determine $\mathbf{E}\left[e^{-j\Theta_\nu(t)}\right]$ let us review some theorems from Appendix A.

Given a random variable X with probability density function $p_x(x)$. Then
$\mathbf{E}\left[g(X)\right] = \int_{-\infty}^{\infty} g(x) p_x(x) dx$ holds, in which $g(...)$ is a given function.

Using this property and prerequisite 3 for Θ_ν, one can find

$$\mathbf{E}\left[e^{-j\Theta_\nu(t)}\right] = \int_{-\infty}^{\infty} e^{-j\Theta_\nu} p_\varphi(\Theta_\nu) d\Theta_\nu = \frac{1}{2\pi} \int_{-\pi}^{\pi} e^{-j\Theta_\nu} d\Theta_\nu = 0 \qquad (8.23)$$

yielding from (8.22)

$$\mathbf{E}\left[e^{-j\Theta_\nu(t)} e^{j\Theta_\mu(t+\zeta)}\right] = \begin{cases} 0 \; ; \; \nu \neq \mu \\ 1 \; ; \; \nu = \mu \end{cases} \qquad (8.24)$$

Then follows from (8.21) with (8.10)

$$R_{q_R q_R}(\zeta) = \sum_{\nu=1}^{N_P} \mathbf{E}\left[\alpha_\nu(t)^2\right] \mathbf{E}\left[e^{-j2\pi f_{D,\nu}\zeta}\right] R_{u_I u_I}(\zeta) \qquad (8.25)$$

With the Doppler frequency in (7.19) and the density function (8.9) for φ_ν, we obtain
$\mathbf{E}\left[e^{-j2\pi f_{D,\nu}\zeta}\right] = \mathbf{E}\left[e^{-j2\pi\zeta f_{D,max}\cos(\varphi_\nu)}\right] = \frac{1}{2\pi}\int_{-\pi}^{\pi} e^{-j2\pi\zeta f_{D,max}\cos(\varphi_\nu)} d\varphi_\nu$. With the
Bessel function $J_0(x) = \frac{1}{2\pi}\int_{-\pi}^{\pi} e^{jx\cos(\beta)} d\beta = \frac{1}{2\pi}\int_{-\pi}^{\pi} \cos\left(x\cos(\beta)\right) d\beta = J_0(-x)$
follows $\mathbf{E}\left[e^{-j2\pi f_{D,\nu}\zeta}\right] = J_0(2\pi f_{D,max}\zeta)$, which is true for all ν. Together with (8.12),
we get finally from (8.25) $R_{q_R q_R}(\zeta) = P_\alpha R_{u_I u_I}(\zeta) J_0(2\pi f_{D,max}\zeta)$ and the proof of
(8.11) ends.

References

1. A. Papoulis, S.U. Pillai, *Probability, Random Variables, and Stochastic Processes*, 4th edn. (McGraw Hill, Boston, 2002)
2. A. Abramowitz, I.A. Stegun, *Handbook of Mathematical Functions* (Dover Publications Inc, New York, 1974)
3. M. Nakagami, The m-distribution - a general formula for intensity distribution of rapid fading, in *Statistical Methods in Radio Wave Propagation*, ed. by W.G. Hoffman (Pergamon, New York, 1960)
4. M. Alouini, J. Goldsmith, A unified approach for calculating the error rates of linearly modulated signals over generalized fading channels. IEEE Trans. Commun. **47**, 1324–1334 (1999)
5. K.M. Simon, M.-S. Alouini, *Digital Communication Over Fading Channels: A Unified Approach to Performance* (Wiley, New York, 2000)
6. G.L. Stueber, *Principles of Mobile Communication* (Kluwer, Dordrecht, 2001)
7. R.H. Clarke, A statistical theory of mobile radio reception. Bell Syst. Tech. J. **47**, 957–1000 (1968)
8. W. Jakes, *Microwave Mobile Communications* (Wiley, New York, 1975)
9. P. Dent, G.E. Bottomley, T. Croft, Jakes fading model revisited. Electron. Lett. **29**, 1162–1163 (1993)
10. A.F. Molisch, *Wireless Communications* (Wiley and IEEE press, New York, 2009)

Part II
Theory of Linear Time-Variant Systems

Chapter 9
Introduction and Some History

Time-variant systems are of general interest, because they play an important role in communications due to the emerging wireless networks for in-house and outdoor applications. As a matter of fact, a wireless channel can change its parameters with time depending on the position of the mobile transmitter, the receiver, and on the change of the surroundings. During the education of electrical engineers, the main focus is on time-invariant systems and the topic of time-variant systems is not always strongly alluded. Thus, also from this perspective a general view of time-variant systems and their mathematical description is favorable.

First contributions to the subject have been made by Carson [1], later by the seminal papers of Zadeh [2], Bello [3], Kailath [4, 5], and Gersho [6]. Zadeh considers electrical circuits which he calls "variable networks" if their circuit elements vary with time. He points out that the fundamental characteristic of "fixed networks", which have constant elements, is the fact that their impulse response $w(t - s)$ is dependent solely upon the "age variable" $t - s$ that is the difference between the instant t of the observation of the response at the output and the instant s of a Dirac impulse at the input. Furthermore, he asserts out that no such property is possessed by variable networks and he concludes that in those the response must be of the general form $w(t, s)$ characterizing a two-dimensional function of the independent variables t and s. For the first time, he defines a time-variant transfer function $H(f, t) = \int_{-\infty}^{\infty} w(t, s) e^{-j2\pi f(t-s)} ds$, where t acts as a parameter and he interprets $H(f, t)$ as the natural extension of the transfer function $H(f) = \int_{-\infty}^{\infty} h(t - s) e^{-j2\pi f(t-s)} ds$, which is the Fourier transform of $h(t)$ of a fixed network. To find the relation between the input $x(t)$ and the output $y(t)$ of variable networks, Zadeh [2] takes the model of a linear electrical circuitry with time-variant circuit elements and describes these networks by an ordinary linear differential equation of higher order with time-variant coefficients. He reports that the general solution of this equation has the form

$$y(t) = \int_{-\infty}^{\infty} w(t, s) x(s) ds \qquad (9.1)$$

© Springer Nature Switzerland AG 2019
J. Speidel, *Introduction to Digital Communications*, Signals and Communication Technology, https://doi.org/10.1007/978-3-030-00548-1_9

where $x(t)$ and $y(t)$ are the input and the output signals, respectively. Furthermore, he shows that $w(t, s)$ is the response at observation instant t to an input Dirac impulse at instant s and calls it impulse response of the time-variant system. The integral in (9.1) is denoted as general superposition integral, generalized convolution integral, or time-variant convolution.

With the Fourier transform $X(f)$ of $x(t)$, he concludes that

$$y(t) = \int_{-\infty}^{\infty} H(f, t) X(f) e^{j2\pi f t} df \qquad (9.2)$$

holds and that many operations can be performed with $H(f, t)$ just like $H(f)$. While the focus of Zadeh [2] is on the description of variable networks in the frequency domain, Bello [3] starts his considerations with the time domain and he builds upon the work of [2, 4], and [5]. Bello added three more input–output relations to (9.1), where two pairs of them turn out to represent Fourier transform pairs and he showed that all are equivalent for the description of linear time-variant systems, [7]. In this chapter, we follow only the first one which is

$$y(t) = \int_{-\infty}^{\infty} K_1(t, s) x(s) ds \qquad (9.3)$$

where $K_1(t, s)$ is identical with $w(t, s)$ in (9.1). Bello denotes $K_1(t, s)$ as kernel system function or time-variant impulse response and as already stated by Zadeh he interprets $K_1(t, s)$ as the response of the system at observation time instant t to an input Dirac impulse $\delta(t - s)$ applied at time instant s. However, he also points out that $K_1(t, s)$ has some drawbacks for system modeling in electrical engineering and therefore takes over the transformation of variables proposed already by Kailath [4, 5]

$$s = t - \tau \qquad (9.4)$$

yielding the "modified impulse response"

$$g(t, \tau) = K_1(t, t - \tau) \qquad (9.5)$$

which Bello [3] calls *(input) delay spread function*.[1] Then (9.3) results in

$$y(t) = \int_{-\infty}^{\infty} K_1(t, t - \tau) x(t - \tau) d\tau \qquad (9.6)$$

Reference [3] gives an interpretation using an analog, densely tapped delay line with an infinite number of taps and the input signal $x(t)$. The differential output signals $K_1(t, t - \tau) x(t - \tau) d\tau$ of the taps are "summed up" by integration yielding the output signal $y(t)$. Later, we will show that the Fourier transform of $g(t, \tau)$ provides

---

[1]Please note that several authors of later literature denote $g(t, \tau)$ bewildered as impulse response.

the time-variant transfer function $H(f, t)$ proposed by Zadeh [2]. Using the previous explanation for $K_1(t, s)$, we can interpret $g(t, \tau)$ as the response of the system at the observation instant t to an input Dirac impulse applied at the instant $t - \tau$, thus the quantity τ earlier, which has also elicited the name *age variable* or *delay time* for τ and *output time* or simply *time* for t. Please note that the time variation of the system parameters is also determined by t.

In the next chapter, we present an alternative derivation for (9.3), which is different from [2, 3]. First, we derive an input–output relation for time-variant systems using a system theoretic approach and next we present the input–output relation of a discrete-time, time-variant system with the help of a discrete-time delay-line filter. Furnished with the latter results, we will construe the discrete-time system as the sampled version of a continuous time one. With these approaches, we can prove (9.3) directly. Then, we discuss several important properties of the time-variant convolution, such as linearity, associativity, commutativity, and apply them for cascading time-variant and time-invariant systems. An important subject will be the Fourier spectrum of the modified impulse response $g(t, \tau)$, which results in the time-variant transfer function and the Doppler spread function. Of great interest is also the Fourier spectrum of the output signal. Two examples will illustrate that the output spectrum of a time-variant system, e.g., a wireless fading channel, can have a larger bandwidth than the input spectrum. This fact is quite in contrast to linear time-invariant systems. Finally, we close with a thorough consideration of important correlation functions for randomly changing time-variant systems and address in particular also stationary random time-variant systems. These considerations allow us to formulate useful parameters for the characterization of time-varying channels, such as coherence time and coherence bandwidth. We are going to describe time-variant systems from the perspective of system theory for communications and do not consider electrical circuits with time-variant elements as in [2] and later in [8].

References

1. J.R. Carson, Theory and calculation of variable systems. Phys. Rev. **17**, (1921)
2. L.A. Zadeh, Frequency analysis of variable networks, in *Proceedings of the IRE* (1950)
3. P.A. Bello, Characterization of randomly time-variant linear channels. IEEE Trans. Commun. Syst. (1963)
4. T. Kailath, Sampling models for linear time-variant filters, Technical report, M.I.T. Research Laboratory of Electronics, Cambridge, Massachusetts, Report 352 (1959)
5. T. Kailath, Channel characterization: time-variant dispersive channels, in *Lectures on Communications System Theory*, ed. by E.J. Baghdady (McGraw Hill, New York, 1961)
6. A.J. Gersho, Characterization of time-varying linear systems, in *Proceedings of the IEEE(Correspondence)* (1963)
7. P.A. Bello, Time-frequency duality. IEEE Trans. Inf. Theory. (1964)
8. Y. Shmaliy, *Continuous-Time Signals* (Springer, Berlin, 2006)

Chapter 10
System Theoretic Approach for the Impulse Response of Linear Time-Variant Systems

10.1 Continuous-Time, Time-Variant Impulse Response

Let $T[\ldots]$ be a linear system operator, which maps the input signal $x(t)$ of a linear dynamic system to an output signal $y(t)$ as

$$y(t) = T[x(t)] \tag{10.1}$$

We assume empty system memories for $t \to -\infty$. First consider the case of a linear time invariant system with impulse response $w(t)$. The Dirac impulse $\delta(t - s)$ yields the output response

$$w(t - s) = T[\delta(t - s)] \tag{10.2}$$

with the property that the shape of w does not change if s is varied because the response $w(t)$ is just shifted by s along the time axis t. No such relation is possessed by a linear time-variant system. The shape of its response w depends also on the time instant s. Thus, as outlined previously, the response of a time-variant system is more general and given by

$$T[\delta(t - s)] = w(t, s) \tag{10.3}$$

as a function of the two independent variables t and s.

To derive the input–output relation of a linear time-variant system we start with the identity using the sifting property of the Dirac impulse

$$x(t) = \int_{-\infty}^{\infty} x(\zeta)\delta(t - \zeta)\, d\zeta \tag{10.4}$$

© Springer Nature Switzerland AG 2019
J. Speidel, *Introduction to Digital Communications*, Signals and Communication
Technology, https://doi.org/10.1007/978-3-030-00548-1_10

In the next step, we apply the linear system operator $\mathcal{T}[...]$ on both sides

$$\mathcal{T}[x(t)] = \mathcal{T}\left[\int_{-\infty}^{\infty} x(\zeta)\delta(t-\zeta)\,d\zeta\right] = \int_{-\infty}^{\infty} \mathcal{T}[x(\zeta)\delta(t-\zeta)]d\zeta = \int_{-\infty}^{\infty} x(\zeta)\mathcal{T}[\delta(t-\zeta)]d\zeta$$

$$(10.5)$$

noting that the system operator acts only with respect to t. With (10.1) and (10.3) follows

$$y(t) = \int_{-\infty}^{\infty} x(\zeta)w(t,\zeta)d\zeta = \int_{-\infty}^{\infty} x(s)w(t,s)ds = x(t) \circledast w(t,s) \qquad (10.6)$$

where we have just replaced ζ by s in the second step. Obviously, (10.6) with (10.3) prove (9.1) in a different way. We denote the operation in (10.6) as the time-variant convolution and indicate it by $\circledast$ to differentiate from the well known time-invariant convolution $*$.

In summary, a linear time-variant system is determined by an impulse response $w(t,s)$, which is a function of the independent variables t and s. Furthermore, $w(t,s)$ is the response observed at time instant t to an input Dirac impulse $\delta(t-s)$, which is active at time instant s. To fully describe a time-variant system an infinite number of responses $w(t,s)$ as functions of t with the parameter $s \in \mathbb{R}$ must be considered compared to a time-invariant system, where only one impulse response $w(t,0) = h(t)$ suffices.

Example 1

Show that $w(t,s)$ is the response to $x(t) = \delta(t-s)$.

Solution:

Using (10.6) and the sifting property of the Dirac impulse yields

$\int_{-\infty}^{\infty} \delta(\zeta-s)\,w(t,\zeta)d\zeta = w(t,s)$.

10.2   Modified Time-Variant Impulse Response—the Delay Spread Function

The time-variant impulse response $w(t,s)$ has some drawbacks in signal processing.

- The causality condition, which is important for a realizable system, requires two variables, t and s. The system is causal, if $w(t,s) = 0$ for $t < s$, because the effect cannot occur before the cause.
- The Fourier transform of $w(t,s)$ does not directly provide meaningful frequency responses, such as the time-variant transfer function or the Doppler spread function of the system, which are introduced in Chap. 12.

- Furthermore, a minor point is as follows. If the system is time-invariant, the input–output relation (10.6) does not turn elegantly into the conventional convolution integral by just dropping one variable of $w(t, s)$, as will be the case for the delay spread function discusses next.

To overcome these drawbacks the transformation of variables (9.4) is proposed. With $s = t - \tau$ follows from (10.6)

$$y(t) = \int_{-\infty}^{\infty} x(t - \tau)w(t, t - \tau)d\tau \tag{10.7}$$

We introduce the *modified time-variant impulse response* also called *delay spread function* $g(t, \tau)$ as

$$w(t, s) = w(t, t - \tau) = g(t, \tau) \tag{10.8}$$

and with $\tau = t - s$ follows on the other hand

$$w(t, s) = g(t, t - s) \tag{10.9}$$

With (10.8) we get from (10.7)

$$y(t) = \int_{-\infty}^{\infty} x(t - \tau)g(t, \tau)d\tau = \int_{-\infty}^{\infty} x(\zeta)g(t, t - \zeta)d\zeta = x(t) \circledast g(t, \tau) \tag{10.10}$$

where the substitution $\zeta = t - \tau$ was used for the second integral. Equations (10.7) and (10.10) are equivalent, because they provide the same output signal $y(t)$.

Important Remark

Please note that the time-variant convolutions $x(t) \circledast w(t, s)$ in (10.6) and $x(t) \circledast g(t, \tau)$ in (10.10) are executed by different integral operators. Nevertheless, we use the same symbol $\circledast$ just to simplify the notation. What operator has to be used will become clear from the context.

Finally, we see from (10.10) that the third drawback listed before can be overcome with $g(t, \tau)$. If the system is time-invariant, we can just skip the first argument t in $g(t, \tau)$. Then the delay spread function turns into the impulse response $g(\tau)$. We can rename τ by t and write $g(t)$. Hence from (10.10) follows the well known input–output relation of the time-invariant convolution

$$y(t) = \int_{-\infty}^{\infty} x(t - \tau)g(\tau)d\tau = \int_{-\infty}^{\infty} x(\zeta)g(t - \zeta)d\zeta = x(t) * g(t) \tag{10.11}$$

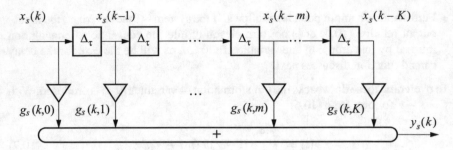

Fig. 10.1 Discrete-time, time-variant finite impulse response filter

10.3   Discrete-Time, Time-Variant System

10.3.1   Discrete-Time Delay Spread Function

We consider the discrete-time system in Fig. 10.1, which is a discrete-time tapped delay line filter also called a finite impulse response (FIR) filter. The discrete-time input and output signals are denoted as $x_s(k)$ and $y_s(k)$, respectively. The index s stands for sampled. The $K + 1$ taps are the filter coefficients indicated by $g_s(k, m)$ with $k, m \in \mathbb{Z}$ and $g_s(k, m) = 0$ for $m > K$ and $m < 0$. Obviously, all parts in the filter are linear namely the delay elements with the delay Δ_τ, the multiplication of the delayed inputs $x_s(k - m)$ with the filter coefficients yielding $g_s(k, m)x_s(k - m)$, and the addition of all component signals. Consequently, we get the output signal

$$y_s(k) = \sum_{m=0}^{K} g_s(k, m)x_s(k - m) = x_s(k) \circledast g_s(k, m) \qquad (10.12)$$

The reader is encouraged to convince oneself of the linearity by using the super-position of two input signals. We denote the relation (10.12) as the discrete-time, time-variant convolution between $g_s(k, m)$ and $x_s(k)$ and allocate the the symbol $\circledast$ to prevent from confusion with the time-invariant convolution operation $*$. For sim-plicity of the notation the same symbol $\circledast$ for the discrete-time and the continuous-time convolution is used. We clearly see that the filter is time-variant because at the time instant k the set of coefficients is $g_s(k, 0), g_s(k, 1), ..., g_s(k, K)$ and at the next instant $k + 1$ the coefficients change to $g_s(k + 1, 0), g_s(k + 1, 1), ..., g_s(k + 1, K)$.

As is well known from signal processing, a discrete-time system can be described in general by a recursive filter with an infinitely long impulse response. With $K \to \infty$ and if we also allow non-causality, $m \to -\infty$, follows from (10.12)

$$y_s(k) = x_s(k) \circledast g_s(k, m) = \sum_{m=-\infty}^{\infty} g_s(k, m)x_s(k - m) \qquad (10.13)$$

By substituting $k - m = \zeta$ and finally replacing ζ by m follows

$$y_s(k) = x_s(k) \circledast g_s(k, m) = \sum_{m=-\infty}^{\infty} g_s(k, k - m)x_s(m) \qquad (10.14)$$

Now we define the unit impulse

$$\delta_k = \begin{cases} 1; & k = 0 \\ 0; & k = \pm 1, \pm 2, \dots \end{cases} \qquad (10.15)$$

and apply at the input of the filter a unit impulse active at time instant $k_0 - m$, i.e., $x_s(k) = \delta_{k-(k_0-m)}$. The observed output $y_s(k)$ at time instant $k \geq k_0 - m$ follows from (10.13)

$$y_s(k) = \sum_{\nu=-\infty}^{\infty} g_s(k, \nu)x_s(k - \nu) = \sum_{\nu=-\infty}^{\infty} g_s(k, \nu)\delta_{k-(k_0-m)-\nu} = g_s(k, k - k_0 + m)$$
$$(10.16)$$

For $k_0 \to k$, the unit impulse at the input is active at time instant $k - m$ and the response observed at instant k follows from (10.16)

$$y_s(k) = g_s(k, m) \qquad (10.17)$$

Hence, $g_s(k, m)$ can be considered as the response of the discrete-time, time-variant system at observation time k to a unit impulse applied at instant $k - m$, i.e. m time intervals earlier. According to our previous nomenclature for continuous-time systems, we can call $g_s(k, m)$ as *discrete-time modified impulse response* or *discrete-time delay spread function*. A causal system cannot react before the instant of the unit impulse, i.e., at $k < k - m$, which is equivalent to $m < 0$. Thus

$$g_s(k, m) = 0 \, \forall m < 0 \qquad (10.18)$$

must hold for a causal time-variant system.

We see that the discrete-time, time-variant system is fully defined by its delay spread function $g_s(k, m)$. The two variables $k \in \mathbb{Z}$ and $m \in \mathbb{Z}$ specify two different time coordinates. k characterizes the time variable of the input and output signal, in addition also the temporal change of the system parameters $g_s(k, m)$. The variable m is called delay time or just delay because the observation instant is m time intervals later than the active unit impulse. Sometimes m is also referred to as integration time because the summation in (10.13) and (10.14) is taken over m.

To grasp the mechanism of a time-variant system more closely assume that we would like to measure the coefficients $g_s(k, m)$ of the FIR filter in Fig. 10.1. In principle they can be found with a series of measurements using (10.17). To get $g_s(k, m)$ as a function of m, we can measure the output at a fixed instant k and change the instant $k - m$, for which $\delta_{k-m} = 1$, by varying $m = 0, 1, 2, \ldots, K$. The result are output samples $y_s(k) = g_s(k, 0)$ up to $g_s(k, K)$. In the next round the same procedure has to be executed for $k + 1$ with the measurement of $y_s(k + 1) = g_s(k + 1, 0)$ up to $g_s(k + 1, K)$ and so on.

This is quite in contrast to a time-invariant filter with coefficients $g_s(k, m) = g'_s(m)$, which are independent of k. As a consequence (10.12) turns into

$$y_s(k) = \sum_{m=0}^{K} g'_s(m) x_s(k - m) = x_s(k) * g'_s(k) \tag{10.19}$$

which is the well known convolution operation for linear, discrete-time, time-invariant systems. An example is a wire-line channel or a static wireless channel, where the transmitter and the receiver are nonmoving and all scatterings and reflections of the electromagnetic waves do not change with time k. To find the time-invariant coefficients $g'_s(k)$ only one unit impulse $x_s(k) = \delta_k$ suffices, yielding from (10.19) the output sequence

$$y_s(k) = g'_s(k) \tag{10.20}$$

which is called the impulse response.

If we apply $x_s(k) = \delta_k$ to the time-variant system, we readily deduce from (10.13) that $y_s(k) = g_s(k, k)$, which does not completely determine the system because $g_s(k, k)$ is just the delay spread function evaluated along the axis $m = k$. Therefore, $g_s(k, m)$ should not be called impulse response.

10.3.2   Transition to Continuous-Time Delay Spread Function

In the following, we extend our considerations to continuous-time, time-variant systems. For that purpose, we interpret $g_s(k, m)$ as the result of a two-dimensional sampling of the continuous function $g(t, \tau)$ at $t = k\Delta_t$ and $\tau = m\Delta_\tau$ with $m, k \in \mathbb{Z}$. Δ_t and Δ_τ are the sampling intervals for the time t and delay time τ, respectively. Hence

$$g(k\Delta_t, m\Delta_\tau) = g_s(k, m) \tag{10.21}$$

and $g(t, \tau)$ can be ideally reconstructed from $g_s(k, m)$, if the sampling theorem is fulfilled. The continuous-time input signal of the continuous-time system shall be $x(t)$ and the relation with the discrete-time version is

$$x(k\Delta_t) = x_s(k) \tag{10.22}$$

and similarly for the output signal $y(t)$

$$y(k\Delta_t) = y_s(k) \tag{10.23}$$

Roughly and without going into all mathematical details for $\Delta_t \to 0$ and $\Delta_\tau \to 0$ the discrete-time variables k and m are approaching the continuous-time variables t and τ, respectively. Moreover, the summations in (10.13) and (10.14) turn into integrals. Thus, the discrete-time, time-variant convolution becomes the continuous-time, time-variant convolution integral

$$y(t) = x(t) \circledast g(t, \tau) = \int_{-\infty}^{\infty} g(t, \tau)x(t - \tau)d\tau \tag{10.24}$$

where the same symbol $\circledast$ as in (10.13) is used to simplify notation. By changing the integration variable $\tau = t - \zeta$ we obtain alternatively

$$y(t) = x(t) \circledast g(t, \tau) = \int_{-\infty}^{\infty} g(t, t - \zeta)x(\zeta)d\zeta \tag{10.25}$$

From these equations it is quickly determined that $g(t, \tau)$ is the response at observation time instant t to an input Dirac impulse at time instant $t - \tau$.

For the proof, we apply a Dirac impulse at $t_0 - \tau$, i.e., the input signal is $x(t) = \delta(t - (t_0 - \tau))$, and obtain from (10.24)

$$y(t) = \int_{-\infty}^{\infty} g(t, \eta)\delta(t - (t_0 - \tau) - \eta)\,d\eta = g(t, t - t_0 + \tau) \tag{10.26}$$

For $t_0 \to t$, the Dirac impulse is active at $t - \tau$, which yields the output signal

$$y(t) = g(t, \tau) \tag{10.27}$$

and the statement is proven.

Referring to the substitution $s = t - \tau$ of variables in (9.4) $g(t, \tau)$ is found by cutting the surface of the two-dimensional impulse response $w(t, s)$ with the plane $\tau = t - s$. Hence, $g(t, \tau)$ is a two-dimensional function of t and the new axis $\tau = t - s$.

We still owe some explanation how to find $w(t, s)$ practically. To this end, let us stimulate the system with a Dirac impulse at $t = s$, consider the response $w(t, s)$ as a function of t and continue this procedure for a large variety of parameters s, $-\infty < s < \infty$. Here again, we recognize the main difference to an impulse response of a time-invariant system, for which an input Dirac impulse at a single s, in particular $s = 0$, suffices. Please note that according to (10.8) a Dirac impulse with $s = 0$ at the input of a time-variant system would just result in an output response $y(t) = w(t, 0) = g(t, t)$ and the determination of the system would not be complete.

Chapter 11
Properties of Time-Variant Convolution

11.1 Relation Between Time-Variant and Time-Invariant Convolution

It is interesting to check whether the time-variant convolution encompasses the time-invariant convolution. By definition, the impulse response $h(t)$ of a time-invariant system is the response to an input Dirac impulse $\delta(t)$. If the system is excited by the time-shifted Dirac impulse $\delta(t-s)$, the response $h(t-s)$ also incorporates a shift without changing its original shape. This holds for any s. Therefore, $w(t,s)$ turns into a one-dimensional function of t

$$w(t,s) = w(t-s) = h(t-s) \tag{11.1}$$

where only one parameter value $s = 0$ suffices resulting in the impulse response $h(t)$. Consequently, the time-variant convolution (10.6) migrates to the time-invariant convolution as

$$
\begin{aligned}
y(t) &= \qquad x(t) \circledast w(t,s) = \int_{-\infty}^{\infty} x(s)w(t,s)ds \\
&= \int_{-\infty}^{\infty} x(s)w(t-s)ds = \int_{-\infty}^{\infty} x(s)h(t-s)ds = x(t) * h(t)
\end{aligned} \tag{11.2}
$$

and we recognize that we can just replace $w(t,s)$ by $w(t-s)$ in (10.6), if the system is time-invariant. Thus, the name "general convolution" mentioned earlier for (9.1) and (10.6) is justified.

The relation (11.2) can also be formulated with the delay spread function $g(t,\tau) = h(\tau)$ using the substitution $t - s = \tau$

$$x(t) \circledast g(t,\tau) = x(t) \circledast h(\tau) = x(t) * h(t) \tag{11.3}$$

© Springer Nature Switzerland AG 2019
J. Speidel, *Introduction to Digital Communications*, Signals and Communication
Technology, https://doi.org/10.1007/978-3-030-00548-1_11

Proof of (11.3)

The time-invariant system shall be described in general by the delay spread function $g(t, \tau) = h(t - s) = h(\tau)$. The time-variant convolution (10.10) yields $x(t) \circledast g(t, \tau) = \int_{-\infty}^{\infty} g(t, \tau)x(t - \tau)d\tau$. As is well known, the time-invariant convolution is defined as $x(t) * h(t) = \int_{-\infty}^{\infty} h(\tau)x(t - \tau)d\tau$. Both integrals are identical, if $g(t, \tau) = h(\tau)$. As any one-dimensional function $h(x)$ can be defined with any variable, $x = \tau$ or $x = t$, the proof of (11.3) is complete.

11.2 Properties

Linearity

The time-variant convolution (10.6) is *linear*. Let

$$x(t) = a_1 x_1(t) + a_2 x_2(t) \tag{11.4}$$

and

$$y_i(t) = x_i(t) \circledast w(t, s) \; ; \; i = 1, 2 \tag{11.5}$$

then

$$y(t) = y_1(t) + y_2(t) \tag{11.6}$$

The proof is straightforward because the convolution integral represents a linear operation.

Convolution of Two Time-Variant Impulse Responses

As depicted in Fig. 11.1a, we can model the convolution of two time-variant impulse responses $w_1(t, s)$ and $w_2(t, s)$ by a cascade of two time-variant systems. The overall impulse response shall be $w_{12}(t, s)$. In the upper line, we have indicated the input signal $x_0(t)$ as well as the output signals $x_1(t)$ and $x_2(t)$. Using (10.6) we get

$$x_i(t) = x_{i-1}(t) \circledast w_i(t, s) \; ; \; i = 1, 2 \tag{11.7}$$

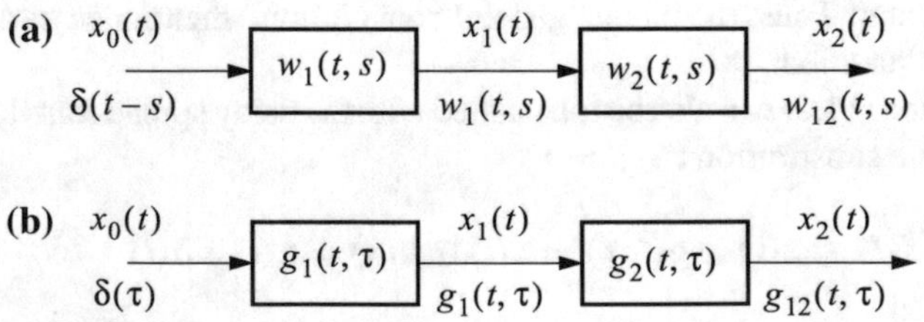

Fig. 11.1 Cascade of two time-variant systems **a** with impulse responses $w_1(t, s)$ and $w_2(t, s)$ **b** with delay spread functions $g_1(t, \tau)$ and $g_2(t, \tau)$

In the lower line of Fig. 11.1a, we indicate the input $x_0(t) = \delta(t - s)$ resulting in the impulse responses $w_1(t, s)$ and $w_{12}(t, s)$. We proof that the system possesses the overall impulse response

$$w_{12}(t, s) = w_1(t, s) \circledast w_2(t, s) = \int_{-\infty}^{\infty} w_1(\zeta, s) w_2(t, \zeta) d\zeta \qquad (11.8)$$

which is the time-variant convolution of the individual impulse responses. It is readily appreciated that the convolution integral (10.6) can be directly applied by interpreting $w_1(t, s)$ as a function of t with a parameter s, thus taking over the role of $x(\zeta)$ in (10.6).

Proof of (11.8)

Let $w_{12}(s, t)$ denote the overall impulse response. Then the response to the input signal $x_0(t)$ is according to (10.6)

$$x_2(t) = \int_{-\infty}^{\infty} x_0(s) w_{12}(t, s) ds \qquad (11.9)$$

Similarly, we obtain $x_1(t) = \int_{-\infty}^{\infty} x_0(s) w_1(t, s) ds$ and $x_2(t) = \int_{-\infty}^{\infty} x_1(\zeta) w_2 (t, \zeta) d\zeta$. Inserting $x_1(t)$ yields $x_2(t) = \int_{-\infty}^{\infty} \int_{-\infty}^{\infty} x_0(s) w_1(\zeta, s) w_2(t, \zeta) ds d\zeta$. Assuming that the integrals converge absolutely, we can interchange the order of integration and get $x_2(t) = \int_{-\infty}^{\infty} x_0(s) \left[\int_{-\infty}^{\infty} w_1(\zeta, s) w_2(t, \zeta) d\zeta \right] ds$.
From the comparison with (11.9), we conclude that the inner integral is $w_{12}(t, s)$ yielding the result $w_{12}(t, s) = \int_{-\infty}^{\infty} w_1(\zeta, s) w_2(t, \zeta) d\zeta$ and the proof ends.

Convolution of Two Delay Spread Functions

In Fig. 11.1b the two time-variant systems are characterized by their delay spread functions

$$g_i(t, \tau) = w_i(t, t - \tau); \ i = 1, 2 \ ; \ g_{12}(t, \tau) = w_{12}(t, t - \tau) \qquad (11.10)$$

We proof that the overall delay spread function is

$$g_{12}(t, \tau) = g_1(t, \tau) \circledast g_2(t, \tau) = \int_{-\infty}^{\infty} g_1(t - \eta, \tau - \eta) g_2(t, \eta) d\eta \qquad (11.11)$$

Please note, to simplify notation we use the same symbol $\circledast$ for the convolution of the impulse responses in (11.8) and the delay spread functions in (11.11) although the integral notations are different.

Proof of (11.11)

We start from (11.8) and obtain with $s = t - \tau$ and (11.10) for the left-hand side $w_{12}(t, t - \tau) = g_{12}(t, \tau) = g_1(t, \tau) \circledast g_2(t, \tau)$. For the integral in (11.8) the substitution $\zeta = t - \eta$, $d\zeta = -d\eta$ yields $g_{12}(t, \tau) = \int_{-\infty}^{\infty} w_1(t - \eta, t - \tau) w_2(t, t - \eta) d\eta$. The term $w_1(t - \eta, t - \tau)$ has to be changed to the form $w_1(x, x - u) = g_1(x, u)$ as follows, $w_1(t - \eta, t - \tau) = w_1(t - \eta, t - \eta - \tau + \eta) = w_1(t - \eta, t - \eta - (\tau - \eta))$ resulting in

$$w_1(t - \eta, t - \tau) = g_1(t - \eta, \tau - \eta) \tag{11.12}$$

Then follows with $w_2(t, t - \eta) = g_2(t, \eta)$ the result
$g_{12}(t, \tau) = \int_{-\infty}^{\infty} g_1(t - \eta, \tau - \eta) g_2(t, \eta) d\eta$ and the proof is finished.

Commutativity

- The time-variant convolution of two time-variant impulse responses $w_1(t, s)$ and $w_2(t, s)$ is in general *noncommutative*.

$$w_1(t, s) \circledast w_2(t, s) \neq w_2(t, s) \circledast w_1(t, s) \tag{11.13}$$

- The time-variant convolution of two delay spread functions $g_1(t, \tau)$ and $g_2(t, \tau)$ is in general *noncommutative*.

$$g_1(t, \tau) \circledast g_2(t, \tau) \neq g_2(t, \tau) \circledast g_1(t, \tau) \tag{11.14}$$

- From an engineering point of view, we conclude that a change of the sequential arrangement of the two time-variant systems in Fig. 11.1a, b will result in different overall responses in general.
- The time-variant convolution of two impulse responses $w_1(t, s) = h_1(t - s)$ and $w_2(t, s) = h_2(t - s)$ of *time-invariant* systems is *commutative*.

$$h_1(t - s) \circledast h_2(t - s) = h_2(t - s) \circledast h_1(t - s) \tag{11.15}$$

As can be seen, the time-variant convolution boils down to the time-invariant convolution, where $s = 0$ suffices.

$$h_1(t) * h_2(t) = h_2(t) * h_1(t) \tag{11.16}$$

With $s = t - \tau$ we can also conclude from (11.15)

$$h_1(\tau) \circledast h_2(\tau) = h_2(\tau) \circledast h_1(\tau) \tag{11.17}$$

- Please note that $h(\tau)$ characterizes a delay spread function of a time-invariant system and not a signal such as $x(t)$. Furthermore, in case of a time-invariant system the impulse response $h(t)$ can be treated just as a signal and the convolution operation is commutative, $x(t) * h(t) = h(t) * x(t)$. For time-variant systems a differentiation between a signal and an impulse response or delay spread function is required for executing the convolution.

Proof of (11.13)

For the left-hand side we obtain $w_{12}(t, s) = \int_{-\infty}^{\infty} w_1(\zeta, s) w_2(t, \zeta) d\zeta$ according to (11.8). For the right-hand side follows $w_{21}(t, s) = \int_{-\infty}^{\infty} w_2(\eta, s) w_1(t, \eta) d\eta$. We see that in general both integrals are different. For example, in $w_{12}(t, s)$ the integration is done over the first variable of $w_1(t, s)$ whereas in $w_{21}(t, s)$ the integration is executed with respect to the second variable of $w_1(t, s)$.

Proof of (11.14)

The proof is similar as before. For the left-hand side of (11.14) we obtain $g_{12}(t, \tau) = \int_{-\infty}^{\infty} g_1(t - \eta, \tau - \eta) g_2(t, \eta) d\eta$ according to (11.11). For the right-hand side follows $g_{21}(t, \tau) = \int_{-\infty}^{\infty} g_2(t - \eta, \tau - \eta) g_1(t, \eta) d\zeta$. We see that in general both integrals are different.

Proof of (11.15) **and** (11.16)

From (11.8) follows for the left-hand side $h_{12}(t - s) = h_1(t - s) \circledast h_2(t - s) = \int_{-\infty}^{\infty} h_1(\zeta - s) h_2(t - \zeta) d\zeta = \int_{-\infty}^{\infty} h_1(\eta) h_2(t - s - \eta) d\eta$, where we have used the substitution $\zeta - s = \eta$. The right-hand side of (11.15) is given by $h_{21}(t - s) = h_2(t - s) \circledast h_1(t - s) = \int_{-\infty}^{\infty} h_2(\zeta - s) h_1(t - \zeta) d\zeta = \int_{-\infty}^{\infty} h_1(\eta) h_2(t - s - \eta) d\eta$, where the substitution $t - \zeta = \eta$ was used. Obviously, $h_{12}(t - s) = h_{21}(t - s)$ is true. We also see that we get an overall impulse response which is just delayed by s, which is not surprising, because at the input of the system the delayed Dirac impulse $\delta(t - s)$ is active. For $s = 0$ we directly obtain the time-invariant convolution (11.16). This finalizes the proof.

Associativity of Time-Variant Convolution

- The time-variant convolution of a signal $x(t)$ with a time-variant impulse response is *associative*,

$$[x(t) \circledast w_1(t, s)] \circledast w_2(t, s) = x(t) \circledast [w_1(t, s) \circledast w_2(t, s)] \qquad (11.18)$$

- The time-variant convolution of a signal $x(t)$ with a delay spread functions is *associative*,

$$[x(t) \circledast g_1(t, \tau)] \circledast g_2(t, \tau) = x(t) \circledast [g_1(t, \tau) \circledast g_2(t, \tau)] \qquad (11.19)$$

- The time-variant convolution of time-variant impulse responses is *associative*,

$$[w_0(t, s) \circledast w_1(t, s)] \circledast w_2(t, s) = w_0(t, s) \circledast [w_1(t, s) \circledast w_2(t, s)] \qquad (11.20)$$

- The time-variant convolution of delay spread functions is *associative*,

$$[g_0(t, \tau) \circledast g_1(t, \tau)] \circledast g_2(t, \tau) = g_0(t, \tau) \circledast [g_1(t, \tau) \circledast g_2(t, \tau)] \qquad (11.21)$$

- The associativity also holds, if one or more systems are time-invariant characterized by $w_i(t, s) = w_i(t - s)$ and $g_i(t, \tau) = g_i(\tau)$, $i = 1, 2$, respectively.

Proof of (11.18) **and** (11.19)

For the left-hand side of (11.18), we get in a first step

$x(t) \circledast w_1(t, s) = \int_{-\infty}^{\infty} x(\zeta) w_1(t, \zeta) d\zeta = y_1(t)$ and in a second step

$u_1(t) = y_1(t) \circledast w_2(t, s) = \int_{-\infty}^{\infty} y_1(\eta) w_2(t, \eta) d\eta = \iint_{-\infty}^{\infty} x(\zeta) w_1(\eta, \zeta) w_2(t, \eta)$
$d\eta d\zeta$. Similarly for the right-hand side we obtain

$w_1(t, s) \circledast w_2(t, s) = \int_{-\infty}^{\infty} w_1(\eta, s) w_2(t, \eta) d\eta = w_{12}(t, s)$ and

$u_2(t) = x(t) \circledast w_{12}(t, s) = \int_{-\infty}^{\infty} x(\zeta) w_{12}(t, \zeta) d\zeta = \iint_{-\infty}^{\infty} x(\zeta) w_1(\eta, \zeta) w_2(t, \eta)$
$d\eta d\zeta$. A comparison shows that $u_1(t) = u_2(t)$ and the proof of (11.18) is finished.

With $s = t - \tau$ we obtain $w_i(t, s) = w_i(t, t - \tau) = g_i(t, \tau)$; $i = 1, 2$. Then from (11.18) directly follows (11.19), which finalizes the proof of (11.19).

Proof of (11.20) **and** (11.21)

For the left-hand side of (11.20), we get

$w_0(t, s) \circledast w_1(t, s) = \int_{-\infty}^{\infty} w_0(\zeta, s) w_1(t, \zeta) d\zeta = y_1(t, s)$ and $u_1(t, s) = y_1(t, s) \circledast$
$w_2(t, s) = \int_{-\infty}^{\infty} y_1(\eta, s) w_2(t, \eta) d\eta = \iint_{-\infty}^{\infty} w_0(\zeta, s) w_1(\eta, \zeta) w_2(t, \eta) d\eta d\zeta$. Similarly the right-hand side yields with $w_{12}(t, s)$ from proof (11.18) $u_2(t, s) = w_0(t, s)$
$\circledast w_{12}(t, s) = \int_{-\infty}^{\infty} w_0(\zeta, s) w_{12}(t, \zeta) d\zeta = \iint_{-\infty}^{\infty} w_0(\zeta, s) w_1(\eta, \zeta) w_2(t, \eta) d\eta d\zeta$.
Obviously, $u_1(t, s) = u_2(t, s)$ holds and we conclude that the left and the right-hand side of (11.20) are identical, which finalizes the proof.
The proof of (11.21) is similar to (11.19).

Distributivity of Time-Variant Convolution

The time-variant convolution is *distributive*,

$$w_0(t, s) \circledast [w_1(t, s) + w_2(t, s)] = [w_0(t, s) \circledast w_1(t, s)] + [w_0(t, s) \circledast w_2(t, s)] \tag{11.22}$$

$$g_0(t, \tau) \circledast [g_1(t, \tau) + g_2(t, \tau)] = [g_0(t, \tau) \circledast g_1(t, \tau)] + [g_0(t, \tau) \circledast g_2(t, \tau)] \tag{11.23}$$

The proof is straightforward because $\circledast$ is a linear operator.

11.3   Summary

Table 11.1 summarizes important convolution operations and properties between impulse responses of time-variant and/or time-invariant systems. With the substitution $s = t - \tau$ of (9.4) the corresponding relations for the delay spread functions of the system are summarized in Table 11.2.

The convolution integrals in Table 11.2 for $g_1(\tau) \circledast g_2(t, \tau)$ and $g_1(t, \tau) \circledast g_2(\tau)$ are proven together with their Fourier transforms in Sect. 12.5. We recognize that we can just drop t in the delay spread function, if the system is not time-variant. Please note that the same symbol $\circledast$ is used for the time-variant convolution of impulse responses w_i and of delay spread functions g_i, although the integral notations differ.

Table 11.1 Summary of time-variant convolution and properties. Impulse response $w_i(t, s)$ of time-variant system, impulse response $w_i(t - s)$ of time-invariant system, signals $x(t)$ and $y(t)$, $i = 0, 1, 2$

$$y(t) = x(t) \circledast w_i(t, s) \quad = \quad \int_{-\infty}^{\infty} x(s) w_i(t, s) ds$$
$$y(t) = x(t) \circledast w_i(t - s) \quad = \quad \int_{-\infty}^{\infty} x(s) w_i(t - s) ds$$

$$w_1(t, s) \circledast w_2(t, s) \quad = \quad \int_{-\infty}^{\infty} w_1(\zeta, s) w_2(t, \zeta) d\zeta$$
$$w_1(t - s) \circledast w_2(t, s) \quad = \int_{-\infty}^{\infty} w_1(\zeta - s) w_2(t, \zeta) d\zeta = \int_{-\infty}^{\infty} w_1(\eta) w_2(t, \eta + s) d\eta$$
$$w_1(t, s) \circledast w_2(t - s) \quad = \int_{-\infty}^{\infty} w_1(\zeta, s) w_2(t - \zeta) d\zeta = \int_{-\infty}^{\infty} w_1(t - \eta, s) w_2(\eta) d\eta$$
$$w_1(t - s) \circledast w_2(t - s) \quad = \quad \int_{-\infty}^{\infty} w_1(\zeta - s) w_2(t - \zeta) d\zeta$$

$$w_1(t, s) \circledast w_2(t, s) \quad \neq \quad w_2(t, s) \circledast w_1(t, s)$$
$$w_1(t - s) \circledast w_2(t - s) \quad = \quad w_2(t - s) \circledast w_1(t - s)$$

$$[x(t) \circledast w_1(t, s)] \circledast w_2(t, s) \quad = \quad x(t) \circledast [w_1(t, s) \circledast w_2(t, s)]$$
$$[w_0(t, s) \circledast w_1(t, s)] \circledast w_2(t, s) = \quad w_0(t, s) \circledast [w_1(t, s) \circledast w_2(t, s)]$$

$$w_0(t, s) \circledast [w_1(t, s) + w_2(t, s)] = \quad [w_0(t, s) \circledast w_1(t, s)] + [w_0(t, s) \circledast w_2(t, s)]$$

Table 11.2 Summary of time-variant convolution and properties. Delay spread functions $g_i(t, \tau) = w_i(t, t - \tau)$ of time-variant system, delay spread function $g_i(\tau) = w_i(t - s)$ of time-invariant system, signals $x(t)$ and $y(t)$, $i = 0, 1, 2$

$$x(t) \circledast g_i(t, \tau) \quad = \quad \int_{-\infty}^{\infty} x(t - \eta) g_i(t, \eta) d\eta = \int_{-\infty}^{\infty} x(\zeta) g_i(t, t - \zeta) d\zeta$$
$$x(t) \circledast g_i(\tau) \quad = \quad \int_{-\infty}^{\infty} x(t - \tau) g_i(\tau) d\tau$$

$$g_1(t, \tau) \circledast g_2(t, \tau) \quad = \quad \int_{-\infty}^{\infty} g_1(t - \zeta, \tau - \zeta) g_2(t, \zeta) d\zeta$$
$$g_1(\tau) \circledast g_2(t, \tau) \quad = \int_{-\infty}^{\infty} g_1(\tau - \zeta) g_2(t, \zeta) d\zeta = \int_{-\infty}^{\infty} g_1(\eta) g_2(t, \tau - \eta) d\eta$$
$$g_1(t, \tau) \circledast g_2(\tau) \quad = \quad \int_{-\infty}^{\infty} g_1(t - \zeta, \tau - \zeta) g_2(\zeta) d\zeta$$
$$g_1(\tau) \circledast g_2(\tau) \quad = \quad \int_{-\infty}^{\infty} g_1(\tau - \zeta) g_2(\zeta) d\zeta$$

$$g_1(t, \tau) \circledast g_2(t, \tau) \quad \neq \quad g_2(t, \tau) \circledast g_1(t, \tau)$$
$$g_1(\tau) \circledast g_2(\tau) \quad = \quad g_2(\tau) \circledast g_1(\tau)$$

$$[x(t) \circledast g_1(t, \tau)] \circledast g_2(t, \tau) \quad = \quad x(t) \circledast [g_1(t, \tau) \circledast g_2(t, \tau)]$$
$$[g_0(t, \tau) \circledast g_1(t, \tau)] \circledast g_2(t, \tau) = \quad g_0(t, \tau) \circledast [g_1(t, \tau) \circledast g_2(t, \tau)]$$

$$g_0(t, \tau) \circledast [g_1(t, \tau) + g_2(t, \tau)] = \quad [g_0(t, \tau) \circledast g_1(t, \tau)] + [g_0(t, \tau) \circledast g_2(t, \tau)]$$

11.4   Examples

Furnished with the theory of time-variant systems we are now going to discuss some applications for signal transmission.

Example 2

A frequent application in communications is the cascade of a time-invariant transmit filter $w_1(t - s)$, a time-variant channel $w_2(t, s)$, and a time-invariant receive filter $w_3(t - s)$. We are interested in the overall impulse response $w_{13}(t, s)$, the corresponding delay spread function $g_{13}(t, \tau)$ and the response $y(t)$ to an input signal $x(t)$.

Solution:

First please note, if time-variant and time-invariant systems are combined, the overall impulse response is the response to the Dirac impulse $\delta(t - s)$. Consequently, all the impulse responses should be defined with two variables t and s including $w_i(t, s) = w_i(t - s)$, $i = 1, 3$ of the time-invariant systems, in which we just interpret the the comma as the minus sign.

Regarding $w_2(t, s)$ in (11.8) as $w_2(t, s) \circledast w_3(t - s)$ and with $w_1(t, s) = w_1(t - s)$ follows the overall impulse response of the cascade

$$w_{13}(t, s) = w_1(t - s) \circledast w_2(t, s) \circledast w_3(t - s) \qquad (11.24)$$

which obviously characterizes an equivalent time-variant system.

We obtain the overall delay spread function from (11.24) by inserting $s = t - \tau$

$$g_{13}(t, \tau) = g_1(\tau) \circledast g_2(t, \tau) \circledast g_3(\tau) \qquad (11.25)$$

Next we are interested in the output signal $y(t)$ for a given input signal $x(t)$. With (10.6) and equivalently with (10.10) we obtain

$$y(t) = x(t) \circledast w_{13}(t, s) = x(t) \circledast g_{13}(t, \tau) \qquad (11.26)$$

Example 3

Figure 11.2 shows a modulator, a time-variant channel, and a demodulator. Determine the time-variant impulse response $w_1(t, s)$ of the modulator, which multiplies

Fig. 11.2 Example of a time-variant channel with impulse response $w_2(t, s)$ embedded between a modulator with $w_1(t, s)$ and a demodulator with $w_3(t, s)$

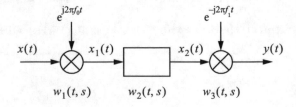

its input signal $x(t)$ with the complex carrier $e^{j2\pi f_0 t}$. Then find the overall impulse response $w_{13}(t, s)$ of the cascade of the modulator, the channel with impulse response $w_2(t, s)$, the demodulator with complex carrier $e^{-j2\pi f_1 t}$, and impulse response $w_3(t, s)$. Discuss the important case that the demodulator is synchronous with the modulator, $f_1 = f_0$, and determine the output signal $y(t)$.

Solution:

The output signal of the modulator is

$$x_1(t) = x(t)e^{j2\pi f_0 t} \tag{11.27}$$

With the input Dirac impulse $x(t) = \delta(t - s)$ the time-variant impulse response can be determined as

$$w_1(t, s) = \delta(t - s)e^{j2\pi f_0 t} \tag{11.28}$$

Similarly, the impulse response of the demodulator follows as

$$w_3(t, s) = \delta(t - s)e^{-j2\pi f_1 t} \tag{11.29}$$

Then we get according to (11.8) the impulse response of the cascade

$$
\begin{aligned}
w_{13}(t, s) &= w_1(t, s) \circledast w_2(t, s) \circledast w_3(t, s) \\
&= \left(\delta(t - s)e^{j2\pi f_0 t}\right) \circledast w_2(t, s) \circledast \left(\delta(t - s)e^{-j2\pi f_1 t}\right)
\end{aligned} \tag{11.30}
$$

Making use of the associativity (11.20) one can first calculate with (11.8) $w_{12}(t, s) = w_1(t, s) \circledast w_2(t, s) = \int_{-\infty}^{\infty} \delta(\zeta - s)e^{j2\pi f_0 \zeta} w_2(t, \zeta)d\zeta$, which represents the cascade of the modulator and the time-variant channel, resulting in

$$w_{12}(t, s) = e^{j2\pi f_0 s} w_2(t, s) \tag{11.31}$$

Next we determine

$$
\begin{aligned}
w_{13}(t, s) &= w_{12}(t, s) \circledast w_3(t, s) \\
&= \int_{-\infty}^{\infty} e^{j2\pi f_0 s} w_2(\eta, s)\delta(t - \eta)e^{-j2\pi f_1 t} d\eta
\end{aligned}
$$

and get the final result

$$w_{13}(t, s) = e^{j2\pi f_0 s} w_2(t, s)e^{-j2\pi f_1 t} \tag{11.32}$$

which is a time-variant impulse response, as expected. We can interpret this result as a modulation of $w_2(t, s)$ with respect to both coordinates s and t.

In the following, we are going to write $w_{13}(t, s)$ in terms of the delay spread function $g_{13}(t, \tau)$. With $s = t - \tau$, $w_{13}(t, t - \tau) = g_{13}(t, \tau)$, and $w_2(t, t - \tau) = g_2(t, \tau)$ follows from (11.32)

$$g_{13}(t, \tau) = g_2(t, \tau)e^{-j2\pi f_0 \tau}e^{-j2\pi(f_1 - f_0)t} \tag{11.33}$$

The output signal of the cascade is obtained with (11.2) as $y(t) = x(t) \circledast w_{13}(t, s)$ and after some manipulation with the convolution integral follows

$$y(t) = \left[x(t) \circledast \left(w_2(t, s)e^{j2\pi f_0 s}\right)\right]e^{-j2\pi f_1 t} = [x_1(t) \circledast w_2(t, s)]e^{-j2\pi f_1 t} \tag{11.34}$$

with $x_1(t) = x(t)e^{j2\pi f_0 t}$. This result is plausible directly from Fig. 11.2 because $x_1(t)$ and $x_2(t) = x_1(t) \circledast w_2(t, s)$ are the output signals of the modulator and the channel, respectively.

Finally, for synchronous demodulation, $f_0 = f_1$, we obtain

$$w_{13}(t, s) = w_2(t, s)e^{-j2\pi f_0(t-s)} \tag{11.35}$$

With $s = t - \tau$ follows the overall delay spread function

$$g_{13}(t, \tau) = g_2(t, \tau)e^{-j2\pi f_0 \tau} \tag{11.36}$$

in which the modulation shows up only with respect to the delay variable τ.

Verification of Noncommutativity: This example is also a good opportunity to demonstrate that the time-variant convolution is noncommutative in general. Assume that we have a cascade of two systems $e^{j2\pi f_3 t}$ and $w_2(t, s)$ in Fig. 11.2. Then we obtain from (11.32) with $f_0 = f_3$ and $f_1 = 0$ the overall impulse response $w_{13}(t, s) = e^{j2\pi f_3 s}w_2(t, s)$. If we interchange the sequential order, the overall impulse response is given by (11.32) with $f_0 = 0$ and $f_1 = -f_3$ yielding $w'_{13}(t, s) = w_2(t, s)e^{j2\pi f_3 t}$, which is clearly different. In a similar way we obtain the delay spread functions from (11.33) as $g_{13}(t, \tau) = g_2(t, \tau)e^{j2\pi f_3(t-\tau)}$ and $g'_{13}(t, \tau) = g_2(t, \tau)e^{j2\pi f_3 t}$.

Chapter 12
System Functions and Fourier Transform

We have already discussed the time-variant impulse response $w(t, s)$ and the delay spread function $g(t, \tau)$, which characterize a linear time-variant system completely and hence are called system functions. Now we will see that the latter provides meaningful Fourier transforms for applications in electrical engineering, and thus system functions in the frequency domain. In the following sections, we will apply the Fourier transform with respect to the variables t and/or τ. Therefore, we define the corresponding variables as

$$t \leftrightarrow f_t; \quad \tau \leftrightarrow f_\tau$$

and use the symbol $\rightarrowtail$ for the transform. On top of the arrow we indicate in what respect the Fourier transform is executed. In wireless communications f_t is called Doppler frequency and f_τ is the "natural" frequency also used for the ordinary frequency response of a time-invariant system or the spectrum of a signal. The system functions as well as cascades of different systems are summarized in Table 11.2 of Sect. 11.3 and discussed in the following in quite some detail.

12.1 Time-Variant Transfer Function

The *time-variant transfer function* $G_t(t, f_\tau)$ is defined as the Fourier transform of the delay spread function $g(t, \tau)$ with respect to τ

$$g(t, \tau) \overset{\tau}{\rightarrowtail} G_t(t, f_\tau) = \int_{-\infty}^{\infty} g(t, \tau) e^{-j2\pi f_\tau \tau} d\tau \qquad (12.1)$$

It represents the transfer function of a time-variant system at a fixed observation instant t, where t has to be regarded as a parameter. So $G_t(t, f_\tau)$ can help to understand the temporal change of the transfer function of a time-variant system. If the

The original version of this chapter was revised: a few typographical errors were corrected. The correction to this chapter can be found at https://doi.org/10.1007/978-3-030-00548-1_25

© Springer Nature Switzerland AG 2019
J. Speidel, *Introduction to Digital Communications*, Signals and Communication Technology, https://doi.org/10.1007/978-3-030-00548-1_12

system parameters are slowly varying with time t, then $G_t(t, f_\tau)$ is approximately
constant with respect to t during a short time interval Δt_{Coh}, which is called coherence time. In this time interval, the transfer function $G_t(t, f_\tau)$ could be measured as
a function of f_τ similar to a time-invariant system.

For the output signal $y(t)$ we proof that in general

$$y(t) = \int_{-\infty}^{\infty} G_t(t, \zeta) X(\zeta) e^{j2\pi\zeta t} d\zeta \qquad (12.2)$$

holds, where

$$x(t) \overset{t}{\rightarrowtail} X(f_t) = \int_{-\infty}^{\infty} x(t) e^{-j2\pi f_t t} dt \qquad (12.3)$$

is the Fourier transform of the input signal $x(t)$.[1] It should be pointed out that (12.2)
does not represent the inverse Fourier transform of $G(t, \zeta) X(\zeta)$, although it looks
similar on the first glance. Consequently, $G_t(t, f_t) X(f_t) \neq Y(f_t)$, where $Y(f_t)$ is
the Fourier transform of $y(t)$.

Proof of (12.2)

From (12.3) follows by inverse Fourier transform

$$x(t) = \int_{-\infty}^{\infty} X(\zeta) e^{j2\pi\zeta t} d\zeta \qquad (12.4)$$

which will be inserted into (10.24) yielding $y(t) = \int_{-\infty}^{\infty} g(t, \tau) \int_{-\infty}^{\infty} X(\zeta) e^{j2\pi\zeta(t-\tau)} d\zeta d\tau$. For this proof and all the following, we assume that the integrals are absolutely
convergent and thus the interchange of the integration order is allowed. Then follows
$y(t) = \int_{-\infty}^{\infty} X(\zeta) \int_{-\infty}^{\infty} g(t, \tau) e^{-j2\pi\zeta\tau} d\tau e^{j2\pi\zeta t} d\zeta$. According to (12.1) the inner integral is $G_t(t, \zeta)$ resulting in $y(t) = \int_{-\infty}^{\infty} G_t(t, \zeta) X(\zeta) e^{j2\pi\zeta t} d\zeta$, which finalizes the
proof.

12.2   Delay Doppler Spread Function

The *delay Doppler spread function* $G_\tau(f_t, \tau)$ also called *Doppler-variant impulse
response* is determined by the Fourier transform of the delay spread function $g(t, \tau)$
with respect to t

$$g(t, \tau) \overset{t}{\rightarrowtail} G_\tau(f_t, \tau) = \int_{-\infty}^{\infty} g(t, \tau) e^{-j2\pi f_t t} dt \qquad (12.5)$$

It represents the frequency response along the f_t-axis for a fixed delay τ as a parameter. In the field of wireless communications $G_\tau(f_t, \tau)$ is also called *spreading function*.

---

[1]Please note, for the one-dimensional Fourier transform of a one-dimensional function it does not
matter whether the pair t, f_t or τ, f_τ is used, because $x(t)$ and $x(\tau)$ are the same functions and
also the spectra $X(f_t)$ and $X(f_\tau)$ are mathematically the same.

12.3 Doppler Spread Function

The *Doppler spread function* $G(f_t, f_\tau)$ also called *Doppler-variant transfer function* is defined as the two-dimensional Fourier transform of the delay spread function

$$g(t, \tau) \overset{t,\tau}{\rightarrowtail} G(f_t, f_\tau) = \iint\limits_{-\infty}^{\infty} g(t, \tau) \mathrm{e}^{-\mathrm{j}2\pi(f_t t + f_\tau \tau)} dt d\tau \qquad (12.6)$$

$G(f_t, f_\tau)$ can also be obtained by the Fourier transformation of the time-variant transfer function $G_t(t, f_\tau)$ with respect to t

$$G_t(t, f_\tau) \overset{t}{\rightarrowtail} G(f_t, f_\tau) = \int_{-\infty}^{\infty} G_t(t, f_\tau) \mathrm{e}^{-\mathrm{j}2\pi f_t t} dt \qquad (12.7)$$

or by the transformation of the delay Doppler spread function $G_\tau(f_t, \tau)$ with respect to τ

$$G_\tau(f_t, \tau) \overset{\tau}{\rightarrowtail} G(f_t, f_\tau) = \int_{-\infty}^{\infty} G_\tau(f_t, \tau) \mathrm{e}^{-\mathrm{j}2\pi f_\tau \tau} d\tau \qquad (12.8)$$

12.4 Spectrum of the Output Signal

Below we prove that the Fourier spectrum

$$Y(f_t) = \int_{-\infty}^{\infty} y(t) \mathrm{e}^{-\mathrm{j}2\pi f_t t} dt \qquad (12.9)$$

of the output signal $y(t)$ of a time-variant system with Doppler spread function $G(f_t, f_\tau)$ is

$$y(t) \overset{t}{\rightarrowtail} Y(f_t) = \int_{-\infty}^{\infty} X(f_\tau) G(f_t - f_\tau, f_\tau) df_\tau = \int_{-\infty}^{\infty} X(\varsigma) G(f_t - \varsigma, \varsigma) d\varsigma \qquad (12.10)$$

By substituting $f_t - f_\tau = \eta$ we get the alternative form

$$Y(f_t) = \int_{-\infty}^{\infty} X(f_t - \eta) G(\eta, f_t - \eta) d\eta \qquad (12.11)$$

If we replace η by f_τ we can write (12.11) as

$$Y(f_t) = \int_{-\infty}^{\infty} X(f_t - f_\tau) G(f_\tau, f_t - f_\tau) df_\tau \qquad (12.12)$$

The interesting Eqs. (12.10)–(12.12) show the relation between the input and the output spectrum of a time-variant system determined by its Doppler spread function. It is well known that $Y(f) = H(f)X(f)$ holds for time-invariant systems with the transfer function $H(f)$. In contrast, the relation (12.10) for time-variant systems contains an integral which shows some similarities to a convolution with respect to one frequency variable. From (12.10) and (12.11), where "neutral" variables ζ and η are used for integration, it is quite clear that the spectrum for any one-dimensional signal $y(t)$ can be given as a function of f_t or f_τ, as it makes no difference mathematically.

Proof of (12.10)

Inserting (12.2) into (12.9) yields $Y(f_t) = \int_{-\infty}^{\infty} \int_{-\infty}^{\infty} G_t(t, \zeta) X(\zeta) e^{j2\pi\zeta t} e^{-j2\pi f_t t} d\zeta dt$. Provided that the integrals converge absolutely, the order of integration can be changed resulting in $Y(f_t) = \int_{-\infty}^{\infty} X(\zeta) \left(\int_{-\infty}^{\infty} G_t(t, \zeta) e^{-j2\pi(f_t - \zeta)t} dt \right) d\zeta$. The inner integral is related to the Doppler spread function given in (12.7) and recognized as $G(f_t - \zeta, \zeta)$. Then we obtain the result $Y(f_t) = \int_{-\infty}^{\infty} X(\zeta) G(f_t - \zeta, \zeta) d\zeta$ and the proof is finished.

12.5   Cascades of Time-Variant and Time-Invariant Systems

12.5.1   Cascade of Time-Invariant $g_1(\tau)$ and Time-Variant System $g_2(t, \tau)$

We consider the cascade in Fig. 11.1b of Sect. 11.2 with the prerequisite that the first system is time-invariant with $g_1(t, \tau) = g_1(\tau)$.

Delay Spread Function

$$g_{12}(t, \tau) = g_1(\tau) \circledast g_2(t, \tau) = \int_{-\infty}^{\infty} g_1(\tau - \zeta) g_2(t, \zeta) d\zeta = \int_{-\infty}^{\infty} g_1(\eta) g_2(t, \tau - \eta) d\eta$$

$$(12.13)$$

Fourier Transform with Respect to τ

$$g_{12}(t, \tau) = g_1(\tau) \circledast g_2(t, \tau) \overset{\tau}{\longmapsto} G_1(f_\tau) G_{2t}(t, f_\tau) \qquad (12.14)$$

where $G_1(f_\tau)$ is the transfer function of the time-invariant system $g_1(\tau)$ and $G_{2t}(t, f_\tau)$ represents the time-variant transfer function of the time-variant system $g_2(t, \tau)$ according to (12.1).

Fourier Transform with Respect to t and τ

$$g_{12}(t,\tau) = g_1(\tau) \circledast g_2(t,\tau) \overset{t,\tau}{\longmapsto} G_1(f_\tau)G_2(f_t,f_\tau) \tag{12.15}$$

where $G_2(f_t, f_\tau)$ is the Doppler spread function of the time-variant system $g_2(t,\tau)$ defined in (12.6) and $G_1(f_\tau)$ is the transfer function of the time-invariant system $g_1(\tau)$.

Proof of (12.13)

In the following, we are going to build upon the basic time-variant impulse responses. We start with (11.8), define $w_1(t,s) = w_1(t-s)$ and obtain $w_1(t-s) \circledast w_2(t,s) = \int_{-\infty}^{\infty} w_1(\eta-s)w_2(t,\eta)d\eta$. With the substitution $\eta = t - \zeta$ follows $w_1(t-s) \circledast w_2(t,s) = \int_{-\infty}^{\infty} w_1(t-\zeta-s)w_2(t,t-\zeta)d\zeta$. The transformation of variables (9.4), $s = t - \tau$, yields $w_1(t-s) = g_1(\tau)$, $w_2(t,s) = g_2(t,\tau)$, and $w_2(t,t-\zeta) = g_2(t,\zeta)$. Then we find $w_1(t-s) \circledast w_2(t,s) = g_1(\tau) \circledast g_2(t,\tau) = \int_{-\infty}^{\infty} g_1(\tau-\zeta) g_2(t,\zeta)d\zeta$. The last term in (12.13) follows directly with the substitution $\tau - \zeta = \eta$ and the proof is finished.

Proof of (12.14)

Let $g_1(\tau) \overset{\tau}{\longmapsto} G_1(f_\tau)$, then $g_1(\tau-\zeta) \overset{\tau}{\longmapsto} G_1(f_\tau)e^{-j2\pi f_\tau \zeta}$. From (12.13) follows $\int_{-\infty}^{\infty} g_1(\tau-\zeta)g_2(t,\zeta)d\zeta \overset{\tau}{\longmapsto} \int_{-\infty}^{\infty} G_1(f_\tau)e^{-j2\pi f_\tau \zeta}g_2(t,\zeta)d\zeta = G_1(f_\tau)\int_{-\infty}^{\infty} g_2(t,\zeta) e^{-j2\pi f_\tau \zeta}d\zeta = G_1(f_\tau)G_{2t}(t,f_\tau)$, where $G_{2t}(t,f_\tau) = \int_{-\infty}^{\infty} g_2(t,\zeta)e^{-j2\pi f_\tau \zeta}d\zeta$ is the Fourier transform of $g_2(t,\tau)$ with respect to τ and represents the time-variant transfer function of the time-variant system according to (12.1). This finalizes the proof of (12.14).

Proof of (12.15)

We take the Fourier spectrum in (12.14) and execute a transform with respect to t yielding with $G_{2t}(t,f_\tau) \overset{t}{\longmapsto} G_2(f_t,f_\tau)$ the result (12.15) and the proof is finished.

12.5.2 Cascade of Time-Variant $g_1(t,\tau)$ and Time-Invariant system $g_2(\tau)$

Again we consider the cascade in Fig. 11.1b of Sect. 11.2, however, with the prerequisite that the second system is time-invariant with $g_2(t,\tau) = g_2(\tau)$.

Delay Spread Function

$$g_{12}(t,\tau) = g_1(t,\tau) \circledast g_2(\tau) = \int_{-\infty}^{\infty} g_1(t-\zeta,\tau-\zeta)g_2(\zeta)d\zeta \tag{12.16}$$

Fourier Transform with Respect to t and τ

$$g_{12}(t, \tau) = g_1(t, \tau) \circledast g_2(\tau) \xrightarrow{t, \tau} G_1(f_t, f_\tau)G_2(f_t + f_\tau) \tag{12.17}$$

where $G_1(f_t, f_\tau)$ represents the Doppler spread function of the time-variant system $g_1(t, \tau)$ defined in (12.6) and $G_2(f_\tau)$ is the transfer function of the time-invariant system $g_2(\tau)$.

Proof of (12.16)

A quick proof is obtained by just skipping the argument t in $g_2(t, \tau)$ and to directly apply the time-variant convolution integral (11.11). However, we will start with (11.8) and obtain $w_1(t, s) \circledast w_2(t - s) = \int_{-\infty}^{\infty} w_1(\eta, s)w_2(t - \eta)d\eta$. With the substitution $\eta = t - \zeta$ follows $w_1(t, s) \circledast w_2(t - s) = \int_{-\infty}^{\infty} w_1(t - \zeta, s)w_2(\zeta)d\zeta$. The transformation of variables (9.4), $s = t - \tau$, yields with (11.12)
$$w_1(t - \zeta, t - \tau) = g_1(t - \zeta, \tau - \zeta)$$
and $w_2(\zeta) = g_2(\zeta)$. Then we find

$$w_1(t, s) \circledast w_2(t - s) = g_1(t, \tau) \circledast g_2(\tau) = \int_{-\infty}^{\infty} g_1(t - \zeta, \tau - \zeta)g_2(\zeta)d\zeta.$$

Proof of (12.17)

Let

$$g_1(t - \zeta, \tau - \zeta) \xrightarrow{t, \tau} G_1(f_t, f_\tau)e^{-j2\pi(f_t + f_\tau)\zeta} \tag{12.18}$$

then follows the transform of (12.16) $g_{12}(t, \tau) \xrightarrow{t, \tau} G_1(f_t, f_\tau) \int_{-\infty}^{\infty} g_2(\zeta)e^{-j2\pi(f_t + f_\tau)\zeta} d\zeta = G_1(f_t, f_\tau)G_2(f_t + f_\tau)$, where the integral represents $G_2(f_t + f_\tau)$ and the proof is finished.

12.5.3 Cascade of Two Time-Variant Systems $g_1(t, \tau)$ and $g_2(t, \tau)$

Finally, both systems in Fig. 11.1b of Sect. 11.2 shall be time-variant. We prove the following:

Fourier Transform with Respect to τ

$$g_{12}(t, \tau) = g_1(t, \tau) \circledast g_2(t, \tau) \xrightarrow{\tau} G_{12t}(t, f_\tau) = \int_{-\infty}^{\infty} G_{1t}(t - \eta, f_\tau)g_2(t, \eta)e^{-j2\pi f_\tau \eta}d\eta \tag{12.19}$$

where $G_{12t}(t, f_\tau)$ and $G_{1t}(t, f_\tau)$ are time-variant transfer functions.

Fourier Transform with Respect to t and τ

$$g_{12}(t, \tau) = g_1(t, \tau) \circledast g_2(t, \tau) \overset{t,\tau}{\mapsto} G_{12}(f_t, f_\tau) = \int_{-\infty}^{\infty} G_1(\zeta, f_\tau) G_2(f_t - \zeta, f_\tau + \zeta) d\zeta$$

(12.20)

Here we can see the difference between the time-variant convolution (11.11) and the two-dimensional convolution between $g_1(t, \tau)$ and $g_2(t, \tau)$. For the latter, the Fourier transform is $G_1(f_t, f_\tau)G_2(f_t, f_\tau)$.

Proof of (12.19) and (12.20)

Table 12.1 Summary of Fourier transforms of delay spread functions $g(t, \tau) = w(t, t - \tau)$ of time-variant system, delay spread function $g(\tau) = w(t - s)$ of time-invariant system, signals $x(t)$ and $y(t)$

$x(t)$	$\overset{t}{\mapsto}$	$X(f_t) = \int_{-\infty}^{\infty} x(t) e^{-j2\pi f_t t} dt$
$g(\tau)$	$\overset{\tau}{\mapsto}$	$G(f_\tau) = \int_{-\infty}^{\infty} g(\tau) e^{-j2\pi f_\tau \tau} d\tau$
$g(t, \tau)$	$\overset{\tau}{\mapsto}$	$G_t(t, f_\tau) = \int_{-\infty}^{\infty} g(t, \tau) e^{-j2\pi f_\tau \tau} d\tau$
$g(t, \tau)$	$\overset{t}{\mapsto}$	$G_\tau(f_t, \tau) = \int_{-\infty}^{\infty} g(t, \tau) e^{-j2\pi f_t t} dt$
$g(t, \tau)$	$\overset{t,\tau}{\mapsto}$	$G(f_t, f_\tau) = \iint_{-\infty}^{\infty} g(t, \tau) e^{-j2\pi(f_t t + f_\tau \tau)} dt d\tau$
$g(t - t_0, \tau - \tau_0)$	$\overset{t,\tau}{\mapsto}$	$G(f_t, f_\tau) e^{-j2\pi(f_t t_0 + f_\tau \tau_0)}$
$g(t, \tau) e^{j2\pi(f_t t_0 + f_\tau \tau_0)}$	$\overset{t,\tau}{\mapsto}$	$G(f_t - f_{t_0}, f_\tau - f_{\tau_0})$
$g_1(\tau) \circledast g_2(t, \tau)$	$\overset{\tau}{\mapsto}$	$G_1(f_\tau) G_{2t}(t, f_\tau)$
$g_1(\tau) \circledast g_2(t, \tau)$	$\overset{t,\tau}{\mapsto}$	$G_1(f_\tau) G_2(f_t, f_\tau)$
$g_1(t, \tau) \circledast g_2(\tau)$	$\overset{t,\tau}{\mapsto}$	$G_1(f_t, f_\tau) G_2(f_t + f_\tau)$
$g_1(t, \tau) \circledast g_2(t, \tau)$	$\overset{\tau}{\mapsto}$	$\int_{-\infty}^{\infty} G_{1t}(t - \eta, f_\tau) g_2(t, \eta) e^{-j2\pi f_\tau \eta} d\eta$
$g_1(t, \tau) \circledast g_2(t, \tau)$	$\overset{t,\tau}{\mapsto}$	$\int_{-\infty}^{\infty} G_1(\zeta, f_\tau) G_2(f_t - \zeta, f_\tau + \zeta) d\zeta$
$y(t) = x(t) \circledast g(t, \tau)$	$\overset{t}{\mapsto}$	$Y(f_t) = \int_{-\infty}^{\infty} X(\zeta) G(f_t - \zeta, \zeta) d\zeta$ $= \int_{-\infty}^{\infty} X(f_t - \eta) G(\eta, f_t - \eta) d\eta$
$y(t) = \int_{-\infty}^{\infty} G_t(t, \zeta) X(\zeta) e^{j2\pi \zeta t} d\zeta$	$\overset{t}{\mapsto}$	$Y(f_t) \neq G_t(t, \zeta) X(\zeta)$

$g_{12}(t,\tau)$ is defined in (11.11) as $g_{12}(t,\tau) = \int_{-\infty}^{\infty} g_1(t-\eta, \tau-\eta)g_2(t,\eta)d\eta$. We will execute the transform with respect to t and τ in two steps. First, we transform with respect to τ and obtain $g_{12}(t,\tau) \overset{\tau}{\longmapsto} G_{12t}(t,f_\tau) = \int_{-\infty}^{\infty} G_{1t}(t-\eta, f_\tau)e^{-j2\pi f_\tau \eta}$ $g_2(t,\eta)d\eta$, where $g_1(t,\tau) \overset{\tau}{\longmapsto} G_{1t}(t,f_\tau)$ and the proof of (12.19) is finished.

Next, the transformation is done with respect to t. As we recognize, the integrand in (12.19) is the product of two functions of t, namely $G_{1t}(t-\eta, f_\tau)$ and $g_2(t,\eta)$. Therefore we use the property that a product of functions in the time domain results in a convolution of their spectra after Fourier transform. Hence, with $G_{1t}(t-\eta, f_\tau) \overset{t}{\longmapsto}$ $G_1(f_t, f_\tau)e^{-j2\pi f_t \eta}$ and $g_2(t,\eta) \overset{t}{\longmapsto} G_{2\tau}(f_t, \eta)$ we obtain $G_{1t}(t-\eta, f_\tau)g_2(t,\eta) \overset{t}{\longmapsto}$ $G_1(f_t, f_\tau)e^{-j2\pi f_t \eta} *_{f_t} G_{2\tau}(f_t, \eta) = \int_{-\infty}^{\infty} G_1(\zeta, f_\tau)e^{-j2\pi\zeta\eta}G_{2\tau}(f_t-\zeta, \eta)d\zeta$. Then follows from (12.19) $g_{12}(t,\tau) \overset{t,\tau}{\longmapsto} G_{12}(f_t, f_\tau) = \int_{-\infty}^{\infty} G_1(\zeta, f_\tau)\int_{-\infty}^{\infty} e^{-j2\pi(\zeta+f_\tau)\eta}$ $G_{2\tau}(f_t-\zeta, \eta)d\eta d\zeta$. The inner integral is $G_2(f_t-\zeta, f_\tau+\zeta)$ and we get the final result $G_{12}(f_t, f_\tau) = \int_{-\infty}^{\infty} G_1(\zeta, f_\tau)G_2(f_t-\zeta, f_\tau+\zeta)d\zeta$, which finalizes the proof of (12.20).

12.6 Summary

Table 12.1 summarizes the Fourier transforms of delay spread functions and for some cascades. Please note that $G_1(f_\tau)$ and $G_2(f_\tau)$ represent transfer functions of time-invariant systems. If we consider these functions isolated, we can use any frequency variable, such as f_t and f_τ that is true for any one-dimensional function. However, in the context of cascading systems the frequency variable makes the difference and indicates in what direction the filtering takes place.

Chapter 13
Applications

In the following, we consider some examples to show the impact of the Doppler spread function $G(f_t, f_\tau)$ of a time-variant system on the output spectrum $Y(f_t)$ for a given input spectrum $X(f_t)$. As already pointed out, $G(f_t, f_\tau)$ is not just limiting the input spectrum in f_t- and f_τ-direction. Due to the integral in (12.10), we expect that the output spectrum $Y(f_t)$ can even have a larger bandwidth compared to the input signal, which does not occur for time-invariant systems. To show the details in principle, we take simple rectangular shaped spectra as examples.

Example 4

In Fig. 13.1a, the top view of the input spectrum $X(f_\tau)$ and the Doppler spread function

$$G(f_t, f_\tau) = \begin{cases} 1; & |f_t| \le f_{t,G}, \ |f_\tau| \le f_{\tau,G} \\ 0; & else \end{cases} \tag{13.1}$$

is given as a function of f_t and f_τ. Inside the shaded area, the functions are equal to one and outside equal to zero. All spectra shall be real-valued. $G(f_t, f_\tau)$ is equipped with the cut-off frequencies $f_{\tau,G}$ and $f_{t,G}$. For a wireless channel $f_{t,G}$ represents the impact of the temporal fading and is normally much smaller than the transmission bandwidth $f_{\tau,G}$ of the channel. The spectrum

$$X(f_\tau) = \begin{cases} 1; & |f_\tau| \le f_{\tau,X} \\ 0; & else \end{cases} \tag{13.2}$$

owns the cut-off frequency $f_{\tau,X}$. Please note that $X(f_\tau)$ is constant with respect to f_t. We also have assumed that $f_{\tau,G} > f_{\tau,X}$ so that the system is not limiting the input spectrum. Figure 13.1b shows $G(0, f_\tau)$ and $X(f_\tau)$ as a function of f_τ. In Fig. 13.1c the term $G(f_t - u, u)$ of the output spectrum $Y(f_t) = \int_{-\infty}^{\infty} X(u)G(f_t - u, u)du$ in (12.10) is illustrated. Similar to a convolution operation $G(f_t - u, u)$ "moves" along

© Springer Nature Switzerland AG 2019

J. Speidel, *Introduction to Digital Communications*, Signals and Communication Technology, https://doi.org/10.1007/978-3-030-00548-1_13

the u-axis for various f_t and covers $X(u)$ to compose the integrand

$$X(u)G(f_t - u, u).$$

Please note the dashed lines indicate that $G(f_t - u, u) = 0$ for $|u| < f_{\tau,G}$. For the indicated f_t the integrand starts to be unequal to zero and $Y(f_t)$ in Fig. 13.1d begins to rise. For $f_t = -(f_{\tau,X} - f_{t,G})$ both spectra completely overlap and consequently $Y(f_t)$ is maximal. Obviously, $Y(f_t)$ is an even function. The spectral parts of $G(f_t, f_\tau)$ in f_t-direction increase the cut-off frequency $f_{t,Y}$ of the output spectrum and make it larger than $f_{\tau,X}$ of the input signal, in our example $f_{t,Y} = f_{\tau,X} + f_{t,G}$. In wireless communications, this effect is called Doppler spread. Thus, for a wireless, fading channel, the larger the bandwidth $f_{t,G}$ of the temporal fading is the more the cut-off frequency of the output spectrum is increased compared to the input. Note, we can use any frequency variable for a one-dimensional spectrum. Hence, the abscissa in Fig. 13.1d may alternatively be named as f_τ.

The reader can assure oneself that the solution is valid also, if the cut-off frequency $f_{\tau,G}$ of the system is equal to $f_{\tau,X}$.

Example 5

We are now considering the example in Fig. 13.2a, where the cut-off frequency $f_{\tau,G}$ of the Doppler spread function $G(f_t, f_\tau)$ of the system limits the input spectrum and thus $f_{\tau,G} < f_{\tau,X}$ holds compared to Fig. 13.1a. The cut-off frequency $f_{t,G}$ of the temporal fading is unchanged. In Fig. 13.2b $X(f_\tau)$ and $G(0, f_\tau)$ are shown. Figure 13.2c depicts the shifted spectrum $G(f_t - u, u)$ as part of the integrand in (12.10), where f_τ is renamed as u. Please note the dashed lines indicate that $G(f_t - u, u) = 0$ for $|u| < f_{\tau,G}$. The resulting output spectrum $Y(f_t)$ is shown in Fig. 13.2d. Its cut-off frequency $f_{t,Y} = f_{\tau,G} + f_{t,G}$ is solely given by the parameters of the Doppler spread function of the system and exceeds the transmission bandwidth $f_{\tau,G}$ by the maximal Doppler frequency $f_{t,G}$. Again we recognize that the time variance of the channel is the cause of the excess bandwidth of the output signal. As mentioned before, we can use any frequency variable for a one-dimensional spectrum Y, hence denote the abscissa in Fig. 13.2d also as f_τ.

Example 6

Now we consider a similar arrangement as in Fig. 13.1, however, the Doppler spread function $G(f_t, f_\tau)$ shall have the cut-off frequency $f_{t,G} = 0$. Hence, the system is time-invariant and in case of a transmission channel it would show no temporal fading. Thus an adequate model is

$$G(f_t, f_\tau) = \delta(f_t)G(0, f_\tau) \tag{13.3}$$

Then we obtain from (12.10)

$$Y(f_t) = \int_{-\infty}^{\infty} X(f_\tau)\delta(f_t - f_\tau)G(0, f_\tau)df_\tau = X(f_t)G(0, f_t) \tag{13.4}$$

Fig. 13.1 **a** Top view of input $X(f_\tau)$, Doppler spread function $G(f_t, f_\tau)$, $f_{\tau,G} \geq f_{\tau,X}$ **b** Details of $G(f_t, f_\tau)$ and $X(f_\tau)$ for $f_t = 0$ **c** $G(f_t - u, u)$ of $Y(f_t) = \int_{-\infty}^{\infty} X(u)G(f_t - u, u)du$ in (12.10) **d** Output spectrum $Y(f_t)$

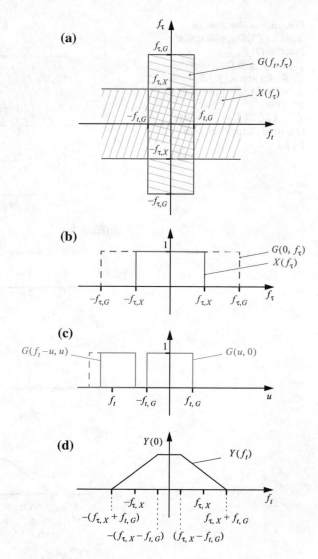

As mentioned earlier for a one-dimensional signal, we can use any variable. To adapt to Fig. 13.1, we rename f_t as f_τ and get

$$Y(f_\tau) = X(f_\tau)G(0, f_\tau) \tag{13.5}$$

which is the well-known frequency response of a time-invariant system. If $f_{\tau,G}$ and $f_{\tau,X}$ are the same as in Fig. 13.1, then $Y(f_\tau) = X(f_\tau)$, because the system is not limiting the input spectrum. Thus, no Doppler spread is present, as expected.

Example 7

Let us now discuss an artificial system, which has zero transmission bandwidth $f_{\tau,G} = 0$. As $G(f_t, f_\tau)$ still has spectral components in f_t direction with cut-off frequency $f_{t,G}$, it is interesting to know the output spectrum of the system. We model the Doppler spread function as

Fig. 13.2 a Top view of input $X(f_t)$, Doppler spread function $G(f_t, f_\tau)$, $f_{\tau,G} < f_{\tau,X}$ **b** Details of $G(f_t, f_\tau)$ and $X(f_\tau)$ for $f_t = 0$ **c** $G(f_t - u, u)$ of $Y(f_t) = \int_{-\infty}^{\infty} X(u)G(f_t - u, u)du$ in (12.10) **d** Output spectrum $Y(f_t)$

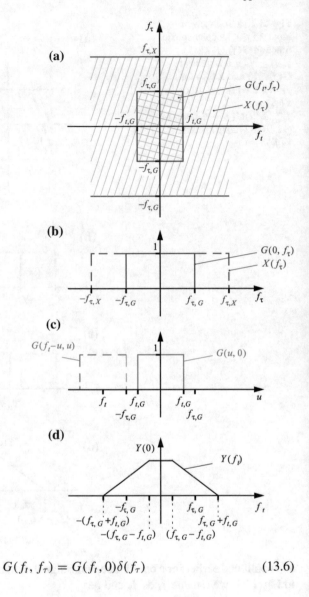

$$G(f_t, f_\tau) = G(f_t, 0)\delta(f_\tau) \tag{13.6}$$

which yields from (12.10)

$$Y(f_t) = \int_{-\infty}^{\infty} X(f_\tau)G(f_t - f_\tau, 0)\delta(f_\tau)df_\tau = X(0)G(f_t, 0) \tag{13.7}$$

Obviously, the output spectrum is nonzero and given by the frequency response of $G(f_t, f_\tau)$ along the f_t-axis multiplied by the mean value $X(0)$ of the input signal. Hence, the output spectrum solely exhibits a Doppler spread. The system is not very practical for signal transmission, as only the spectral component $X(0)$ of the input is transferred to the output.

Chapter 14
Interrelation Between Time-Variant and Two-Dimensional Convolution

14.1 Input–Output Relation

We look back to the input–output relation of a time-variant system in (10.24) and (10.25), where we have recognized that a one-dimensional function $x(t)$ has to be convolved with a two-dimensional function $g(t, \tau)$ using a special form of the convolution operation. One may raise the question, whether we can apply the conventional two-dimensional convolution in some way, well bearing in mind that the input and the out signals finally must be interpreted as one-dimensional signals.

Following this idea, we start to define the input signal as $x(t, \tau)$ and the output signal as $y(t, \tau)$. Then the two-dimensional convolution provides

$$y(t, \tau) = x(t, \tau) * g(t, \tau) = \iint\limits_{-\infty}^{\infty} g(u, v)x(t - u, \tau - v)dudv \qquad (14.1)$$

How can we extend $x(t)$ to a two-dimensional function? A similar question has already been touched upon in the Examples 5 and 6 in Chap. 13. There we had to plot the spectrum $X(f_\tau)$ of $x(t)$ in the two-dimensional frequency space f_t, f_τ together with the Doppler spread function $G(f_t, f_\tau)$. We have interpreted $X(f_\tau)$ as a constant with respect to f_t. Consequently, we can extend $x(t)$ in the following way

$$x(t, \tau) = \delta(t)x(\tau) \qquad (14.2)$$

because then its Fourier transform $X(f_t, f_\tau)$ is constant with respect to f_t. This is easily verified as

$$X(f_t, f_\tau) = \iint\limits_{-\infty}^{\infty} \delta(t)x(\tau)e^{-j2\pi(f_t t + f_\tau \tau)}dtd\tau = \int_{-\infty}^{\infty} x(\tau)e^{-j2\pi f_\tau \tau}d\tau = X(f_\tau) \qquad (14.3)$$

© Springer Nature Switzerland AG 2019
J. Speidel, *Introduction to Digital Communications*, Signals and Communication
Technology, https://doi.org/10.1007/978-3-030-00548-1_14

From (14.1) results with (14.2)

$$y(t, \tau) = \iint\limits_{-\infty}^{\infty} g(u, v)\delta(t - u)x(\tau - v)dudv = \int_{-\infty}^{\infty} g(t, v)x(\tau - v)dv \quad (14.4)$$

which yields for $\tau = t$ the time-variant convolution integral

$$y(t, t) = y(t) = \int_{-\infty}^{\infty} g(t, v)x(t - v)dv \quad (14.5)$$

where we use the short hand notation $y(t)$ for $y(t, t)$.

In summary, we can operate with the two-dimensional input signal (14.2), apply the two-dimensional convolution integral (14.1) and set $\tau = t$ to get the final result (14.5).

14.2 Fourier Spectrum of the Output Signal

Now we are now going to determine the Fourier spectrum $Y(f_t)$ of the output signal $y(t)$ of the time-variant system. To this end we consider (14.1) again. Knowing that the Fourier transform turns the two-dimensional convolution in the original domain into the product of the spectra in the frequency domain we get for the two-dimensional Fourier spectrum of $y(t, \tau)$

$$Y(f_t, f_\tau) = X(f_t, f_\tau)G(f_t, f_\tau) = X(f_\tau)G(f_t, f_\tau) \quad (14.6)$$

To introduce the condition $t = \tau$ used in (14.5), we consider the inverse transform

$$y(t, \tau) = \iint\limits_{-\infty}^{\infty} X(f_\tau)G(f_t, f_\tau)e^{j2\pi(f_t t + f_\tau \tau)}df_t df_\tau \quad (14.7)$$

With $t = \tau$ and the substitution $f_t = \eta - f_\tau$, $df_t = d\eta$ follows $y(t, t) = y(t) = \iint_{-\infty}^{\infty} X(f_\tau)G(\eta - f_\tau, f_\tau)df_\tau e^{j2\pi\eta t} d\eta$. Next, we rename the integration variable η as f_t and get

$$y(t) = \int_{-\infty}^{\infty} \left(\int_{-\infty}^{\infty} X(f_\tau)G(f_t - f_\tau, f_\tau)df_\tau \right) e^{j2\pi f_t t} df_t \quad (14.8)$$

The inner integral must be the Fourier transform $Y(f_t)$ of $y(t)$

$$Y(f_t) = \int_{-\infty}^{\infty} X(f_\tau)G(f_t - f_\tau, f_\tau)df_\tau \quad (14.9)$$

Hence, with (14.9) we have confirmed our earlier result (12.10) in a different way.

Chapter 15
Randomly Changing Time-Variant Systems

15.1 Prerequisites

Hitherto we have considered signals and characteristic functions of time-variant system, in particular the delay spread function $g(t, \tau)$, as deterministic. With the Fourier transform different spectra or transfer functions, such as the Doppler spread function $G(f_t, f_\tau)$, have been defined. In many applications, e.g. wireless communications the time-variant channel can take on a fast of different characteristics depending on the environment, the speed of the transmitter or receiver, and other effects. Hence, there is a need for the introduction of a statistical description for the most important system parameters. The use of multivariate probability density functions could help in principle. However, it would be hard or even prohibitive to determine them exhaustively in many applications. For most practical cases, second-order statistics provide reasonable approximations, at least for the performance comparison of systems and can still be handled, as proposed in [1]. Several functions characterizing the time-variant system, such as the Doppler spread function, can be separated into a deterministic part, identified by its mean value and a pure stochastic part. In the following, we are only interested in latter and, therefore, assume stochastic signals with zero mean. Consequently, we focus on the correlation rather than the covariance functions. For details of stochastic processes, the reader is relegated to the survey given in the Appendix A.

In the following, we start with the definitions of the autocorrelation functions with no restriction whether the processes are stationary or not. In a second step, we focus on stationary processes in the wide sense. Finally, we consider the case that the stochastic processes under consideration are uncorrelated with respect to the delay time τ, which is also called uncorrelated scattering in case of a time-variant wireless channel.

© Springer Nature Switzerland AG 2019

J. Speidel, *Introduction to Digital Communications*, Signals and Communication Technology, https://doi.org/10.1007/978-3-030-00548-1_15

15.2   Correlation Functions of Randomly Changing Time-Variant Systems

We define the autocorrelation functions for various system functions of a randomly changing time-variant system.

Autocorrelation Function of the Delay Spread Function $g(t, \tau)$

$$R_{gg}(t, t'; \tau, \tau') = \mathbf{E}\left[g^*(t, \tau)g(t', \tau')\right] \tag{15.1}$$

If not otherwise noted, the expected value $\mathbf{E}[...]$ is considered with respect to all time and delay instants t, t', τ, and τ', respectively. This holds similarly also for the following.

Autocorrelation Function of the Time-Variant Transfer Function $G_t(t, f_\tau)$

$$R_{G_t G_t}(t, t'; f_\tau, f_\tau') = \mathbf{E}\left[G_t^*(t, f_\tau)G_t(t', f_\tau')\right] \tag{15.2}$$

Autocorrelation Function of the Delay Doppler Spread Function $G_\tau(f_t, \tau)$

$$R_{G_\tau G_\tau}(f_t, f_t'; \tau, \tau') = \mathbf{E}\left[G_\tau^*(f_t, \tau)G_\tau(f_t', \tau')\right] \tag{15.3}$$

Autocorrelation Function of the Doppler Spread Function $G(f_t, f_\tau)$

$$R_{GG}(f_t, f_t'; f_\tau, f_\tau') = \mathbf{E}\left[G^*(f_t, f_\tau)G(f_t', f_\tau')\right] \tag{15.4}$$

With the autocorrelation function $R_{gg}(t, t'; \tau, \tau')$ of $g(t, \tau)$ in (15.1), we show at the end of this section that

$$R_{GG}(f_t, f_t'; f_\tau, f_\tau') = \iiiint\limits_{-\infty}^{\infty} R_{gg}(t, t'; \tau, \tau')e^{-j2\pi(-f_t t + f_t' t' - f_\tau \tau + f_\tau' \tau')}dt\,dt'\,d\tau\,d\tau' \tag{15.5}$$

holds. Hence $R_{GG}(f_t, f_t'; f_\tau, f_\tau')$ can be interpreted as the four-dimensional Fourier spectrum of $R_{gg}(t, t'; \tau, \tau')$, evaluated at the frequency positions $-f_t$, f_t', $-f_\tau$, and f_τ'.

Interrelation Between the Autocorrelation Functions

Because the frequency domain system functions $G_t(t, f_\tau)$, $G_\tau(f_t, \tau)$, and $G(f_t, f_\tau)$ are the result of the one- or two-dimensional Fourier transform of the delay spread function $g(t, \tau)$, it is evident that the autocorrelation functions will also show some interrelations through the Fourier transform. We just give the following two examples.

$$R_{gg}(t, t'; \tau, \tau') \overset{\tau, \tau'}{\multimap} R_{G_t G_t}(t, t'; -f_\tau, f_\tau') \quad \text{with}$$
$$R_{G_t G_t}(t, t'; -f_\tau, f_\tau') = \iint_{-\infty}^{\infty} R_{gg}(t, t'; \tau, \tau')e^{-j2\pi(f_\tau \tau + f_\tau' \tau')}d\tau\,d\tau' \tag{15.6}$$

$$R_{gg}(t, t'; \tau, \tau') \overset{t,t'}{\rightarrowtail} R_{G_\tau G_\tau}(-f_t, f_t'; \tau, \tau') \quad \text{with}$$
$$R_{G_\tau G_\tau}(-f_t, f_t'; \tau, \tau') = \iint_{-\infty}^{\infty} R_{gg}(t, t'; \tau, \tau')e^{-j2\pi(f_t t + f_t' t')}dt dt' \quad (15.7)$$

The proof of (15.6) and (15.7) is given at the end of this section.

Autocorrelation Function of the Output Signal $y(t)$

The autocorrelation function of the output signal $y(t)$ is given by

$$R_{yy}(t, t') = \mathbf{E}\left[y^*(t)y(t')\right] = \iint_{-\infty}^{\infty} R_{gg}(t, t'; \tau, \tau')R_{xx}(t - \tau, t' - \tau')d\tau d\tau' \quad (15.8)$$

with the autocorrelation function of $x(t)$

$$R_{xx}(t, t') = \mathbf{E}\left[x^*(t)x(t')\right] \quad (15.9)$$

We recognize from (15.8) that this relation has some similarities to the Wiener-Lee relation for time-invariant systems outlined in the Appendix A.

Proof of (15.5)

From (15.4) follows with (12.6)
$$R_{GG}(f_t, f_t'; f_\tau, f_\tau') = \mathbf{E}\left[\iint_{-\infty}^{\infty} g^*(t, \tau)e^{j2\pi(f_t t + f_\tau \tau)}dt d\tau \iint_{-\infty}^{\infty} g(t', \tau')e^{-j2\pi(f_t' t' + f_\tau' \tau')}\right.$$
$$\left. dt' d\tau'\right] = \iiiint_{-\infty}^{\infty} \mathbf{E}\left[g^*(t, \tau)g(t', \tau')\right]e^{-j2\pi(f_t' t' + f_\tau' \tau' - f_t t - f_\tau \tau)}dt d\tau dt' d\tau'. \quad \text{Plugging}$$
in $R_{gg}(t, t'; \tau, \tau')$ from (15.1) yields the final result
$$R_{GG}(f_t, f_t'; f_\tau, f_\tau') = \iiiint_{-\infty}^{\infty} R_{gg}(t, t'; \tau, \tau')e^{-j2\pi(-f_t t + f_t' t' - f_\tau \tau + f_\tau' \tau')}dt dt' d\tau d\tau'$$
and the proof is finished.

Proof of (15.6)

With (15.1) we obtain $R_{gg}(t, t'; \tau, \tau') \overset{\tau,\tau'}{\rightarrowtail} \iint_{-\infty}^{\infty} \mathbf{E}\left[g^*(t, \tau)g(t', \tau')\right]e^{-j2\pi(f_\tau \tau + f_\tau' \tau')}$
$d\tau d\tau'$. Due to the linearity, we can exchange the expectation operation and the integration and get $\mathbf{E}\left[\int_{-\infty}^{\infty} g^*(t, \tau)e^{-j2\pi f_\tau \tau}d\tau \int_{-\infty}^{\infty} g(t', \tau')e^{-j2\pi f_\tau' \tau'}d\tau'\right]$. With (12.1), the first integral is identified as $G_t^*(t, -f_\tau)$ and the second one as $G_t(t', f_\tau')$. Hence,
$$R_{gg}(t, t'; \tau, \tau') \overset{\tau,\tau'}{\rightarrowtail} \mathbf{E}\left[G_t^*(t, -f_\tau)G_t(t', f_\tau')\right] = R_{G_t G_t}(t, t'; -f_\tau, f_\tau') \text{ results and the}$$
proof is finished.

Proof of (15.7)

With (15.1) we obtain $R_{gg}(t, t'; \tau, \tau') \overset{t,t'}{\rightarrowtail} \iint_{-\infty}^{\infty} \mathbf{E}\left[g^*(t, \tau)g(t', \tau')\right]e^{-j2\pi(f_t t + f_t' t')}$
$dt dt'$. Due to the linearity, we can exchange the expectation operation and the integration and get $\mathbf{E}\left[\int_{-\infty}^{\infty} g^*(t, \tau)e^{-j2\pi f_t t}dt \int_{-\infty}^{\infty} g(t', \tau')e^{-j2\pi f_t' t'}dt'\right]$. With (12.5) the first integral is identified as $G_\tau^*(-f_t, \tau)$ and the second one as $G_\tau(f_t', \tau')$. Hence,
$$R_{gg}(t, t'; \tau, \tau') \overset{t,t'}{\rightarrowtail} \mathbf{E}\left[G_\tau^*(-f_t, \tau)G_\tau(f_t', \tau')\right] = R_{G_\tau G_\tau}(-f_t, f_t'; \tau, \tau') \text{ results and}$$
the proof is finished.

Proof of (15.8)

With (10.24) follows
$R_{yy}(t, t') = \mathbf{E}\left[y^*(t)y(t')\right] = \mathbf{E}\left[\iint_{-\infty}^{\infty} g^*(t, \tau)x^*(t - \tau)g(t', \tau')x(t' - \tau')d\tau d\tau'\right] = \iint_{-\infty}^{\infty} \mathbf{E}\left[g^*(t, \tau)g(t', \tau')x^*(t - \tau)x(t' - \tau')\right]d\tau d\tau'$. We can assume that $g(t, \tau)$ and $x(t)$ are uncorrelated yielding $\mathbf{E}\left[g^*(t, \tau)g(t', \tau')x^*(t - \tau)x(t' - \tau')\right] = \mathbf{E}\left[g^*(t, \tau)g(t', \tau')\right]\mathbf{E}\left[x^*(t - \tau)x(t' - \tau')\right]$. With (15.1) and (15.9) follows
$R_{yy}(t, t') = \iint_{-\infty}^{\infty} R_{gg}(t, t'; \tau, \tau')R_{xx}(t - \tau, t' - \tau')d\tau d\tau'$ and the proof is finished.

15.3 Wide Sense Stationary Time-Variant Systems

15.3.1 Wide Sense Stationarity

We are now specializing on the case that the system function $g(t, \tau)$ is a wide sense stationary (WSS) stochastic process with respect to the time t. The same shall hold for the input and output signal $x(t)$ and $y(t)$ of the system, respectively. In Appendix A, a wide sense stationary process is characterized by the (joint) probability density function and the mean value, both independent of time. Furthermore, the autocorrelation functions depend on t and t' just through the time difference

$$\Delta t = t' - t \tag{15.10}$$

Please note, when the delay variables τ and τ' are present the autocorrelation functions still depend on τ and τ' separately and not automatically on the difference $\Delta \tau = \tau' - \tau$. The latter is only true, if the delay spread function $g(t, \tau)$ is also stationary with respect to the delay variable.

15.3.2 Autocorrelation Functions and Power Spectral Densities

With $t' = t + \Delta t$ from (15.10), we define the following functions for wide sense stationary processes

Autocorrelation Function of the Delay Spread Function $g(t, \tau)$

from (15.1)

$$R_{gg}(\Delta t; \tau, \tau') = \mathbf{E}\left[g^*(t, \tau)g(t + \Delta t, \tau')\right] \tag{15.11}$$

Autocorrelation Function of the Time-Variant Transfer Function $G_t(t, f_\tau)$

from (15.2)

$$R_{G_t G_t}(\Delta t; f_\tau, f'_\tau) = \mathbf{E}\left[G_t^*(t, f_\tau)G_t(t + \Delta t, f'_\tau)\right] \qquad (15.12)$$

Power Spectral Density of the Delay Spread Function

According to the Wiener–Khintchine theorem in Appendix A, the *cross power spectral density* $S_{gc}(f'_t; \tau, \tau')$ between $g(t, \tau)$ and $g(t', \tau')$ can be defined by the Fourier transform of the autocorrelation function $R_{gg}(\Delta t; \tau, \tau')$ of $g(t, \tau)$ with respect to Δt

$$R_{gg}(\Delta t; \tau, \tau') \overset{\Delta t}{\leftrightarrow} S_{gc}(f'_t; \tau, \tau') \text{ with} \qquad (15.13)$$
$$S_{gc}(f'_t; \tau, \tau') = \int_{-\infty}^{\infty} R_{gg}(\Delta t; \tau, \tau')e^{-j2\pi f'_t \Delta t} d(\Delta t)$$

For $\tau = \tau'$, this function just depends on two variables and we use the short hand notation $S_{gc}(f'_t; \tau, \tau) = S_{gc}(f'_t; \tau)$, which can be interpreted as a form of the *power spectral density* of the delay spread function $g(t, \tau)$. Please note that τ and τ' are parameters, which are not touched by the Fourier transform in (15.13).

Relation Between $R_{G_\tau G_\tau}(f_t, f'_t; \tau, \tau')$ and $S_{gc}(f'_t; \tau, \tau')$

We prove that the autocorrelation function of the delay Doppler spread function is given by

$$R_{G_\tau G_\tau}(f_t, f'_t; \tau, \tau') = S_{gc}(f'_t; \tau, \tau')\delta(f'_t - f_t) \qquad (15.14)$$

According to Appendix A, we can conclude from (15.14) that the delay Doppler spread functions $G_\tau(f_t, \tau)$ and $G_\tau(f'_t, \tau')$ are uncorrelated with respect to any of the two Doppler frequencies $f_t \neq f'_t$.

Proof of (15.14)

From (15.7) follows with the WSS property (15.10), $t' = t + \Delta t$, $dt' = d(\Delta t)$, and by replacing $-f_t$ by f_t the autocorrelation function
$R_{G_\tau G_\tau}(f_t, f'_t; \tau, \tau') = \int_{-\infty}^{\infty} e^{-j2\pi(f'_t - f_t)t} dt \int_{-\infty}^{\infty} R_{gg}(\Delta t; \tau, \tau')e^{-j2\pi f'_t \Delta t} d(\Delta t).$
Knowing that $\int_{-\infty}^{\infty} e^{-j2\pi(f'_t - f_t)t} dt = \delta(f'_t - f_t)$ and using (15.13) yields the result
$R_{G_\tau G_\tau}(f_t, f'_t; \tau, \tau') = S_{gc}(f'_t; \tau, \tau')\delta(f'_t - f_t)$, which finalizes the proof.

Input–Output Relation of Correlation Functions and Power Spectral Densities

In the following, we assume that the input signal $x(t)$ and the time-variant system are WSS. Then the autocorrelation function (15.9)

$$R_{xx}(t, t') = R_{xx}(\Delta t) \qquad (15.15)$$

is just a function of $t' - t = \Delta t$ and for the argument of R_{xx} in (15.8) follows $(t' - \tau') - (t - \tau) = \Delta t - (\tau' - \tau)$. Then we obtain from (15.8)

$$R_{yy}(\Delta t) = \iint\limits_{-\infty}^{\infty} R_{gg}(\Delta t; \tau, \tau') R_{xx}\left(\Delta t - (\tau' - \tau)\right) d\tau d\tau' \qquad (15.16)$$

which is the final relation between the input and the output correlation function of a time-variant system. It has some similarities with the Wiener-Lee theorem for time-invariant systems given in Appendix A.

Next we are going to consider the Fourier transforms with respect to Δt

$$R_{xx}(\Delta t) \overset{\Delta t}{\longmapsto} S_{xx}(f_t) \; ; \; R_{yy}(\Delta t) \overset{\Delta t}{\longmapsto} S_{yy}(f_t) \qquad (15.17)$$

which are the power spectral density functions of $x(t)$ and $y(t)$ according to the Wiener–Khintchine theorem in the Appendix A, respectively. With (15.13) and $R_{xx}\left(\Delta t - (\tau' - \tau)\right) \overset{\Delta t}{\longmapsto} S_{xx}(f_t) e^{-j2\pi f_t(\tau' - \tau)}$ follows for the Fourier transform of the integrand in (15.16)

$$R_{gg}(\Delta t; \tau, \tau') R_{xx}\left(\Delta t - (\tau' - \tau)\right) \overset{\Delta t}{\longmapsto} S_{gc}(f_t; \tau, \tau') * \left(S_{xx}(f_t) e^{-j2\pi f_t(\tau' - \tau)}\right)$$
$$(15.18)$$

where $*$ indicates the time-invariant convolution with respect to f_t. Please note that we changed the argument of S_{gc} to f_t, because for a one-dimensional function we can take any variable, f_t or f_t'. Finally we obtain from (15.16) the power spectral density

$$S_{yy}(f_t) = \iint\limits_{-\infty}^{\infty} S_{gc}(f_t; \tau, \tau') * \left(S_{xx}(f_t) e^{-j2\pi f_t(\tau' - \tau)}\right) d\tau d\tau' \qquad (15.19)$$

We recognize that (15.19) is an extension of the Wiener–Khintchine theorem in Appendix A, whereby the time-variance of the system implies the additional integration with respect to τ and τ', as is also the case in (15.16).

15.4   Time-Variant Systems with Uncorrelated Scattering

The term uncorrelated scattering (US) stems from wireless channels and we apply it here for the general time-variant system. The condition defines a system with a delay spread function $g(t, \tau)$ and functions derived thereof, which are uncorrelated with respect to the delay variable τ. As shown in the Appendix A for white noise, the autocorrelation function must feature a Dirac impulse $\delta(\tau)$ in principle. In this Section we review some important properties of uncorrelated scattering. Then in Sect. 15.5 we combine the characteristics of processes showing both, wide sense stationarity and uncorrelated scattering (WSSUS).

15.4.1 Delay Cross Power Spectral Density of $g(t, \tau)$

For a wireless channel with uncorrelated scattering, the delay spread functions $g(t, \tau)$ and $g(t, \tau')$ are by definition uncorrelated for any delay $\tau \neq \tau'$. Thus, the autocorrelation function (15.1) of $g(t, \tau)$ must possess the following form:

$$R_{gg}(t, t'; \tau, \tau') = R_{gg;US}(t, t'; \tau, \tau') = P_g(t, t'; \tau)\delta(\tau' - \tau) \qquad (15.20)$$

$P_g(t, t'; \tau)$ is called *delay cross power spectral density* of $g(t, \tau)$ and obviously can still be a function of time. Next we show that $P_g(t, t'; \tau)$ is represented by the Fourier transform of $R_{gg}(t, t'; \tau, \tau')$ in (15.1) with respect to τ' at $f'_\tau = 0$. This results in

$$\int_{-\infty}^{\infty} R_{gg;US}(t, t'; \tau, \tau')d\tau' = P_g(t, t'; \tau) \qquad (15.21)$$

Proof of (15.20)

We apply the Fourier transform on (15.20) and obtain
$\int_{-\infty}^{\infty} R_{gg}(t, t'; \tau, \tau')e^{-j2\pi f'_\tau \tau'}d\tau' = P_g(t, t'; \tau) \int_{-\infty}^{\infty} \delta(\tau' - \tau)e^{-j2\pi f'_\tau \tau'}d\tau' = P_g(t, t'; \tau)e^{-j2\pi f'_\tau \tau}$. Then we set $f'_\tau = 0$ and get the final result, which finalizes the proof.

15.4.2 Autocorrelation Function of Time-Variant Transfer Function

Now we are going to consider the autocorrelation function $R_{G_t G_t}(t, t'; f_\tau, f'_\tau)$ of the time-variant transfer function $G_t(t, f_\tau)$ in (15.2) for the case of uncorrelated scattering specified by (15.20). In the following, we prove that the autocorrelation function of $G_t(t, f_\tau)$ then depends on the frequency difference

$$\Delta f_\tau = f'_\tau - f_\tau \qquad (15.22)$$

and hence must exhibit the form

$$R_{G_t G_t}(t, t'; f_\tau, f'_\tau) = R_{G_t G_t}(t, t'; \Delta f_\tau) \qquad (15.23)$$

Proof of (15.23)

We first make use of (15.6) and obtain
$R_{G_t G_t}(t, t'; f_\tau, f'_\tau) = \iint_{-\infty}^{\infty} R_{gg}(t, t'; \tau, \tau')e^{-j2\pi(-f_\tau \tau + f'_\tau \tau')}d\tau d\tau'$. Plugging in (15.20) yields $R_{G_t G_t}(t, t'; f_\tau, f'_\tau) = \iint_{-\infty}^{\infty} P_g(t, t'; \tau)\delta(\tau' - \tau)e^{-j2\pi(-f_\tau \tau + f'_\tau \tau')}$
$d\tau d\tau' = \int_{-\infty}^{\infty} P_g(t, t'; \tau)e^{-j2\pi \Delta f_\tau \tau}d\tau$. The last term is just a function of Δf rather

than f_τ and f'_τ. Consequently $R_{G_t G_t}(t, t'; f_\tau, f'_\tau) = R_{G_t G_t}(t, t'; \Delta f_\tau)$, which finalizes the proof.

15.5 Wide Sense Stationary Processes with Uncorrelated Scattering

15.5.1 Delay Cross Power Spectral Density of $g(t, \tau)$

Now we consider a time-variant system with a delay spread function $g(t, \tau)$, which represents a wide sense stationary process (with respect to the time t) and exhibits uncorrelated scattering (with respect to the delay time τ). To this end, the conditions (15.11) for the wide sense stationarity and (15.20) for the uncorrelated scattering must be true simultaneously. Then follows from (15.20) with the replacement of t, t' by Δt the autocorrelation function of $g(t, \tau)$

$$R_{gg;WSSUS}(t, t'; \tau, \tau') = R_{gg}(\Delta t; \tau, \tau') = P_g(\Delta t; \tau)\delta(\tau' - \tau) \qquad (15.24)$$

$P_g(\Delta t; \tau)$ is denoted as the *delay cross power spectral density* of $g(t, \tau)$ that just depends on the time difference Δt and features a Dirac impulse at $\tau' = \tau$.

15.5.2 Doppler Power Spectrum

The Fourier transform of $P_g(\Delta t; \tau)$ with respect to Δt yields another power density spectrum of $g(t, \tau)$, which is called *Doppler power spectrum* $S_{gD}(f_t, \tau)$

$$P_g(\Delta t; \tau) \overset{\Delta t}{\rightharpoonup} S_{gD}(f_t, \tau) = \int_{-\infty}^{\infty} P_g(\Delta t; \tau)e^{-j2\pi f_t \tau \Delta t}d\Delta t \qquad (15.25)$$

15.5.3 Autocorrelation Function of Time-Variant Transfer Function

From (15.23), we can directly obtain the autocorrelation function of the time-variant transfer function $G_t(t, f_\tau)$ for the case of WWSUS, if the argument t, t' is replaced by Δt

$$R_{G_t G_t, WSSUS}(t, t'; f_\tau, f'_\tau) = R_{G_t G_t}(\Delta t; \Delta f_\tau) \qquad (15.26)$$

and is also called *time frequency correlation function*.

15.6   Simplified Parameters for Time-Variant Systems

In several applications, such as wireless communications, different categories of the time-variance of a system can be distinguished. In case of a slow change of the environment and/or slow movement of the transmitter or receiver, the time-variant system can be considered as quasi static during time periods or blocks, which are much larger than the duration of a transmit symbol. The time-variance is then denoted as block fading. For other use cases, such as fast moving terminals in cars, trains or even airplanes, the channel characteristics are changing rapidly. Consequently, there is a need to differentiate between these applications and use dedicated, yet simple parameters. In many applications, the *coherence bandwidth* $\Delta f_{\tau,coh}$ and the *coherence time* Δt_{coh} turn out to be useful indicators and, therefore, frequently used. In the following, we discuss these parameters for deterministic and stochastic system functions.

15.6.1   Coherence Bandwidth

We define the coherence bandwidth as the frequency interval $\Delta f_{\tau,coh}$, in which the magnitude of the time-variant transfer function $G_t(t, f_\tau)$ does not change significantly. Thus, in case of a deterministic system we can just take the 3 dB cut-off frequency and define

$$\left| G_t(t, \Delta f_{\tau,coh}) \right|^2 = \frac{1}{2} |G_t(t, 0)|^2 \qquad (15.27)$$

for a fixed time instant t, or a small time interval. Alternatively, the coherence bandwidth can also be defined by means of the Doppler spread function $G(f_t, f_\tau)$ along the f_τ-axis.

For a randomly changing time-variant system, we have to apply statistical measures. Presuming a channel with wide sense stationary uncorrelated scattering we can make use of the autocorrelation function $R_{G_t G_t}(\Delta t; \Delta f_\tau)$ of the time-variant transfer function given in (15.26) at $\Delta t = 0$ and define a 3 dB cut-off frequency as coherence bandwidth $\Delta f_{\tau,coh}$

$$\left| R_{G_t G_t}(\Delta t; \Delta f_{\tau,coh}) \right| = \frac{1}{2} \left| R_{G_t G_t}(\Delta t; 0) \right| \qquad (15.28)$$

for a fixed time, e.g., $\Delta t = 0$. Please note that $\Delta t = 0$ reflects the time instant $t = t'$ according to (15.10). Often $R_{G_t G_t}(\Delta t; \Delta f_\tau)$ is frequency selective with many ripples rather than monotonic with respect to Δf_τ. In this case adequate root mean square values are useful [2]. Inside the coherence bandwidth the spectral components of $R_{G_t G_t}(\Delta t; \Delta f_\tau)$ are strongly correlated. Recall that f_τ characterizes the frequency for the transmission spectrum. Hence, the larger $f_{\tau,coh}$ is the more spectral components can be conveyed from the input to the output of the time-variant system.

15.6.2 Coherence Time

The coherence time Δt_{coh} can be defined as that time interval, in which the delay spread function $g(t, \tau)$ does not change significantly with respect to time t for any fixed delay time τ. For a deterministic system we may determine a 3 dB cut-off frequency $f_{t,coh}$ using the Doppler spread function $G(f_t, f_\tau)$ as

$$\left|G(f_{t,coh}, f_\tau)\right|^2 = \frac{1}{2}\,|G(0, f_\tau)|^2 \tag{15.29}$$

for any fixed f_τ. For illustration, let us consider the following example.

Example 8

Given a Doppler spread function $G(f_t, f_\tau)$, which shall exhibits the ideal lowpass spectrum

$$G(f_t, f_\tau) = \begin{cases} 1 : |f_t| \le f_{t,c} \, , \, |f_\tau| \le f_{\tau,c} \\ 0 \, ; \qquad\qquad else \end{cases} \tag{15.30}$$

with the cut-off frequencies $f_{t,c}$ and $f_{\tau,c}$. Of course due to the step-wise transition there is no 3 dB point. Therefore, we alternatively take $f_{t,c}$. If we apply the two-dimensional inverse Fourier transform we obtain

$$g(t, \tau) = 4 f_{t,c} f_{\tau,c} \,\mathrm{sinc}(2\pi f_{t,c} t) \,\mathrm{sinc}(2\pi f_{\tau,c} \tau) \tag{15.31}$$

Roughly speaking, we consider the function $\mathrm{sinc}(2\pi f_{t,c} t)$ approximately as constant up to the first zero at $t_0 = \frac{1}{2 f_{t,c}}$ and can find the coherence time as

$$\Delta t_{coh} = t_0 = \frac{1}{2 f_{t,c}} \tag{15.32}$$

Apparently, Δt_{coh} is inversely proportional to the maximal Doppler frequency $f_{t,c}$.

As already mentioned for the coherence bandwidth, in case of a randomly time-variant system we have to work on the basis of the autocorrelation and the power spectral density functions. Then we define the coherence time as the time interval Δt_{coh}, in which the delay spread function $g(t, \tau)$ is strongly correlated. For a wide sense stationary system with uncorrelated scattering the delay cross power spectral density $P_g(\Delta t; \tau)$ given in (15.24) is of significance and should be large in the interval Δt_{coh}. Equivalently in the frequency domain we can consider the 3 dB cut-off frequency $f_{t,c}$ of the Doppler power spectrum $S_{gD}(f_t, \tau)$ in (15.25) as a measure

$$\left|S_{gD}(f_{t,c}, \tau)\right| = \frac{1}{2} \left|S_{gD}(0, \tau)\right| \tag{15.33}$$

for any fixed delay τ, mostly taken at $\tau = 0$. Similar to the example above, if $S_{gD}(f_t, \tau)$ is a rectangular lowpass as a function of f_t and with cut-off frequency $f_{t,c}$, then the coherence time Δt_{coh} can be defined likewise as in (15.32).

References

1. P.A. Bello, Characterization of randomly time-variant linear channels. IEEE Trans. Commun. Syst. (1963)
2. A.F. Molisch, H. Asplund, R. Heddergott, M. Steinbauer, T. Zwick, The COST259 directional channel model - part i: overview and methodology. IEEE Trans. Wirel. Commun. **5** (2006)

Part III
Multiple Input Multiple Output Wireless Transmission

Chapter 16
Background

After the first demonstrations of electromagnetic waves in the year 1887 by the physicist Heinrich Hertz at the Technical University of Karlsruhe in Germany, wireless telegraphy transmission was demonstrated at the end of the nineteenth century by the radio pioneer and founder of the later company Guglielmo Marconi. Besides quite some important developments of different antenna technologies, the early ideas for multiple input multiple output (MIMO) schemes using multiple antennas trace back to Kaye and George (1970), Branderburg and Wyner (1974), and van Etten (1975), [1–3]. A concise survey is given in [4]. Later in 1984 and 1986, Winters and Salz considered beamforming techniques at Bell Laboratories, [5]. In 1994, Paulraj and Kailath introduced a patent on the concept of spatial multiplexing using multiple antennas. Raleigh and Cioffi investigated the transmission of multiple data streams using spatial–temporal coding, [6]. In the same year, Foschini introduced the concept of Bell Labs Layered Space-Time (BLAST), [7], which was refined and implemented later in 1999 by Golden et al. [8]. The digital cellular system GSM (Global System for Mobile Communications) put into operation around 1992 in Europe did not yet use the MIMO principle. However, later standards such as the 3.5 Generation (3.5G, UMTS advanced, IMT 2000), the 4G, and the 5G cellular systems adopt this key technology. Similar developments and standards prevailed for the wireless local area network WLAN and WIMAX IEEE 802.11.

Starting around the year 2000, ideas came up to introduce the MIMO principle not only in the area of wireless broadcasting but also in the field of wire-line digital transmission. There the multiple antennas are replaced in principle by wire-line multiple transceivers. Today, applications and standards are present in the field of digital transmission on power lines [9, 10], and digital subscriber lines (vectoring), [11]. A survey on advances in wireless MIMO research and technology is also found in [12].

© Springer Nature Switzerland AG 2019
J. Speidel, *Introduction to Digital Communications*, Signals and Communication Technology, https://doi.org/10.1007/978-3-030-00548-1_16

References

1. A. Kaye, D. George, Transmission of multiplexed PAM signals over multiple channel and diversity systems. IEEE Trans. Commun. Technol. (1970)
2. L. Brandenburg, A. Wyner, Capacity of the Gauss' ian channel with memory: the multivariate case. Bell Syst. Tech. J. **53** (1974)
3. W. Van Etten, Maximum likelihood receiver for multiple channel transmission systems. IEEE Trans. Commun. **24** (1976)
4. Wikipedia, MIMO, https://en.wikipedia.org/wiki/MIMO
5. J. Salz, Digital transmission over cross-coupled linear channels. AT&T Tech. J. **64** (1985)
6. G. Raleigh, J. Cioffi, Spatio-temporal coding for wireless communications, in *IEEE Global Telecommunications Conference* (1996)
7. G. Foschini, Layered space-time architecture for wireless communication in a fading environment when using multi-element antennas. Bell Syst. Tech. J. (1996)
8. G. Golden, G. Foschini, R. Valenzuela, P. Wolniansky, Detection algorithm and initial laboratory results using V-BLAST space-time communication architecture. Electron. Lett. **35** (1999)
9. D. Schneider, J. Speidel, L. Stadelmeier, D. Schill, A. Schwager, MIMO for inhome power line communications, in *International Conference on Source and Channel Coding (SCC), ITG Fachberichte* (2008)
10. L.T. Berger, A. Schwager, P. Pagani, D. Schneider, *MIMO Power Line Communications - Narrow and Broadband Standards, EMC and Advanced Processing* (CRC Press, Florida, 2014)
11. G.993.5: Self-FEXT cancellation (vectoring) for use with VDSL2 transceivers, ITU-T Std
12. F. Khalid, J. Speidel, Advances in MIMO techniques for mobile communications - a survey. Int. J. Commun. Netw. Sys. Sci. **3** (2010)

Chapter 17
Principles of Multiple Input Multiple Output Transmission

17.1 Introduction

In the following sections, we introduce the principles of signal transmission over a Multiple Input Multiple Output (MIMO) channel and emphasize on wireless links. We will derive a block diagram of a MIMO system and compact it using the concept of the equivalent baseband channel, which is described in Part I for single input single output (SISO) channels. We characterize the physical channels between the various inputs and outputs of the MIMO system by the delay spread functions, which are functions of two variables t and τ to prepare for the case of time-variant wireless channels. For time-invariant MIMO channels, such as cables composed of many two-wire electrical lines or optical fibers, the delay spread function turns into the impulse response and just depends on the delay variable τ, which is then renamed as t.

17.2 MIMO Transmission System with Quadrature Amplitude Modulation

17.2.1 System Model

Figure 17.1 shows the principle block diagram of a MIMO transmission system. The transmitter and the receiver are equipped with M and N parallel branches, respectively. In principle, the branches on one side are composed of the same building blocks. Only transmit branch j and receive branch i are shown in more detail.

Single Input Single Output Link

Assume for the moment that only the transmit branch j is active and that all other transmit signals are zero, $s_p(k) = 0 \; \forall \; p \neq j$. Then the link from transmit branch j to

---

The original version of this chapter was revised: a few typographical errors were corrected. The correction to this chapter can be found at https://doi.org/10.1007/978-3-030-00548-1_25

© Springer Nature Switzerland AG 2019

J. Speidel, *Introduction to Digital Communications*, Signals and Communication Technology, https://doi.org/10.1007/978-3-030-00548-1_17

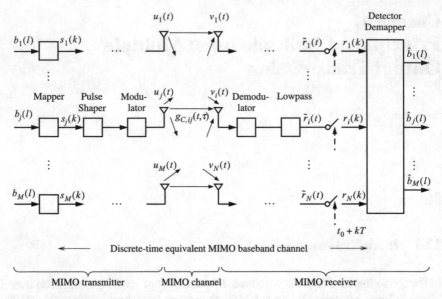

Fig. 17.1 Principle block diagram of a digital MIMO transmission system with single carrier QAM

receive branch i represents a single input single output scheme, which is described in quite some detail in Part I. Briefly, its operation is as follows. The binary bit stream $b_j(l)$, which contains the information bits and in most cases redundancy bits for forward error correction (FEC), is fed into a mapper, which periodically allocates κ_j consecutive bits to a complex symbol $s_j(k)$. We denote $l \in \mathbb{Z}$ and $k \in \mathbb{Z}$ as the discrete-time of the bit and the symbol sequence, respectively. Thus, the symbol alphabet $\mathcal{B}$ consists of $L_j = 2^{\kappa_j}$ symbols, which can be portrait in the complex domain. Together with the complex modulation carrier $e^{j2\pi f_0 t}$ the scheme constitutes a quadrature amplitude modulation (QAM) transmitter. The pulse shaper, which is a lowpass filter, limits the infinite bandwidth of the symbol sequence $s_j(k)$ to at most half the symbol rate $v_S = \frac{1}{T}$. We call T the symbol interval, i.e., every T a symbol leaves the mapper. The modulator with the carrier frequency f_0 shifts the spectrum to the channel passband with the center frequency f_0. As all physical channels own real-valued impulse responses, only the real part of the complex modulator output signal can be transmitted. Consequently, the modulator in Fig. 17.1 also contains a unit, which selects this part to create the real-valued output signal $u_j(t)$. The channel will be discussed later in more detail. For the moment it shall be characterized by the delay spread function $g_{C,i,j}(t, \tau)$.

In the receiver branch i, a bandpass filter may be employed to limit the noise spectrum to the channel passband, yet leaving the signal part of the receive signal unchanged. After synchronous demodulation with the complex carrier $e^{-j2\pi f_0 t}$ the demodulated signal is lowpass filtered yielding the complex-valued signal $\tilde{r}_i(t)$, which is sampled with symbol rate $\frac{1}{T}$ at $t = t_0 + kT$, resulting in

$$r_i(k) = \tilde{r}_i(t_0 + kT) \; ; \; i = 1, 2, \ldots, N \tag{17.1}$$

where the delay t_0 between the transmitter and the receiver has to be estimated at the receiver.

Physical Single Input Single Output Channel

Electrical cables or optical fibers are in general time-invariant and will be described by an impulse response $g_{C,ij}(t)$. A wireless and mobile channel is time-variant due to the temporal change of the environment and the movement of the transmitter or receiver. As is known from Part I and II, such a channel can be characterized by a time-variant impulse response $w_{C,ij}(t, s)$, which is the response at observation time t to a Dirac impulse active at time instant $s \leq t$, [1, 2]. The variables t and s are independent. The use of $w_{C,ij}(t, s)$ has some drawbacks for signal processing, in particular it does not provide a meaningful Fourier spectrum. Therefore, the transformation of variables $s = t - \tau$ is applied yielding the *modified impulse response* or *delay spread function*

$$w_{C,ij}(t, t - \tau) = g_{C,ij}(t, \tau) \tag{17.2}$$

which can then be interpreted as the response at observation time t to an input Dirac impulse at $t - \tau \leq t$. We call t the observation time or just time and τ is referred to as delay time or age variable. The input–output relation between $u_j(t)$ and $v_i(t)$ in Fig. 17.1 is given by the time-variant convolution

$$v_i(t) = \int_{-\infty}^{\infty} u_j(s) w_{C,ij}(t, s) ds = \int_{-\infty}^{\infty} u_j(t - \tau) g_{C,ij}(t, \tau) d\tau \tag{17.3}$$

where we have used $s = t - \tau$ and (17.2). In case of time-invariance, the delay spread function turns into the impulse response, $g_{C,ij}(t, \tau) = g_{C,ij}(\tau)$. It just depends on the variable τ, which is then renamed as t yielding $g_{C,ij}(t)$. Then the time-variant convolution boils down to the well known convolution for time-invariant systems

$$v_i(t) = \int_{-\infty}^{\infty} u_j(t - \tau) g_{C,ij}(\tau) d\tau \tag{17.4}$$

Equivalent Baseband System Model

To obtain a compact mathematical description without details of impulse shaping, modulation, demodulation, etc, the described SISO link from transmitter j to receiver i in Fig. 17.1 can be simplified by allocating between the input symbol sequence $s_j(k)$ of the impulse shaper and the output signal $\tilde{r}_i(t)$ of the receive lowpass an equivalent baseband model with overall impulse response $\tilde{w}_{ij}(t, s)$ or with the delay spread function $\tilde{h}_{ij}(t, \tau) = \tilde{w}_{ij}(t, t - \tau)$ using $s = t - \tau$. Similarly, $\tilde{h}_{ij}(t, \tau)$ is the response observed at time instant t to a Dirac impulse at instant $t - \tau \leq t$. After sampling at $t = t_0 + kT$ and $\tau = \tau_0 + mT$ we define $\tilde{h}_{ij}(t_0 + kT, \tau_0 + mT) = h_{ij}(k, m)$, which we call in the following the *discrete-time modified impulse response of the equivalent baseband system* or *discrete-time delay spread function of the equivalent baseband system*. In Fig. 17.1 $h_{ij}(k, m)$ characterizes the SISO link from the input sequence

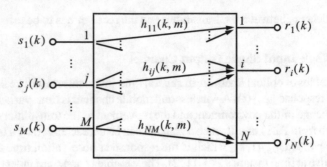

Fig. 17.2 Principle block diagram of a MIMO system with discrete-time equivalent baseband channels characterized by the delay spread functions $h_{ij}(k, m)$

$s_j(k)$ at the transmitter to the output sequence $r_i(k)$ of the sampling device at the receiver. Similarly, it can be interpreted as the response observed at discrete-time instant k to a unit impulse at instant $k - m \leq k$. This leads to the simplified block diagram in Fig. 17.2 of the MIMO system with discrete-time equivalent baseband channels.

In case of a time-invariant channel, such as a cable, the equivalent baseband system impulse response $\tilde{h}_{ij}(t)$ and its sampled version $\tilde{h}_{ij}(t_0 + kT) = h(k)$ are representing the SISO link. To simplify the notation in the remainder we always use the general term $h_{i,j}(k, m)$ and whether we mean $h_{ij}(m)$ of a time-invariant link will become clear from the context. Please note that due to modulation and demodulation with a complex carrier, $h_{ij}(k, m)$ is complex in general.

If all M transmit signals $s_j(k)$ in Fig. 17.2 are active, the receive signal $r_i(k)$ is the superposition. This holds for all receive signals $r_1(k), \ldots, r_N(k)$. They are "combined" by the building block detector and demapper in Fig. 17.1, which provides the estimates $\hat{b}_j(l)$ of the transmit bit sequences $b_j(l)$; $j = 1, 2, \ldots, M$. Various signal combiners and detection methods for MIMO receivers are described in the Chaps. 18 and 19 in detail.

17.2.2 Input–Output Relation of MIMO System with Time-Variant Channel

Single Input Single Output (SISO)

Let us assume that only the transmit branch j is active. As mentioned before, we characterize the SISO channel composed of the transmit branch j , the physical channel, and the receive branch i by the equivalent delay spread function $h_{ij}(k, m)$. Then we obtain at the output of the receive branch i

$$r_i(k) = s_j(k) \circledast h_{ij}(k, m) + n_i(k) \tag{17.5}$$

where $n_i(k)$ is the additive noise at the receive branch i and $\circledast$ denotes the discrete-time, time-variant convolution defined as

$$s_j(k) \circledast h_{ij}(k, m) = \sum_{m=-\infty}^{\infty} h_{ij}(k, m) s_j(k - m) \tag{17.6}$$

We have introduced the operator $\circledast$ to prevent from confusion with the time-invariant convolution, which we indicate by $*$. The prove of (17.6) is evident from (17.12), which is discussed later. The time-variant and the time-invariant convolution look similar, however, they differ in several properties. The two variables k and m specify two different time variables. m is called delay time, age variable, or integration time, as the summation in (17.6) is executed over m, and k characterizes the time variable of the output signal of the system and in addition also the temporal change of the system parameters. If the channel is time-invariant, then $h_{ij}(k, m)$ does not depend on k. Hence, we can just skip k in $h_{ij}(k, m)$ of (17.5) as well as (17.6) and write $h_{ij}(k, m) = h(m)$, as is shown in Part II. Consequently, $\circledast$ turns into the well-known convolution operation $*$ for linear time-invariant systems. An example for $h_{ij}(k)$ is a wire-line channel or a static wireless channel, where the transmitter and the receiver are nonmoving and all scatterers or reflectors of the electromagnetic waves do not vary with time k.

Multiple Input Single Output (MISO)

Now we consider that all transmit branches $j = 1, 2, \ldots, M$ and one receive branch i are active. Then we obtain for the receive branch i with (17.5)

$$r_i(k) = \sum_{j=1}^{M} s_j(k) \circledast h_{ij}(k, m) + n_i(k) \tag{17.7}$$

Multiple Input Multiple Output (MIMO)

Finally all transmit and all receive branches are active. Consequently, (17.7) holds for $i = 1, 2, \ldots, N$, which gives rise to a matrix notation, in which (17.7) represents the ith equation

$$
\underbrace{\begin{pmatrix} r_1(k) \\ \vdots \\ r_i(k) \\ \vdots \\ r_N(k) \end{pmatrix}}_{\mathbf{r}(k)} = \underbrace{\begin{pmatrix} h_{11}(k, m) & \cdots & h_{1j}(k, m) & \cdots & h_{1M}(k, m) \\ & & \vdots & & \\ h_{i1}(k, m) & \cdots & h_{ij}(k, m) & \cdots & h_{1M}(k, m) \\ & & \vdots & & \\ h_{N1}(k, m) & \cdots & h_{Nj}(k, m) & \cdots & h_{NM}(k, m) \end{pmatrix}}_{\mathbf{H}(k,m)} \circledast \underbrace{\begin{pmatrix} s_1(k) \\ \vdots \\ s_j(k) \\ \vdots \\ s_M(k) \end{pmatrix}}_{\mathbf{s}(k)} + \underbrace{\begin{pmatrix} n_1(k) \\ \vdots \\ n_i(k) \\ \vdots \\ n_N(k) \end{pmatrix}}_{\mathbf{n}(k)}
$$
$$\tag{17.8}$$

which can be written as

$$\mathbf{r}(k) = \mathbf{H}(k, m) \circledast \mathbf{s}(k) + \mathbf{n}(k) \tag{17.9}$$

$\mathbf{r}(k)$ is the receive vector, $\mathbf{H}(k, m)$ the matrix of the delay spread functions of the discrete-time equivalent MIMO channel model, $\mathbf{s}(k)$ the transmit signal vector, and $\mathbf{n}(k)$ the vector of the additive noise. Please note that $\circledast$ turns the matrix multiplication into the convolution operation defined in (17.6), where the order of operation is $s_j(k) \circledast h_{ij}(k, m)$ and has to be respected because $\circledast$ is in general noncommutative.

17.3   Deterministic Models for Wireless MIMO Channels

In Part I, on digital wireless communications various models for SISO channels and their properties are outlined in quite some detail. There are deterministic and stochastic models. All these findings can be applied for any link between the transmit branch j and the receive branch i in Figs. 17.1 and 17.2. The significance needs no detailed explanation and we restrict ourselves to a brief introduction of the finite impulse response channel model. Furthermore, we emphasize on the spatial interrelation of these SISO channels by considering their correlation properties.

17.3.1   Uniform Linear and Uniform Circular Antenna Arrays

Figure 17.3 shows in principle the configurations of MIMO antennas, which are frequently used, the uniform linear and the uniform circular array antenna. In both antennas, the distance between adjacent elements is equal. These antennas can be applied for transmission and reception. From the theory of electromagnetic waves, we know that the signals emitted by the antenna elements can interact and then are correlated, if approximately $l \leq \frac{\lambda_0}{2}$ holds, where $\lambda_0 = \frac{c}{f_0}$ is the operating wavelength, c is the speed of light in the air, and f_0 is the carrier frequency, [3]. The closer the antenna elements are, the higher the correlation of the signals can be, because in case of a transmit MIMO antenna the waves of neighboring elements overlap. Correlation also allows the shaping of the emitted electromagnetic beam into the direction of a hot spot of users by appropriate antenna feeding signals generated by a precoding vector. This technique is called transmit beamforming. In case of a receive MIMO antenna adjacent narrow elements are excited by signals, which are quite similar and thus receive correlation is introduced. If the distance between adjacent antenna elements is increased, the correlation among the signals decreases. For $l \approx \lambda_0$ microdiversity starts and for larger distances such as $l > 3\lambda_0$ the correlation almost vanishes. This mode of operation is used for spatial diversity, which allows the transmission or reception of almost independent signals. Depending on the applications MIMO antennas with narrow and far spaced elements can also be combined.

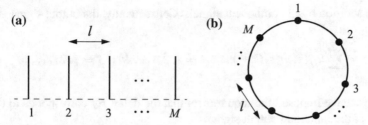

Fig. 17.3 Arrays with M antenna elements spaced by l, **a** uniform linear array, **b** uniform circular array

17.3.2 Finite Impulse Response Channel Model

As is well known from signal processing a linear discrete-time system is characterized by an impulse response with infinite duration (infinite impulse response, IIR). In addition, if the system is noncausal and time-invariant the input–output relation for the SISO link between transmitter j and receiver i is given by $r_i(k) = \sum_{m=-\infty}^{\infty} h_{ij}(m)s_j(k-m)$. Typically, the magnitude of the impulse response $h_{ij}(m)$ of a time-invariant equivalent baseband channel declines with increasing m. Therefore, an adequate approximation can be obtained by pruning the response at $m = K'_{ij}$ resulting in a finite impulse response (FIR) filter model.

Similarly, the magnitude of the delay spread function $h_{ij}(k,m)$ of the equivalent baseband wireless channel approaches zero for increasing m and thus we can discard the samples $h_{ij}(k,m)$ for $m > K_{ij}$ to obtain an approximation with respect to m. This simplified channel model is depicted in Fig. 17.4 and shows the structure of a finite impulse response (FIR) filter, which is also called tapped delay-line or transversal filter. The $K_{ij}+1$ channel coefficients are given by $h_{ij}(k,m)$, where $m = 0, 1, \ldots, K_{ij}$. In contrast to a time-invariant filter the coefficients (taps) can change their values at every time instant k.

All operations in the filter are linear namely the delay Δ_τ of the input signal $s_j(k-m)$, the multiplication with the tap $h_{ij}(k,m)$ yielding $h_{ij}(k,m)s_j(k-m)$,

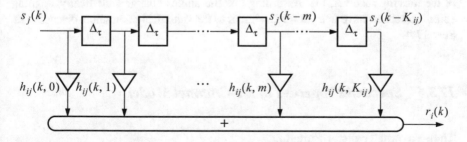

Fig. 17.4 Time-variant transversal filter for modeling the equivalent baseband SISO channel between transmitter j and receiver i

and the addition of all component signals. Consequently, the output signal is

$$r_i(k) = \sum_{m=0}^{K_{ij}} h_{ij}(k, m)s_j(k - m) \; ; \; i = 1, 2, \ldots, N \; ; \; j = 1, 2, \ldots, M \quad (17.10)$$

where we have imposed the requirement that the delay Δ_τ corresponds to the time base T of the input and output signal,

$$\Delta_\tau = T \tag{17.11}$$

The reader is encouraged to convince oneself of the linearity by using the superposition of two input signals. Please note that the considered FIR filter is causal because $h_{ij}(k, m) = 0 \, \forall \, m < 0$. A physical interpretation of (17.10) is a wireless channel where the transmit signal $s_j(k)$ is spread over $K_{ij} + 1$ different paths on its way to the receiver. Each path m introduces a dedicated delay mT and a path loss indicated by the complex channel coefficient $h_{ij}(k, m)$. Such a channel is called a time-variant multipath channel. It should be mentioned that the described modeling is done on the basis that the delay of each physical path is quantized by multiples of $\Delta_\tau = T$, as indicated by (17.11), where T is the symbol interval.

As portrait in Fig. 17.2, a wireless MIMO system is characterized by $N \cdot M$ of such multipath SISO channel models, where $j = 1, 2, \ldots, M$ and $i = 1, 2, \ldots, N$.

17.3.3   Spatial Channel Models

As outlined in Chap. 7, in most cases the transmit signal propagates along a multitude of paths with reflections and scattering. For MIMO systems, several studies have been made to find models, which incorporate spatial parameters of the delay spread function. Among others each propagation path is divided into a large number of sub-paths and their signals are characterized by the angle of departure from the transmit antenna, the angle of arrival at the receive antenna, the distances of the antenna elements, the phase difference of the waves, and the direction of the velocity of the moving receiver, [4]. Assuming that the angles change statistically, angular spread functions have been defined, similar to the statistical parameters discussed in Sect. 17.4.

17.3.4   Spectral Properties of the Channel Model

Time-Variant Transfer Function

We consider again the SISO channel model between the transmitter j and the receiver i with its delay spread function $h_{ij}(k, m)$. As described in Part II, a time-variant

transfer function $\tilde{H}_{ij}(k, e^{j2\pi f_\tau T})$ can be defined by the Fourier transform of $h_{ij}(k, m)$
with respect to m. To this end, we first apply the z-transform on $h_{ij}(k, m)$ with respect
to m resulting in $\tilde{H}_{ij}(k, z) = \sum_{m=0}^{\infty} h_{ij}(k, m)z^{-m}$ and then substitute $z = e^{j2\pi f_\tau T}$
yielding

$$h_{ij}(k, m) \overset{m}{\longmapsto} \tilde{H}_{ij}(k, e^{j2\pi f_\tau T}) = \sum_{m=0}^{\infty} h_{ij}(k, m)e^{-j2\pi f_\tau T m} \qquad (17.12)$$

where the discrete delay variable mT corresponds to the frequency variable f_τ.
Obviously, $\tilde{H}_{ij}(k, e^{j2\pi f_\tau T})$ is periodic with period $f_\tau = \frac{1}{T}$. Its baseband is located
in $|f_\tau| \le \frac{1}{2T}$. If $\left| \tilde{H}_{ij}(k, e^{j2\pi f_\tau T}) \right|$ varies in this frequency interval significantly with
f_τ for any fixed k, then we call this channel *frequency selective* and the effect as
frequency selective fading. Given the fact that this property is owned by at least one
SISO channel of the MIMO scheme in Fig. 17.2, the MIMO channel is referred to as
frequency selective.

In contrast, if $\left| \tilde{H}_{ij}(k, e^{j2\pi f_\tau T}) \right|$ is approximately constant with respect to f_τ in the
aforesaid interval up to a certain cut-off frequency for any fixed k, then the SISO
channel is referred to as non-frequency selective or *frequency flat*. Then we conclude
from (17.12) that all channel coefficients must be zero except for the first one,

$$h_{ij}(k, 0) = h_{ij}(k) \qquad (17.13)$$

where $h_{ij}(k)$ has to be understood as a short hand notation. Hence, the SISO channel
is just time-variant and from (17.5) follows with (17.6) the input–output relation

$$r_i(k) = s_j(k)h_{ij}(k) + n_i(k) \qquad (17.14)$$

Here we can study easily the effect of a time-variant channel. Assume there is no
noise. Even if we sent a constant signal s_j, we get from (17.14) the response $r_i(k) = s_j h_{ij}(k)$ which varies with time.

If all SISO channels of the MIMO system are frequency flat, at least approximately,
then the MIMO scheme is called frequency flat and from (17.9) follows

$$\mathbf{r}(k) = \mathbf{H}(k)\mathbf{s}(k) + \mathbf{n}(k) \qquad (17.15)$$

with $\mathbf{H}(k, m) = \mathbf{H}(k, 0) = \mathbf{H}(k)$ as a short hand notation. Obviously, the time-variant
convolution turns into the multiplication.

Delay Doppler Spread Function

As is known from the theory of time-variant systems in Part II, we obtain the delay
Doppler spread function also called Doppler-variant impulse response, if we consider
the Fourier spectrum of the delay spread function with respect to the time t or k. We
start with the z-transform with respect to k yielding $\tilde{H}_{ij}(z, m) = \sum_{k=0}^{\infty} h_{ij}(k, m)z^{-k}$.

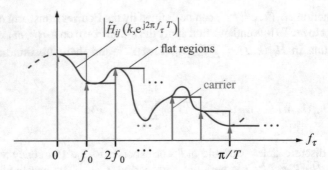

Fig. 17.5 Dividing the magnitude of the time-variant transfer function $\left|\tilde{H}_{ij}(k, \mathrm{e}^{\mathrm{j}2\pi f_\tau T})\right|$ of the equivalent baseband channel into small regions of approximately flat segments using multi-carrier modulation with carrier frequencies $f_0, 2f_0,...$

Substituting $z = \mathrm{e}^{\mathrm{j}2\pi f_t T}$ results in

$$h_{ij}(k, m) \overset{k}{\multimap} \bar{H}_{ij}(\mathrm{e}^{\mathrm{j}2\pi f_t T}, m) = \sum_{k=0}^{\infty} h_{ij}(k, m)e^{-\mathrm{j}2\pi f_t Tk} \qquad (17.16)$$

where k of the original domain corresponds to the Doppler frequency variable f_t in the frequency domain. We see that $\bar{H}_{ij}(\mathrm{e}^{\mathrm{j}2\pi f_t T}, m)$ is periodic with period $f_t = \frac{1}{T}$ and its baseband is located in the interval $|f_t| \leq \frac{1}{2T}$. If the cut-off frequency of $\left|\bar{H}_{ij}(\mathrm{e}^{\mathrm{j}2\pi f_t T}, m)\right|$ in this interval is small, then the variations of $h_{ij}(k, m)$ with respect to k are slow. In this case, the channel exhibits a weak temporal fading. The contrary is true, if the cut-off frequency is large resulting in fast changes.

Step-Wise Frequency Flat Channel Transfer Function

Frequency flat regions of the channel transfer function can be achieved, if multi-carrier modulation such as orthogonal frequency division multiplexing (OFDM) is applied. As illustrated in Fig. 17.5, this modulation technique subdivides the total transmission band into approximately flat segments between adjacent carriers $e^{\mathrm{j}2\pi f_0 t}$, $e^{\mathrm{j}2\pi 2 f_0 t}$, $e^{\mathrm{j}2\pi 3 f_0 t}$, ..., which are in case of OFDM generated by an inverse discrete Fourier transform (IDFT) as a modulator and by a discrete Fourier transform (DFT) as a demodulator.

To this end, each transmit and receive branch in Fig. 17.1 is composed of an IDFT and a DFT, respectively. In each flat frequency region an input–output relation similar to (17.14) holds per SISO channel. We will focus for all further considerations on MIMO systems with frequency flat fading channels described by (17.15).

17.4 Statistical Models for MIMO Channels

17.4.1 I.I.D. Gaussian MIMO Channel Model

As outlined in [3] and summarized in Part I on digital communications over single input single output links, multipath propagation, rich scattering, and reflections lead to a multitude of scenarios which in most cases cannot be described by detailed deterministic models. Suitable are statistical models, which will be reviewed in the following. As already stated, all SISO channels of the MIMO system shall exhibit a flat frequency response, $\left| \tilde{H}_{ij}(k, e^{j2\pi f_\tau T}) \right| = const.$ with respect to f_τ, for any k. Therefore, only the temporal fading of the complex-valued delay spread functions

$$h_{ij}(k) = \text{Re}\left[h_{ij}(k)\right] + j\text{Im}\left[h_{ij}(k)\right] ; \; j = 1, 2, \ldots, M ; \; i = 1, 2, \ldots, N \quad (17.17)$$

has to be considered, which are now stochastic processes. A summary on the definitions and properties of random variables and stochastic processes is given in the Appendix A.

Assume that the output signal of an unidirectional transmit antenna element j undergoes rich scattering and reflections resulting in a multipath propagation with an infinite number of statistically independent paths carrying signals, which superimpose at the receive antenna element i resulting in the stochastic process $h_{ij}(k)$. Then the conditions of the central limit theorem are fulfilled, which states that $h_{ij}(k)$ exhibits a complex Gaussian probability density function. If there is no line of sight between the transmitter and the receiver, the receive signal has zero mean. In detail we can model each delay spread function in (17.17) as follows:

1. All $h_{ij}(k)$ exhibit zero mean,

$$\mathbf{E}\left[h_{ij}(k)\right] = 0 \quad (17.18)$$

   from which follows that real and imaginary part of $h_{ij}(k)$ have zero mean. This prerequisite is fulfilled, if there is no line of sight between transmitter and receiver.
2. All $h_{ij}(k)$ are independent and identically distributed (i.i.d.) according to a Gaussian probability density function given by

$$p_x(x) = \frac{1}{\sqrt{2\pi}\sigma_x} e^{-\frac{x^2}{2\sigma_x^2}} \quad (17.19)$$

   where x stands for $\text{Re}\left[h_{ij}(k)\right]$ and $\text{Im}\left[h_{ij}(k)\right]$ for any fixed k. The corresponding variances are $var(x) = \mathbf{E}\left[(x - \mathbf{E}[x])^2\right] = \sigma_x^2$.
3. All $h_{ij}(k)$ are circular symmetric Gaussian processes, i.e.,

$$var\left(\text{Re}\left[h_{ij}(k)\right]\right) = var\left(\text{Im}\left[h_{ij}(k)\right]\right) \quad (17.20)$$

4. All $h_{ij}(k)$ are pairwise spatially uncorrelated,

$$
\mathbf{E}\left[h_{pq}h_{rs}^*\right] = \begin{cases} 1 & ; \; p = r \; ; \; q = s \\ \mathbf{E}\left[h_{pq}(k)\right]\mathbf{E}\left[h_{rs}^*(k)\right] = 0 & ; \; else \end{cases} \tag{17.21}
$$

The channel matrix $\mathbf{H}(k)$ is, therefore, denoted as $\mathbf{H}_w(k)$, where the index w
stand for white and shall indicate the uncorrelated entries $h_{ij}(k)$. The second line
in (17.21) is zero, because of (17.18). As the $h_{ij}(k)$ are Gaussian, they are even
statistically independent.

The requirements above characterize a wide sense stationary (WSS) Gaussian process. The prerequisite (17.21) is often denoted as uncorrelated scattering. Therefore,
the considered channel model is also named *wide sense stationary uncorrelated
scattering (WSSUS)* model.

With these prerequisites, it can be shown that all $\left|h_{ij}\right|$ exhibit a Rayleigh proba-
bility density function and that $\arg\left[h_{ij}\right]$ is equally distributed in the interval $[-\pi, \pi]$.
If there is a line of sight between transmitter and receiver, then $\mathbf{E}\left[h_{ij}(k)\right] \neq 0$ and
$\left|h_{ij}\right|$ shows a Rician distribution.

17.4.2   Covariance Matrix of the MIMO Channel

If we use the term correlation, we always mean spatial correlation in the following.
First, we define the covariance matrix $\mathbf{R}_{HH}$ of our channel matrix $\mathbf{H}(k)$. As a covariance matrix is defined on vectors, we have to rearrange the matrix $\mathbf{H}(k)$ as a vector
by stacking all column vectors of the matrix one over the other. In the following, we
drop k to simplify the notation. Let

$$
\mathbf{H} = \left(\mathbf{h}_1 \; \mathbf{h}_2 \; \cdots \; \mathbf{h}_M\right) \tag{17.22}
$$

be the channel matrix decomposed into column vectors $\mathbf{h}_j \in \mathbb{C}^{N\times1}$; $j = 1, 2, \ldots, M$.
Then the stacking operation is defined as

$$
\mathrm{vec}\left(\mathbf{H}\right) = \begin{pmatrix} \mathbf{h}_1 \\ \mathbf{h}_2 \\ \vdots \\ \mathbf{h}_M \end{pmatrix} \tag{17.23}
$$

and the covariance matrix of $\mathbf{H}$ is

$$
\mathbf{R}_{HH} = \mathbf{E}\left[\mathrm{vec}\left(\mathbf{H}\right)\left(\mathrm{vec}\left(\mathbf{H}\right)\right)^H\right] \tag{17.24}
$$

where the superscript H denotes the Hermiteian operation also called conjugate transposition. The reader can find a summary of useful theorems on matrix calculus in the Appendix B. Given the matrix $\mathbf{X}$ then $\mathbf{X}^H = \left(\mathbf{X}^T\right)^* = \left(\mathbf{X}^*\right)^T$, where the superscript T stand for the transposition. In a similar way the Hermiteian operation can also be applied on vectors $\mathbf{v}$. With a column vector $\mathbf{v}$ the product $\mathbf{v}\mathbf{v}^H$ defines a matrix whereas $\mathbf{v}^H\mathbf{v}$ is the scalar product of the two vectors resulting in a scalar. A matrix $\mathbf{X}$ is said to be Hermiteian, if $\mathbf{X}^H = \mathbf{X}$. One convinces easily that $\mathbf{R}_{HH}$ is a Hermiteian matrix.

Example 1

As an example, we calculate the covariance matrix for a 2x2 channel matrix

$$\mathbf{H} = \begin{pmatrix} h_{11} & h_{12} \\ h_{21} & h_{22} \end{pmatrix} = \left(\mathbf{h}_1 \; \mathbf{h}_2\right) \tag{17.25}$$

and get from (17.24)

$$\mathbf{R}_{HH} = \mathbf{E}\left[\begin{pmatrix} \mathbf{h}_1 \\ \mathbf{h}_2 \end{pmatrix} \left(\mathbf{h}_1^H \; \mathbf{h}_2^H\right)\right] = \mathbf{E}\left[\begin{pmatrix} h_{11} \\ h_{21} \\ h_{12} \\ h_{22} \end{pmatrix} \left(h_{11}^* \; h_{21}^* \; h_{12}^* \; h_{22}^*\right)\right] \tag{17.26}$$

finally resulting in

$$\mathbf{R}_{HH} = \begin{pmatrix} \mathbf{E}\left[|h_{11}|^2\right] & \mathbf{E}\left[h_{11}h_{21}^*\right] & \mathbf{E}\left[h_{11}h_{12}^*\right] & \mathbf{E}\left[h_{11}h_{22}^*\right] \\ \mathbf{E}\left[h_{21}h_{11}^*\right] & \mathbf{E}\left[|h_{21}|^2\right] & \mathbf{E}\left[h_{21}h_{12}^*\right] & \mathbf{E}\left[h_{21}h_{22}^*\right] \\ \mathbf{E}\left[h_{12}h_{11}^*\right] & \mathbf{E}\left[h_{12}h_{21}^*\right] & \mathbf{E}\left[|h_{12}|^2\right] & \mathbf{E}\left[h_{12}h_{22}^*\right] \\ \mathbf{E}\left[h_{22}h_{11}^*\right] & \mathbf{E}\left[h_{22}h_{21}^*\right] & \mathbf{E}\left[h_{22}h_{12}^*\right] & \mathbf{E}\left[|h_{22}|^2\right] \end{pmatrix} \tag{17.27}$$

A square matrix $\mathbf{X}$ is called Hermiteian, if $\mathbf{X}^H = \mathbf{X}$. Apparently, $\mathbf{R}_{HH}$ is a Hermiteian matrix. Moreover we generalize from (17.27) that $\mathbf{R}_{HH} \in \mathbb{C}^{NM \times NM}$ holds, if $\mathbf{H} \in \mathbb{C}^{N \times M}$. For an uncorrelated MIMO channel with channel matrix $\mathbf{H} = \mathbf{H}_w$ the property (17.21) results in the the covariance matrix

$$\mathbf{R}_{H_w H_w} = \begin{pmatrix} 1 & 0 & 0 & 0 \\ 0 & 1 & 0 & 0 \\ 0 & 0 & 1 & 0 \\ 0 & 0 & 0 & 1 \end{pmatrix} = \mathbf{I}_4 \tag{17.28}$$

which is the 4x4 identity matrix $\mathbf{I}_4$. Please note, if the stochastic process $h_{ij}(k)$ is at least wide sense stationary, then $\mathbf{R}_{HH}$ does not depend on time k.

17.4.3   MIMO Channel Model with Correlation

With $\mathbf{H}_w$ and its covariance matrix $\mathbf{R}_{H_w H_w} = \mathbf{I}_{NM}$ in (17.28) we have introduced a spatially uncorrelated MIMO channel. In the following we prove that a channel matrix $\mathbf{H} \in \mathbb{C}^{N \times M}$ defined with the stacked column vectors

$$\text{vec}\,(\mathbf{H}) = \mathbf{G}^H \text{vec}\,(\mathbf{H}_w) \tag{17.29}$$

has the covariance matrix

$$\mathbf{R}_{HH} = \mathbf{G}^H \mathbf{G} \tag{17.30}$$

where $\mathbf{G} = \mathbf{R}_{HH}^{\frac{1}{2}}$ is called the square root matrix of $\mathbf{R}_{HH}$.

Proof of (17.30)

From (17.24) follows with (17.29) $\mathbf{R}_{HH} = \mathbf{E}\left[\mathbf{G}^H \text{vec}\,(\mathbf{H}_w)\,(\text{vec}\,(\mathbf{H}_w))^H\,\mathbf{G}\right]$. As $\mathbf{G}$ shall be nonrandom, this matrix is not subject to the expected value. Consequently, $\mathbf{R}_{HH} = \mathbf{G}^H \mathbf{E}\left[\text{vec}\,(\mathbf{H}_w)\,(\text{vec}\,(\mathbf{H}_w))^H\right]\mathbf{G}$ follows and $\mathbf{E}\left[\text{vec}\,(\mathbf{H}_w)\,(\text{vec}\,(\mathbf{H}_w))^H\right] = \mathbf{R}_{H_w H_w} = \mathbf{I}_{NM}$ results in $\mathbf{R}_{HH} = \mathbf{G}^H \mathbf{G}$, which finalizes the proof.

Example 2

Given a single input multiple output (SIMO) system with $M = 1$ transmit and $N = 2$ receive antennas. The channel matrix shall be $\mathbf{H} = \left(h_{11}\ h_{21}\right)^T = \left(1\ \frac{j}{2}\right)^T$ and approximately constant in a considered time interval. We are looking for the decomposition of the covariance matrix $\mathbf{R}_{HH} = \mathbf{G}^H \mathbf{G}$ with the square root matrix $\mathbf{G} = \mathbf{R}_{HH}^{\frac{1}{2}}$.

First, we determine the covariance matrix

$$\mathbf{R}_{HH} = \mathbf{E}\left[\begin{pmatrix} h_{11} \\ h_{21} \end{pmatrix}\begin{pmatrix} h_{11}^* & h_{21}^* \end{pmatrix}\right] = \mathbf{E}\left[\begin{pmatrix} 1 \\ \frac{j}{2} \end{pmatrix}\begin{pmatrix} 1 & -\frac{j}{2} \end{pmatrix}\right] = \begin{pmatrix} 1 & -\frac{j}{2} \\ \frac{j}{2} & \frac{1}{4} \end{pmatrix} \tag{17.31}$$

As $\mathbf{R}_{HH}$ is a Hermiteian matrix, we know from Appendix B that this matrix is unitarily diagonalizable with the eigenvalues $\lambda_i \geq 0$; $i = 1, 2$ and the unitary transfom matrix $\mathbf{V}$ with $\mathbf{V}^{-1} = \mathbf{V}^H$

$$\mathbf{R}_{HH} = \mathbf{V}\Lambda\mathbf{V}^H = \mathbf{V}\Lambda^{\frac{1}{2}}\left(\Lambda^{\frac{1}{2}}\right)^H \mathbf{V}^H \tag{17.32}$$

From the comparison with (17.30), we conclude $\mathbf{V}\Lambda^{\frac{1}{2}} = \mathbf{G}^H$ and get the square root matrix $\mathbf{G} = \Lambda^{\frac{1}{2}}\mathbf{V}^H = \mathbf{R}_{HH}^{\frac{1}{2}}$ knowing that the entries of the diagonal matrix $\Lambda^{\frac{1}{2}}$ are real. Apparently, $\mathbf{R}_{HH}^{\frac{1}{2}}$ is not a Hermiteian matrix in general. The diagonal matrix $\Lambda = \text{diag}\,(\lambda_1, \lambda_2)$ contains the eigenvalues of $\mathbf{R}_{HH}$. The characteristic equation for the eigenvalues is

$$\begin{vmatrix} 1 - \lambda & -\frac{j}{2} \\ \frac{j}{2} & \frac{1}{4} - \lambda \end{vmatrix} = 0 \tag{17.33}$$

which yields $\lambda_1 = 0$, $\lambda_2 = \frac{5}{4}$ and thus

$$\Lambda = \begin{pmatrix} 0 & 0 \\ 0 & \frac{5}{4} \end{pmatrix} \; ; \; \Lambda^{\frac{1}{2}} = \begin{pmatrix} 0 & 0 \\ 0 & \frac{\sqrt{5}}{2} \end{pmatrix} \tag{17.34}$$

The matrix $\mathbf{V}$ is composed of the eigenvectors $\mathbf{v}_i = \begin{pmatrix} v_{1i} & v_{2i} \end{pmatrix}^T$ corresponding to the two eigenvalues and determined by

$$\begin{pmatrix} 1 - \lambda_i & -\frac{j}{2} \\ \frac{j}{2} & \frac{1}{4} - \lambda_i \end{pmatrix} \begin{pmatrix} v_{1i} \\ v_{2i} \end{pmatrix} = \mathbf{0} \tag{17.35}$$

For λ_1 we obtain $\mathbf{v}_1 = \begin{pmatrix} \frac{j}{2} v_{21} & v_{21} \end{pmatrix}^T$. In the same way we get $\mathbf{v}_2 = \begin{pmatrix} -j2v_{22} & v_{22} \end{pmatrix}^T$. The entries $v_{21} \in \mathbb{C}$ and $v_{22} \in \mathbb{C}$ are free parameters, however, as $\mathbf{R}_{HH}$ is a Hermiteian matrix, the two eigenvectors must be orthogonal. An orthogonalization and a normalization of these vectors are not required at this point. Thus we obtain the transform matrix

$$\mathbf{V} = \begin{pmatrix} \mathbf{v}_1 & \mathbf{v}_2 \end{pmatrix} = \begin{pmatrix} \frac{j}{2} v_{21} & -j2v_{22} \\ v_{21} & v_{22} \end{pmatrix} \tag{17.36}$$

Then follows with (17.34)

$$\mathbf{G} = \mathbf{R}_{HH}^{\frac{1}{2}} = \Lambda^{\frac{1}{2}} \mathbf{V}^H = \begin{pmatrix} 0 & 0 \\ j\sqrt{5} v_{22}^* & \frac{\sqrt{5}}{2} v_{22}^* \end{pmatrix} \tag{17.37}$$

$\mathbf{R}_{HH}^{\frac{1}{2}}$ is not a Hermiteian matrix, as $\mathbf{R}_{HH}^{\frac{1}{2}} \neq \left(\mathbf{R}_{HH}^{\frac{1}{2}} \right)^H$ holds. Finally, with the condition $\mathbf{R}_{HH} = \mathbf{G}^H \mathbf{G}$ follows for the free parameter $|v_{22}|^2 = \frac{1}{5}$ yielding $v_{22} = \pm\frac{1}{\sqrt{5}}$ and $v_{22} = \pm\frac{j}{\sqrt{5}}$. Then the results for the matrix decomposition are

$$\mathbf{G} = \mathbf{R}_{HH}^{\frac{1}{2}} = (\pm 1) \begin{pmatrix} 0 & 0 \\ j & \frac{1}{2} \end{pmatrix} \quad \text{and} \quad \mathbf{G} = \mathbf{R}_{HH}^{\frac{1}{2}} = (\pm j) \begin{pmatrix} 0 & 0 \\ j & \frac{1}{2} \end{pmatrix} \tag{17.38}$$

The reader convinces oneself easily that $\mathbf{G}^H \mathbf{G} = \mathbf{R}_{HH}$ is true.

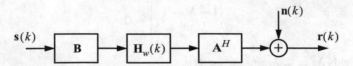

Fig. 17.6 Wireless link with an uncorrelated channel $\mathbf{H}_w(k)$ enhanced by matrices $\mathbf{B} = \mathbf{R}_{tx}^{1/2}$ and $\mathbf{A} = \mathbf{R}_{rx}^{1/2}$ introducing transmit and receive correlation, respectively

17.4.4   MIMO Channel Model with Transmit and Receive Correlation (Kronecker Model)

The wireless transmitter and receiver may operate in environments with different scattering and reflections at the transmitter and the receiver side. Then a description of the correlation close to the antenna elements by dedicated transmit and receive covariance matrices, $\mathbf{R}_{tx}$ and $\mathbf{R}_{rx}$, rather than by only one channel correlation matrix $\mathbf{R}_{HH}$ is reasonable.

The resulting channel model is depicted in Fig. 17.6 as a block diagram, where the matrices $\mathbf{B} \in \mathbb{C}^{M \times M}$ and $\mathbf{A} \in \mathbb{C}^{N \times N}$ shall enhance the uncorrelated MIMO channel with matrix $\mathbf{H}_w(k)$ by the square root matrices $\mathbf{B} = \mathbf{R}_{tx}^{\frac{1}{2}}$ and $\mathbf{A} = \mathbf{R}_{rx}^{\frac{1}{2}}$, where

$$\mathbf{R}_{tx} = \mathbf{B}^H \mathbf{B} \tag{17.39}$$

and

$$\mathbf{R}_{rx} = \mathbf{A}^H \mathbf{A} \tag{17.40}$$

holds.

Obviously, the channel matrix for this model is then described by

$$\mathbf{H}(k) = \mathbf{A}^H \mathbf{H}_w(k) \mathbf{B} \tag{17.41}$$

The correlation matrices $\mathbf{R}_{tx}$ and $\mathbf{R}_{rx}$ are determined from the channel matrix

$$\mathbf{H} = \begin{pmatrix} h_{11} & \cdots & h_{1j} & \cdots & h_{1M} \\ & & \vdots & & \\ h_{i1} & \cdots & h_{ij} & \cdots & h_{iM} \\ & & \vdots & & \\ h_{N1} & \cdots & h_{Nj} & \cdots & h_{NM} \end{pmatrix} = \left(\mathbf{h}_1 \; \cdots \; \mathbf{h}_j \; \cdots \; \mathbf{h}_M \right) = \begin{pmatrix} \mathbf{g}_1^T \\ \vdots \\ \mathbf{g}_i^T \\ \vdots \\ \mathbf{g}_N^T \end{pmatrix} \tag{17.42}$$

using the column vector

$$\mathbf{h}_j = \left(h_{1j} \; \cdots \; h_{ij} \; \cdots \; h_{Nj} \right)^T \tag{17.43}$$

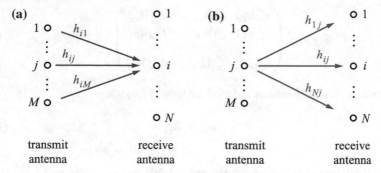

Fig. 17.7 (a) Multiple input single output (MISO) channel determined by row vector $\mathbf{g}_i^T = \left(h_{i1} \cdots h_{ij} \cdots h_{iM} \right)$, (b) Single input multiple output (SIMO) channel determined by column vector $\mathbf{h}_j = \left(h_{1j} \cdots h_{ij} \cdots h_{Nj} \right)^T$

and the row vector

$$\mathbf{g}_i^T = \left(h_{i1} \cdots h_{ij} \cdots h_{iM} \right) \tag{17.44}$$

The channel vector $\mathbf{g}_i^T$ determines the multiple input single output (MISO) channel depicted in Fig. 17.7a and $\mathbf{h}_j$ describes the single input multiple output (SIMO) scheme in Fig. 17.7b.

Then the transpose of the transmit covariance matrix is defined as

$$\mathbf{R}_{tr}^T = \mathbf{E}\left[\mathbf{g}_i \mathbf{g}_i^H \right] \; ; \; i = 1, 2, \ldots, N \tag{17.45}$$

which shall be unchanged irrespective of what antenna element i is receiving. Similarly, the receive covariance matrix is determined as

$$\mathbf{R}_{rx} = \mathbf{E}\left[\mathbf{h}_j \mathbf{h}_j^H \right] \; ; \; j = 1, 2, \ldots, M \tag{17.46}$$

and shall be the same irrespective of what transmit antenna element j is emitting. Furthermore, we assume that all main diagonal entries t_{ii} of $\mathbf{R}_{tx}$ and r_{jj} of $\mathbf{R}_{rx}$ are normalized equal to one.

Below we show that from the channel model (17.41) the stacked vector of the channel matrix can be derived as

$$\text{vec}\left(\mathbf{H} \right) = \left(\mathbf{B}^T \otimes \mathbf{A}^H \right) \text{vec}\left(\mathbf{H}_w \right) \tag{17.47}$$

With the covariance matrices (17.45) and (17.46) the covariance matrix $\mathbf{R}_{HH}$ of the channel follows as:

$$\mathbf{R}_{HH} = \mathbf{R}_{tx}^* \otimes \mathbf{R}_{rx} \tag{17.48}$$

where $\otimes$ symbolizes the Kronecker product. The proof is also found below. Given the matrix $\mathbf{R}_{tx}^* \in \mathbb{C}^{M \times M}$ with elements $t_{lm}^* \; ; \; l, m = 1, 2, \ldots, M$ and $\mathbf{R}_{rx} \in \mathbb{C}^{N \times N}$ then the Kronecker product is defined as

$$\mathbf{R}_{HH} = \begin{pmatrix} t_{11}^* \mathbf{R}_{rx} & t_{12}^* \mathbf{R}_{rx} & \cdots & t_{1M}^* \mathbf{R}_{rx} \\ t_{21}^* \mathbf{R}_{rx} & t_{22}^* \mathbf{R}_{rx} & \cdots & t_{2M}^* \mathbf{R}_{rx} \\ & & \cdots & \\ t_{M1}^* \mathbf{R}_{rx} & t_{M2}^* \mathbf{R}_{rx} & \cdots & t_{MM}^* \mathbf{R}_{rx} \end{pmatrix} \in \mathbb{C}^{MN \times MN} \qquad (17.49)$$

Furthermore, it is straightforward to show with (17.45) that

$$\mathbf{R}_{tx}^* = \mathbf{R}_{tx}^T \qquad (17.50)$$

holds.

In the following proofs of (17.47) and (17.48), we use three Lemmas of linear algebra, [5]: Let $\mathbf{Q}$, $\mathbf{R}$, $\mathbf{S}$, and $\mathbf{T}$ be matrices of compatible dimensions. Then

$$\text{vec} \left(\mathbf{QRS} \right) = \left(\mathbf{S}^T \otimes \mathbf{Q} \right) \text{vec} \left(\mathbf{R} \right) \qquad (17.51)$$

$$\left(\mathbf{Q} \otimes \mathbf{R} \right)^H = \mathbf{Q}^H \otimes \mathbf{R}^H \qquad (17.52)$$

$$\left(\mathbf{Q} \otimes \mathbf{R} \right) \left(\mathbf{S} \otimes \mathbf{T} \right) = \mathbf{QS} \otimes \mathbf{RT} \qquad (17.53)$$

Proof of (17.47)

With (17.51) follows directly from the channel model defined by (17.41) $\text{vec} \left(\mathbf{H} \right) = \text{vec} \left(\mathbf{A}^H \mathbf{H}_w \mathbf{B} \right) = \left(\mathbf{B}^T \otimes \mathbf{A}^H \right) \text{vec} \left(\mathbf{H}_w \right)$ and the proof is finished.

Proof of (17.48)

From (17.24), we obtain with (17.47) $\mathbf{R}_{HH} = \mathbf{E} \left[\left(\left(\mathbf{B}^T \otimes \mathbf{A}^H \right) \text{vec} \left(\mathbf{H}_w \right) \right) \left(\left(\mathbf{B}^T \otimes \mathbf{A}^H \right) \text{vec} \left(\mathbf{H}_w \right) \right)^H \right]$ and using (17.52) $\mathbf{R}_{HH} = \mathbf{E} \left[\left(\left(\mathbf{B}^T \otimes \mathbf{A}^H \right) \text{vec} \left(\mathbf{H}_w \right) \right) \left(\text{vec} \left(\mathbf{H}_w \right) \right)^H \left(\mathbf{B}^* \otimes \mathbf{A} \right) \right] = \left(\mathbf{B}^T \otimes \mathbf{A}^H \right) \mathbf{E} \left[\left(\text{vec} \left(\mathbf{H}_w \right) \right) \left(\text{vec} \left(\mathbf{H}_w \right) \right)^H \right] \left(\mathbf{B}^* \otimes \mathbf{A} \right)$.
With $\mathbf{E} \left[\left(\text{vec} \left(\mathbf{H}_w \right) \right) \left(\text{vec} \left(\mathbf{H}_w \right) \right)^H \right] = \mathbf{I}_{MN}$ and (17.53) follows $\mathbf{R}_{HH} = \mathbf{B}^T \mathbf{B}^* \otimes \mathbf{A}^H \mathbf{A}$. Inserting (17.39) and (17.40), eventually yields the result $\mathbf{R}_{HH} = \mathbf{R}_{tx}^* \otimes \mathbf{R}_{rx}$ and the proof is finished.

Proof of (17.46)

We prove in the following that the channel model determined by (17.41) results in a receive correlation matrix $\mathbf{R}_{rx}$ given by (17.46), which is identical for all transmit antenna indices j. First, we write $\mathbf{R}_{HH}$ from (17.26) for the general case

$$\mathbf{R}_{HH} = \mathbf{E} \left[\begin{pmatrix} \mathbf{h}_1 \\ \vdots \\ \mathbf{h}_M \end{pmatrix} \left(\mathbf{h}_1^H \cdots \mathbf{h}_M^H \right) \right] = \begin{pmatrix} \mathbf{E} \left[\mathbf{h}_1 \mathbf{h}_1^H \right] & \cdots & \mathbf{E} \left[\mathbf{h}_1 \mathbf{h}_M^H \right] \\ \vdots & \cdots & \vdots \\ \mathbf{E} \left[\mathbf{h}_M \mathbf{h}_1^H \right] & \cdots & \mathbf{E} \left[\mathbf{h}_M \mathbf{h}_M^H \right] \end{pmatrix} \qquad (17.54)$$

When we compare the main diagonal entries of (17.49) and (17.54) we see that $\mathbf{E} \left[\mathbf{h}_1 \mathbf{h}_1^H \right] = \mathbf{R}_{rx}, \ldots, \mathbf{E} \left[\mathbf{h}_M \mathbf{h}_M^H \right] = \mathbf{R}_{rx}$ are independent of j, because we have assumed that the main diagonal elements of $\mathbf{R}_{tx}$ are $t_{11} = \cdots = t_{MM} = 1$. This finalizes the proof.

Proof of (17.45)

$\mathbf{R}_{tx}^T$ is defined in (17.45) through the vectors $\mathbf{g}_i$ and we can determine the transpose of the channel matrix (17.42) as

$$\mathbf{H}^T = \begin{pmatrix} \mathbf{g}_1 & \cdots & \mathbf{g}_N \end{pmatrix} \tag{17.55}$$

Please note that $\mathbf{g}_i$ is the column vector associated to $\mathbf{g}_i^T$. Then follows

$$\mathbf{R}_{H^T H^T} = \mathbf{E}\left[\text{vec}\,(\mathbf{H}^T)\,(\text{vec}\,(\mathbf{H}^T))^H\right] = \begin{pmatrix} \mathbf{E}\left[\mathbf{g}_1 \mathbf{g}_1^H\right] & \cdots & \mathbf{E}\left[\mathbf{g}_1 \mathbf{g}_N^H\right] \\ & \cdots & \\ \mathbf{E}\left[\mathbf{g}_N \mathbf{g}_1^H\right] & \cdots & \mathbf{E}\left[\mathbf{g}_N \mathbf{g}_N^H\right] \end{pmatrix} \tag{17.56}$$

Next, we show that from the channel model (17.41) follows:

$$\mathbf{R}_{H^T H^T} = \mathbf{R}_{rx} \otimes \mathbf{R}_{tx}^* \tag{17.57}$$

Equation (17.41) provides $\text{vec}\,(\mathbf{H}^T) = \text{vec}\,(\mathbf{B}^T \mathbf{H}_w^T \mathbf{A}^*)$ and with Lemma (17.51) follows $\text{vec}\,(\mathbf{H}^T) = \left(\mathbf{A}^H \otimes \mathbf{B}^T\right)\text{vec}\,(\mathbf{H}_w^T)$. Then we obtain with Lemma (17.52) $\left(\text{vec}\,(\mathbf{H}^T)\right)^H = \left(\text{vec}\,(\mathbf{H}_w^T)\right)^H (\mathbf{A} \otimes \mathbf{B}^*)$, and for the covariance matrix of $\mathbf{H}^T$ follows $\mathbf{R}_{H^T H^T} = \mathbf{E}\left[\left(\text{vec}\,(\mathbf{H}^T)\right)\left(\text{vec}\,(\mathbf{H}^T)\right)^H\right] = \left(\mathbf{A}^H \otimes \mathbf{B}^T\right)\mathbf{I}_{MN}\,(\mathbf{A} \otimes \mathbf{B}^*) = \mathbf{A}^H \mathbf{A} \otimes \mathbf{B}^T \mathbf{B}^*$, where we have used Lemma (17.53) in the last step. With (17.39) and (17.40) we obtain eventually $\mathbf{R}_{H^T H^T} = \mathbf{R}_{rx} \otimes \mathbf{R}_{tx}^*$ and the proof of (17.57) ends.

Let $\mathbf{R}_{rx} \in \mathbb{C}^{N \times N}$ be equipped with the entries r_{pq} ; $p, q = 1, 2, \ldots, N$, then the Kronecker product in (17.57) yields

$$\mathbf{R}_{H^T H^T} = \begin{pmatrix} r_{11}\mathbf{R}_{tx}^* & r_{12}\mathbf{R}_{tx}^* & \cdots & r_{1N}\mathbf{R}_{tx}^* \\ r_{21}\mathbf{R}_{tx}^* & \mathbf{R}_{tx}^* & \cdots & r_{2N}\mathbf{R}_{tx}^* \\ & & \cdots & \\ r_{N1}\mathbf{R}_{tx}^* & r_{N2}\mathbf{R}_{tx}^* & \cdots & r_{NN}\mathbf{R}_{tx}^* \end{pmatrix} \in \mathbb{C}^{MN \times MN} \tag{17.58}$$

When comparing the main diagonals of (17.56) and (17.58) taking the prerequisite $r_{11} = \ldots = r_{NN} = 1$ into account we conclude $\mathbf{R}_{tx}^* = \mathbf{E}\left[\mathbf{g}_1 \mathbf{g}_1^H\right] = \cdots = \mathbf{E}\left[\mathbf{g}_N \mathbf{g}_N^H\right]$, which is independent of the index i of the receive antenna element. With $\mathbf{R}_{tx}^* = \mathbf{R}_{tx}^T$ from (17.50) the proof is finished.

Example 3

We come back to the Example 2 in Sect. 17.4.3 and ask whether the channel meets the conditions of the Kronecker model. We find from (17.46) for the receive covariance matrix $\mathbf{R}_{rx} = \mathbf{E}\left[\mathbf{h}_1 \mathbf{h}_1^H\right] = \mathbf{R}_{HH}$ given in (17.31). The transmit covariance matrices defined in (17.45) are $\mathbf{R}_{tx,1}^T = \mathbf{E}\left[\mathbf{g}_1 \mathbf{g}_1^H\right] = \mathbf{E}\left[h_{11} h_{11}^*\right] = 1$ and $\mathbf{R}_{tx,2}^T = \mathbf{E}\left[\mathbf{g}_2 \mathbf{g}_2^H\right] = \mathbf{E}\left[h_{21} h_{21}^*\right] = \frac{1}{4}$. Obviously, they are different and we conclude that the channel matrix in Example 2 does not fulfill the conditions of the Kronecker channel model.

17.4.5 Exponential Covariance Matrix Model

This model also follows the idea of separate covariance matrices at the transmitter and the receiver. However, instead of determining these matrices by a MISO and a SIMO scheme, as applied by the Kronecker model, fixed covariance matrices with an exponential decay of the magnitudes of the correlation coefficients outside of the main diagonals is presumed.

Let

$$(\mathbf{R}_{tx})_{pq} = \rho_{tx}^{|p-q|^\beta} \; ; \; p, q = 1, 2, \ldots, M \tag{17.59}$$

be the covariance coefficients with $|\rho_{tx}| \leq 1$ and be the entries of the transmit covariance matrix $\mathbf{R}_{tx}$. Similarly,

$$(\mathbf{R}_{rx})_{pq} = \rho_{rx}^{|p-q|^\beta} \; ; \; p, q = 1, 2, \ldots, N \tag{17.60}$$

are the entries of the receive covariance matrix $\mathbf{R}_{rx}$ with $|\rho_{rx}| \leq 1$ and β typically is in the range from 1 to 2. Only for the main diagonal elements the equal sign holds. Small magnitudes $|\rho_{tx}|$ and $|\rho_{rx}|$ characterize low correlation and values close to 1 indicate a strong correlation of the delay spread functions between the antenna elements p and q. Obviously, with this model the magnitudes of the covariance coefficients decay exponentially with the distance $|p - q|$ between the antenna elements with numbers p and q. This model is motivated from measurements of the local transmit and receive region. $\mathbf{R}_{HH}$ is then given by (17.48).

Finally, the exponential covariance model is also applicable directly to the entries of $\mathbf{R}_{HH}$ in (17.24) without the differentiation between the transmit and the receive correlation, [6].

Example 4

An example of a transmit correlation matrix for a transmit antenna with $M = 4$ elements follows from (17.59) assuming $\beta = 1$,

$$\mathbf{R}_{tx} = \begin{pmatrix} 1 & \rho_{tx} & \rho_{tx}^2 & \rho_{tx}^3 \\ \rho_{tx} & 1 & \rho_{tx} & \rho_{tx}^2 \\ \rho_{tx}^2 & \rho_{tx} & 1 & \rho_{tx} \\ \rho_{tx}^3 & \rho_{tx}^2 & \rho_{tx} & 1 \end{pmatrix} \tag{17.61}$$

References

1. L.A. Zadeh, Frequency analysis of variable networks. Proc. IRE (1950)
2. P.A. Bello, Characterization of randomly time-variant linear channels. IEEE Trans. Commun. Syst. (1963)
3. A.F. Molisch, *Wireless Communications* (Wiley and IEEE Press, New York, 2009)

4. Spatial channel model for MIMO simulations, Technical report, 3GPP TR 25996 V 9.0.0 (2010)
5. R.A. Horn, C.R. Johnson, *Matrix Analysis* (Cambridge University Press, Cambridge, 2013)
6. S. Loyka, Channel capacity of MIMO architecture using the exponential correlation matrix. IEEE Commun. Lett. **5**, 369–371 (2001)

Chapter 18
Principles of Linear MIMO Receivers

18.1 Introduction

As depicted in the block diagram of Fig. 18.1, we consider a MIMO system with frequency flat and in general time-varying channel with channel matrix $\mathbf{H}(k) \in \mathbb{C}^{N \times M}$, input signal vector $\mathbf{s}(k) \in \mathbb{C}^{M \times 1}$, noise vector $\mathbf{n}(k) \in \mathbb{C}^{N \times 1}$, and receive vector $\mathbf{r}(k) \in \mathbb{C}^{N \times 1}$. At the receiver, a linear filter described by a matrix $\mathbf{W}(k) \in \mathbb{C}^{M \times N}$ is employed to obtain at its output a good replica $\mathbf{y}(k)$ of the transmit signal vector $\mathbf{s}(k)$. We assume that a channel estimator not shown in Fig. 18.1 has provided perfect channel state information so that the instantaneous channel matrix $\mathbf{H}(k)$ is known for every discrete-time instant k at the receiver.

From Fig. 18.1, we obtain

$$\mathbf{r}(k) = \mathbf{H}(k)\mathbf{s}(k) + \mathbf{n}(k) \tag{18.1}$$

and

$$\mathbf{y}(k) = \mathbf{W}(k)\mathbf{r}(k) \tag{18.2}$$

By inserting (18.1) into (18.2), we get

$$\mathbf{y}(k) = \mathbf{W}(k)\mathbf{H}(k)\mathbf{s}(k) + \mathbf{W}(k)\mathbf{n}(k) = \mathbf{G}(k)\mathbf{s}(k) + \mathbf{n}'(k) \tag{18.3}$$

with the interference term

$$\mathbf{G}(k) = \mathbf{W}(k)\mathbf{H}(k) \tag{18.4}$$

and the noise

$$\mathbf{n}'(k) = \mathbf{W}(k)\mathbf{n}(k) \tag{18.5}$$

Let $\mathbf{G}(k) = \left(g_{ij}(k)\right)_{M \times M}$ and $\mathbf{W}(k) = \left(w_{ij}(k)\right)_{M \times N}$, then we obtain the ith line of the system of equations (18.3)

© Springer Nature Switzerland AG 2019
J. Speidel, *Introduction to Digital Communications*, Signals and Communication
Technology, https://doi.org/10.1007/978-3-030-00548-1_18

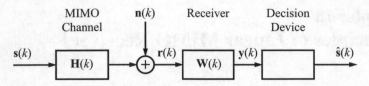

Fig. 18.1 Block diagram of a MIMO system with channel matrix $\mathbf{H}(k)$ and linear receiver with matrix $\mathbf{W}(k)$

$$y_i(k) = g_{ii}(k)s_i(k) + \sum_{\substack{j=1 \\ j \neq i}}^{M} g_{ij}(k)s_j(k) + \sum_{j=1}^{M} w_{ij}(k)n_j(k) \; ; \; i = 1, ..., M$$

(18.6)

The first term in (18.6) is the desired signal $s_i(k)$ of the transmit antenna i with some weighting coefficient $g_{ii}(k)$. The second term is denoted as inter-channel interference composed of the signals from all other transmit antennas and the last contribution is the noise component at the receiver output. Please note that the $g_{ij}(k)$ are functions of the receive matrix elements $w_{ij}(k)$. The task of the linear receiver is to adjust its matrix $\mathbf{W}$ in such a way that the inter-channel interference is reduced or even completely removed.

18.2   Operation Modes for MIMO Systems

Before we are going into more details of the MIMO receivers, we first take the opportunity with (18.1) to get some inside into the options provided by MIMO systems. For that purpose, we first take a simple example with $M = N = 2$ and rewrite (18.1) in detail.

$$r_1(k) = h_{11}(k)s_1(k) + h_{12}(k)s_2(k) + n_1(k)$$

(18.7)

$$r_2(k) = h_{21}(k)s_1(k) + h_{22}(k)s_2(k) + n_2(k)$$

(18.8)

Spatial Multiplexing to Increase Data Rate

In this MIMO transmission mode, two independent signals $s_1(k)$ and $s_2(k)$ from antenna 1 and antenna 2 can be sent, respectively. Consequently, we can increase the symbol rate provided that the receiver is able to recover $s_1(k)$ and $s_2(k)$ from the receive signals $r_1(k)$ and $r_2(k)$, and the spatial correlation of the delay spread functions h_{ij} is small.

Spatial Diversity to Improve Transmission Quality

As a starting point, let us consider that only one antenna at the transmitter and one antenna at receiver are available. Then, the receive signal is

$$r_1(k) = h_{11}(k)s_1(k) + n_1(k) \tag{18.9}$$

If the delay spread function $h_{11}(k)$ is zero due to fading at time instant k, it is impossible to recover $s_1(k)$ because we just have received noise $n_1(k)$. This can be overcome by providing more than one antenna either at the transmitter or at the receiver or at both sides provided that the spatial correlation of the delay spread functions is small.

Simple Transmit Diversity

Let us consider the case of a simple transmit diversity scheme with $M = 2$ transmit and $N = 1$ receive antennas and that on each of the transmit antennas the same signal is sent, thus $s_2(k) = s_1(k)$. The receive signal then is

$$r_1(k) = (h_{11}(k) + h_{12}(k)) s_1(k) + n_1(k) \tag{18.10}$$

We see even if one of the channel coefficients is zero due to fading the transmit signal $s_1(k)$ can still be recovered, owing to the redundant channel path established by the other transmit antenna. The probability that both paths $h_{11}(k)$ and $h_{12}(k)$ are zero at the same time instant k due to fading is rather small, if the two channels are almost independent. Therefore, transmit diversity helps to increase transmission quality by providing spatial redundancy.

Simple Receive Diversity

Consider now the case of a simple receive diversity with $N = 2$ receive and $M = 1$ transmit antennas, which allows us to transmit one signal $s_1(k)$. Then, we obtain at the receiver

$$r_1(k) = h_{11}(k)s_1(k) + n_1(k) \tag{18.11}$$

$$r_2(k) = h_{21}(k)s_1(k) + r_2(k) \tag{18.12}$$

and we can argue in a similar way as before. Therefore receive diversity can also help to improve transmission quality. However, we see that in both cases of simple transmit and receive diversity only one signal $s_1(k)$ can be transmitted and no increase of the symbol rate is achieved.

Beamforming

With more than one antenna at the transmitter the emitted electromagnetic beam can be steered in a certain direction, e.g., to a "hot spot" of many mobile users. The principle is that a matrix $\mathbf{A}$ changes the direction of a vector $\mathbf{x}$ after multiplication $\mathbf{A}\mathbf{x}$. This technique is called beamforming and in most cases only one receive antenna is used.

Combining Spatial Multiplex and Diversity or Beamforming

The general MIMO approach can take advantage of all these features, namely, spatial multiplex to increase data rate, diversity to improve transmission quality, or beam forming in the sense of a compromise. In case of $M = N = 2$, the equations (18.7) and (18.8) describe the input–output relation of the MIMO system.

In the next Sections, we will discuss methods how to recover the transmit from the receive signals. For that purpose, we assume perfect channel knowledge at the receiver, i.e., we impose the prerequisite that the complete channel matrix $\mathbf{H}(k)$ is known at every time instant k. Furthermore, we drop the discrete-time variable k to simplify the notation.

18.3 Zero-Forcing Receiver for Equal Number of Transmit and Receive Antennas

In this section, we derive the matrix $\mathbf{W}$ of a zero-forcing (ZF) receiver for the case that the transmitter and the receiver are equipped with the same number of antennas, i.e.,

$$M = N \tag{18.13}$$

Consequently, $\mathbf{H} \in \mathbb{C}^{M \times M}$ is a square matrix. Furthermore, assume a channel matrix with full rank

$$\text{rank}(\mathbf{H}) = M \tag{18.14}$$

Then, the inverse matrix $\mathbf{H}^{-1}$ exists. Moreover, if we consider very small noise at the receiver, $\mathbf{n} \approx \mathbf{0}$, we get from (18.3) approximately

$$\mathbf{y} = \mathbf{WHr} \tag{18.15}$$

With $\mathbf{WH} = \mathbf{I}_M$, the inverse matrix

$$\mathbf{W} = \mathbf{H}^{-1} \tag{18.16}$$

provides the solution. Then, we obtain the output signal of the receiver from (18.3)

$$\mathbf{y} = \mathbf{s} + \mathbf{H}^{-1}\mathbf{n} \tag{18.17}$$

We see that the inter-channel interference (18.3) is completely removed, $\mathbf{G}(k) = \mathbf{I}_M$ and the receiver output signal $\mathbf{y}$ is just corrupted by the noise $\mathbf{Wn} = \mathbf{H}^{-1}\mathbf{n}$. To check how the noise has changed, we calculate the mean noise power

$$\mathbf{E}\left[\|\mathbf{Wn}\|^2\right] \tag{18.18}$$

and find that in some cases this value can be larger than the original mean noise power $\mathbf{E}\left[\|\mathbf{n}\|^2\right]$ making the detection of the bit sequence in the receive signal more inaccurate.

18.4 Zero-Forcing Receiver for Unequal Number of Transmit and Receive Antennas

18.4.1 Receiver with More Antennas than Transmitter, $N > M$

Solving the System of Linear Equations

Example 1

Let us first consider a simple example with $M = 2$ transmit and $N = 3$ receive antennas and assume small noise $\mathbf{n} \approx \mathbf{0}$. Then we obtain from the basic input–output relation (18.1)

$$r_1 = h_{11}s_1 + h_{12}s_2 \tag{18.19}$$

$$r_2 = h_{21}s_1 + h_{22}s_2 \tag{18.20}$$

$$r_3 = h_{31}s_1 + h_{32}s_2 \tag{18.21}$$

Obviously, this is an over-determined system with $N = 3$ equations for the $M = 2$ unknowns. The channel matrix $\mathbf{H}$ is non-square, and consequently an inverse matrix does not exist. We know from linear algebra that such a system of equations can only be solved exactly, if and only if at least $N - M$ equations are linearly depending on each other. In other words, for the rank of the matrix $\mathbf{H}$, which is the number of linearly independent lines or columns, $\text{rank}(\mathbf{H}) \leq M$ must be true. Consequently, we can cancel $N - M$ superfluous equations. The remaining system with M equations for the M unknowns can be solved conventionally. Whether this remaining system has one unique solution or an infinite manifold of a parametric solution depends on the rank of the remaining $M \times M$ square matrix.

If the three equations are all independent, no solution exists. This fact is easily understood because in our example we could take two equations out of the three and solve them. Then we insert the resulting s_1 and s_2 into the third equation and are faced with a contradiction indicating that the obtained "solution" is not feasible for the original 3x2 system of equations.

This is the mathematical view. But what happens in a real system? Are the three equations linearly depending or are they independent? For the answer, we can think of an experiment, where we sent two known signals $s_1(k)$ and $s_2(k)$ from the two antennas over the channel with the known coefficients h_{11} and h_{12}. Then we measure the received signals $r_1(k), ..., r_3(k)$. Equations (18.19)–(18.21) must hold because every element in these equations is physically valid. Thus, we conclude that in reality the three equations must be linearly depending on each other. However, if additive noises, which are independent of each other and independent of all signals come into play, then the resulting equations may no longer be linearly depending and the over-determined system will have no solution.

Solution with the Pseudo Inverse

The fact that a system of equations can have no solution was somehow disappointing in mathematics. Therefore, in early 1920, the mathematician E. H. Moore came up with an approximate solution, [1]. However, it is reported that his notation was hard to understand because of the telegraphic style and the idiosyncratic notation. Therefore, his work was not recognized enough, until a formulation by R. Penrose [2, 3] appeared. Penrose formulated four axioms for a special kind of an inverse matrix, which is denoted as Moore–Penrose inverse in the literature. It is applicable in general for finding an inverse of rectangular matrices of any dimension approximately and is also called pseudo inverse matrix.

Our problem can be formulated as follows. We are looking for a general not necessarily exact solution, which gives us the transmit signal vector $\mathbf{s}$ or an estimate of it from the receive signal vector $\mathbf{r}$ at every time instant k. As the system of equations (18.1) is over-determined for $N > M$, we can obtain an approximate solution in the sense that the left-hand side of (18.1) is just approximately equal to the right-hand side. As a prerequisite, the channel matrix $\mathbf{H} \in \mathbb{C}^{N \times M}$ shall have full rank, i.e.,

$$\text{rank}\,(\mathbf{H}) = M \tag{18.22}$$

In the following, we also assume small noise, i.e., $\mathbf{n} \approx \mathbf{0}$, and minimize the difference between the left- and the right-hand side of (18.1),

$$\Delta = \mathbf{r} - \mathbf{H}\mathbf{s} \tag{18.23}$$

in the mean squared error sense,

$$\|\Delta\|^2 = \Delta^H \Delta = \min_{\mathbf{s}} \tag{18.24}$$

The result is

$$\mathbf{s} = \left(\mathbf{H}^H \mathbf{H}\right)^{-1} \mathbf{H}^H \mathbf{r} \tag{18.25}$$

and we find the matrix of the zero-forcing receiver as

$$\mathbf{H}^+ = \left(\mathbf{H}^H \mathbf{H}\right)^{-1} \mathbf{H}^H = \mathbf{W} \tag{18.26}$$

One easily checks that $\mathbf{H}^+ \in \mathbb{C}^{M \times N}$. Plugging (18.26) into (18.4) yields

$$\mathbf{G} = \mathbf{H}^+ \mathbf{H} = \left(\mathbf{H}^H \mathbf{H}\right)^{-1} \mathbf{H}^H \mathbf{H} = \mathbf{I}_M \tag{18.27}$$

Hence, $\mathbf{H}^+$ acts as a kind of inverse matrix with respect to $\mathbf{H}$. Therefore, $\mathbf{H}^+$ is called (left-hand sided) *pseudo inverse* or *Moore–Penrose inverse matrix*. From the perspective of linear algebra, we are solving the modified system of equations $\mathbf{H}^+\mathbf{H}\mathbf{s} = \mathbf{H}^+\mathbf{r}$ rather than the original one $\mathbf{H}\mathbf{s} = \mathbf{r}$ given in (18.19)–(18.21). This provides the exact solution $\mathbf{s}$. As a consequence for our MIMO system, we have to equip the receiver with the matrix $\mathbf{W}$ in (18.26) yielding the output signal

$$\mathbf{y} = \mathbf{W}\mathbf{r} = \mathbf{s} + \mathbf{H}^+\mathbf{n} \tag{18.28}$$

Again, we see that the output signal vector $\mathbf{y}$ of the zero-forcing receiver contains the original transmit signal vector, which is corrupted by additive noise. The inter-channel interference is completely canceled. As in the previous case $M = N$, the mean power of the resulting noise

$$\mathbf{n}' = \mathbf{H}^+\mathbf{n}. \tag{18.29}$$

can be increased compared to $\mathbf{n}$ by the receiver matrix $\mathbf{H}^+$.

From our assumption (18.22) follows that $\mathbf{H}^H\mathbf{H}$ has also full rank M and consequently $\left(\mathbf{H}^H\mathbf{H}\right)^{-1}$ exists.

Example 2

Show that the minimum squared error is

$$\|\Delta\|_{\min}^2 = \mathbf{r}^H \left(\mathbf{I}_N - \mathbf{H}\mathbf{H}^+\right)\mathbf{r} \tag{18.30}$$

Solution:

From (18.24) follows with (18.23) $\|\Delta\|^2 = \left(\mathbf{r}^H - \mathbf{s}^H\mathbf{H}^H\right)(\mathbf{r} - \mathbf{H}\mathbf{s})$. Owing to the solution $\mathbf{s} = \mathbf{H}^+\mathbf{r}$ from (18.25) with (18.26) and knowing that $\left(\mathbf{H}\mathbf{H}^+\right)^H = \mathbf{H}\mathbf{H}^+$ follows $\|\Delta\|^2 = \|\Delta\|_{\min}^2 = \mathbf{r}^H\mathbf{r} - \mathbf{r}^H\mathbf{H}\mathbf{H}^+\mathbf{r}$. This is equal to $\|\Delta\|_{\min}^2 = \mathbf{r}^H\left(\mathbf{I}_N - \mathbf{H}\mathbf{H}^+\right)\mathbf{r}$, which finalizes the proof. As expected, the minimal squared error depends on the matrix $\mathbf{H}$, the left-hand side $\mathbf{r}$ of the system of equations, and is in general not zero.

Example 3

To illustrate the problem of an over-determined system of equations let us consider the following example of three equations with two unknowns s_1 and s_2.

$$\begin{pmatrix} 1 & 1 \\ 1 & -1 \\ 2 & 1 \end{pmatrix} \begin{pmatrix} s_1 \\ s_2 \end{pmatrix} = \begin{pmatrix} 0 \\ 1 \\ 1 \end{pmatrix}$$

Obviously, the three equations are linearly independent. We can follow the path outlined above and derive a solution from the first and the second equation yielding

$s_1 = \frac{1}{2}$ and $s_2 = -\frac{1}{2}$. By plugging those into the third equation results in the contradiction $\frac{1}{2} = 1$. Alternatively, we can start with another pair of equations and proceed as before ending up with a contradiction again. Consequently, there is no solution. If we calculate the squared error between the left- and the right-hand side of the system of equations, we obtain $\|\Delta\|^2 = \frac{1}{4}$.

As shown before, the pseudo inverse (18.26) can provide an approximation by minimizing $\|\Delta\|^2$. The channel matrix is $\mathbf{H} = \begin{pmatrix} 1 & 1 \\ 1 & -1 \\ 2 & 1 \end{pmatrix}$. Next we have to calculate

$\mathbf{H}^H\mathbf{H} = \begin{pmatrix} 6 & 2 \\ 2 & 3 \end{pmatrix}$ and $(\mathbf{H}^H\mathbf{H})^{-1} = \frac{1}{14}\begin{pmatrix} 3 & -2 \\ -2 & 6 \end{pmatrix}$. Consequently, the pseudo inverse

follows as $\mathbf{H}^+ = \frac{1}{14}\begin{pmatrix} 1 & 5 & 4 \\ 4 & -8 & 2 \end{pmatrix}$. The minimal squared error is obtained from (18.30) as $\|\Delta\|^2_{\min} = \frac{1}{14}$, which is well below $\|\Delta\|^2 = \frac{1}{4}$.

For a MIMO transmission, the reduction of inter-channel interference is decisive. The reader assures oneself easily that the cascade of the channel and receive matrix

results in $\mathbf{H}^+\mathbf{H} = \frac{1}{14}\begin{pmatrix} 1 & 5 & 4 \\ 4 & -8 & 2 \end{pmatrix}\begin{pmatrix} 1 & 1 \\ 1 & -1 \\ 2 & 1 \end{pmatrix} = \begin{pmatrix} 1 & 0 \\ 0 & 1 \end{pmatrix}$ and the inter-channel interference is completely removed by the pseudo inverse $\mathbf{H}^+$.

This example verifies that the exact solution $\mathbf{s}$ is obtained by multiplying the receive vector $\mathbf{r}$ by the pseudo inverse matrix of the channel. However, the linear system of equations $\mathbf{r} = \mathbf{H}\mathbf{s}$ is solved only approximately with a squared error between left- and right-hand side of $\|\Delta\|^2 = \frac{1}{4}$.

Example 4

Is $\mathbf{W} = \mathbf{H}^+$ also valid for $N = M$?

The answer is yes because for $N = M$, the channel matrix is a square matrix and the prerequisite (18.22) guarantees that $\mathbf{H}^{-1}$ exists. Consequently, we obtain for the pseudo inverse from (18.26)

$$\mathbf{H}^+ = \mathbf{H}^{-1}\left(\mathbf{H}^H\right)^{-1}\mathbf{H}^H = \mathbf{H}^{-1}\mathbf{I}_M = \mathbf{H}^{-1} \tag{18.31}$$

and the pseudo inverse boils down to the inverse matrix (18.16). Furthermore, we conclude from (18.31) $\mathbf{H}\mathbf{H}^+ = \mathbf{I}_N$ and from (18.30) we find that the squared error then is zero, $\|\Delta\|^2_{\min} = 0$, as expected.

Proof of the Pseudo Inverse Matrix $\mathbf{H}^+$ in (18.26)

For the proof, we set the first partial derivative of our target function with respect to the free parameters equal to zero. To follow this line, we use the trace of a matrix, which we differentiate with respect to a matrix or a vector, as outlined in the Appendix B, and rewrite the squared error of (18.24) as

$$\|\Delta\|^2 = \text{tr}\left(\Delta\Delta^H\right) \tag{18.32}$$

Please note that the product of a column vector Δ and a row vector Δ^H results in a matrix. With (18.23) we obtain

$$\Delta\Delta^H = \mathbf{rr}^H - \mathbf{rs}^H\mathbf{H}^H - \mathbf{Hsr}^H + \mathbf{Hss}^H\mathbf{H}^H \tag{18.33}$$

We recognize that all terms are elements of $\mathbb{C}^{N\times N}$. According to the Appendix B, the trace of the sum of matrices is identical to the sum of the traces. To find the minimum, we set the partial derivative equal to zero, $\frac{\partial\,\mathrm{tr}(\Delta\Delta^H)}{\partial \mathbf{s}^*} = 0$, which is a necessary and sufficient condition here, because our quadratic error $\mathrm{tr}\left(\Delta\Delta^H\right)$ is a convex function of $\mathbf{s}$ or $\mathbf{s}^*$. The partial derivation yields

$$\frac{\partial\,\mathrm{tr}\left(\Delta\Delta^H\right)}{\partial \mathbf{s}^*} = \frac{\partial\,\mathrm{tr}\left(\mathbf{rr}^H\right)}{\partial \mathbf{s}^*} - \frac{\partial\,\mathrm{tr}\left(\mathbf{rs}^H\mathbf{H}^H\right)}{\partial \mathbf{s}^*} - \frac{\partial\,\mathrm{tr}\left(\mathbf{Hsr}^H\right)}{\partial \mathbf{s}^*} + \frac{\partial\,\mathrm{tr}\left(\mathbf{Hss}^H\mathbf{H}^H\right)}{\partial \mathbf{s}^*} = 0 \tag{18.34}$$

and with the differentiation Lemmas for traces shown in the Appendix B we obtain for the individual terms

$$\frac{\partial\,\mathrm{tr}\left(\mathbf{rr}^H\right)}{\partial \mathbf{s}^*} = 0 \tag{18.35}$$

$$\frac{\partial\,\mathrm{tr}\left(\mathbf{rs}^H\mathbf{H}^H\right)}{\partial \mathbf{s}^*} = \mathbf{H}^H\mathbf{r} \tag{18.36}$$

$$\frac{\partial\,\mathrm{tr}\left(\mathbf{Hsr}^H\right)}{\partial \mathbf{s}^*} = 0 \tag{18.37}$$

$$\frac{\partial\,\mathrm{tr}\left(\mathbf{Hss}^H\mathbf{H}^H\right)}{\partial \mathbf{s}^*} = \mathbf{H}^H\mathbf{Hs} \tag{18.38}$$

Inserting (18.35)–(18.38) into (18.34) yields

$$\mathbf{H}^H\mathbf{r} = \mathbf{H}^H\mathbf{Hs} \tag{18.39}$$

from which we conclude by multiplication from the left-hand side with $\left(\mathbf{H}^H\mathbf{H}\right)^{-1}$

$$\mathbf{s} = \left(\mathbf{H}^H\mathbf{H}\right)^{-1}\mathbf{H}^H\mathbf{r} \tag{18.40}$$

and consequently $\mathbf{H}^+ = \left(\mathbf{H}^H\mathbf{H}\right)^{-1}\mathbf{H}^H = \mathbf{W}$ in (18.26) follows, which finalizes the proof.

An Alternative Derivation of the Pseudo Inverse Matrix

Motivated by the fact that $\mathbf{H}^H\mathbf{H}$ is a square matrix with full rank, M, we can come to a straightforward derivation of the pseudo inverse as follows. We consider (18.1) with $\mathbf{n} = \mathbf{0}$

$$\mathbf{r} = \mathbf{Hs} \tag{18.41}$$

and multiply from the left with $\mathbf{H}^H$ resulting in

$$\mathbf{H}^H \mathbf{r} = \mathbf{H}^H \mathbf{H} \mathbf{s} \tag{18.42}$$

In mathematics (18.42) is called the Gaussian normal equation. $\mathbf{H}^H \mathbf{H}$ is a square matrix with the dimension $M \times M$. Then, the task is to solve a system of linear equations defined by a non-singular square matrix. The solution is straightforward. We multiply from the left with the inverse matrix $\left(\mathbf{H}^H \mathbf{H}\right)^{-1}$, which exists due to the fact that $\mathbf{H}^H \mathbf{H}$ has full rank, resulting in

$$\left(\mathbf{H}^H \mathbf{H}\right)^{-1} \mathbf{H}^H \mathbf{r} = \left(\mathbf{H}^H \mathbf{H}\right)^{-1} \mathbf{H}^H \mathbf{H} \mathbf{s} \tag{18.43}$$

from which we conclude

$$\mathbf{s} = \left(\mathbf{H}^H \mathbf{H}\right)^{-1} \mathbf{H}^H \mathbf{r} \tag{18.44}$$

and finally the pseudo inverse

$$\mathbf{H}^+ = \left(\mathbf{H}^H \mathbf{H}\right)^{-1} \mathbf{H}^H \tag{18.45}$$

As a final remark, if $\mathbf{H}$ does not have full rank M, a special pseudo inverse can be determined from a singular value decomposition, [4].

18.4.2   Receiver with Less Antennas than Transmitter, $N < M$

Solving the System of Linear Equations

Example 5

Let us consider a simple example with $M = 3$ transmit and $N = 2$ receive antennas. Assuming small noise, $\mathbf{n} \approx \mathbf{0}$, we obtain from the basic input–output equation (18.1)

$$r_1 = h_{11}s_1 + h_{12}s_2 + h_{13}s_3 \tag{18.46}$$

$$r_2 = h_{21}s_1 + h_{22}s_2 + h_{23}s_3 \tag{18.47}$$

and recognize an under-determined system of equations. Mathematically this can be solved by assuming any value for one variable, e.g., s_3. Then we can move $h_{13}s_3$ and $h_{23}s_3$ to the left-hand side and obtain a 2x2 system of equations with a square matrix, which can be solved conventionally. If this matrix has full rank two, the solutions for s_1 and s_2 contain one free parameter s_3, which can be set to any value. However, from a communications point of view the selection of an arbitrary s_3 at the receiver is useless because the actual transmit signal s_3 is not recovered. So, we are only in a position to transmit two information signals s_1 and s_2 instead of three and the third transmit antenna seems to be of no use. For a special selection of the free parameter, such as $s_3 = 0$, the system would effectively operate with $M = 2$ and thus one transmit antenna could be dropped. Another choice would be to keep $M = 3$ and to introduce redundancy by sending one signal s_1 or s_2 also from the third antenna, e.g., allocating $s_3 = s_1$.

Solution with the Pseudo Inverse

Mathematically, the $N \times M$ under-determined system of linear equations can have in general an exact and unique solution, if and only if at least $M - N$ columns of $\mathbf{H}$ are linearly depending on each other. Then the resulting solution contains at least $M - N$ free parameters, which give rise to an $(M - N)$ infinite linear manifold. From a technical point of view, we waste $M - N$ transmit antennas and can drop them. To keep the $M - N$ signals as useful signals, we may give up the idea of an exact solution at the receiver and look for an approximation, as for the over-determined system of equations discussed in the previous section. Consequently, we follow the line of the pseudo inverse here as well. For that purpose, we take the prerequisite that $\mathbf{H}$ has full rank, which is

$$\text{rank}(\mathbf{H}) = N \leq \min\{M, N\} \tag{18.48}$$

As a consequence, the $M \times M$ matrix $\mathbf{H}^H\mathbf{H}$ in (18.26) can only have the rank N with $N < M$. Thus $\left(\mathbf{H}^H\mathbf{H}\right)^{-1}$ does not exist and consequently $\mathbf{H}^+$ given by (18.26) is not applicable. However, a Moore–Penrose pseudo inverse also exists and is given by

$$\mathbf{H}^{++} = \mathbf{H}^H \left(\mathbf{H}\mathbf{H}^H\right)^{-1} \tag{18.49}$$

Then $\mathbf{H}\mathbf{H}^H$ is an $N \times N$ matrix with full rank N, if $\mathbf{H}$ exhibits the full rank N. Consequently, the inverse matrix $\left(\mathbf{H}\mathbf{H}^H\right)^{-1}$ exists and the pseudo inverse $\mathbf{H}^{++}$ can be calculated. It should be noted that $\mathbf{H}^{++}$ does not contain free parameters any more. Equation (18.49) can be proven by checking the four Moore–Penrose axioms for which we refer to [3, 5]. We will give an alternative proof in the next section using some results from the minimum mean squared error receiver. An approach similar to (18.24) is not successful.

As a final remark, if $\mathbf{H}$ does not have full rank N, a special pseudo inverse can be determined from a singular value decomposition, [4].

Receiver Output Signal and $\mathbf{H}^{++}$ as a Transmit Prefilter

We are interested in the output signal of the receiver. With $\mathbf{W} = \mathbf{H}^{++}$, we first calculate the inter-channel interference term from (18.4) resulting in $\mathbf{G} = \mathbf{H}^H \left(\mathbf{H}\mathbf{H}^H\right)^{-1} \mathbf{H}$ and see that in general $\mathbf{G} \neq \mathbf{I}_M$ holds, which means that the receive matrix $\mathbf{H}^{++}$ cannot remove inter-channel interference in general, as is the case for $\mathbf{H}^+$.

However, if we allocate a filter with matrix $\mathbf{H}^{++}$ at the transmitter as a prefilter, we obtain the transmit signal $\mathbf{s} = \mathbf{H}^{++}\mathbf{c}$ and the receive signal from (18.1) as

$\mathbf{r} = \mathbf{H}\mathbf{H}^{++}\mathbf{c} + \mathbf{n}$, where $\mathbf{c}$ is the vector of symbols to be transmitted. We directly conclude with (18.49) that $\mathbf{H}\mathbf{H}^{++} = \mathbf{I}_N$ holds resulting in

$$\mathbf{r} = \mathbf{c} + \mathbf{n} \tag{18.50}$$

Obviously, the inter-channel interference is completely reduced by $\mathbf{H}^{++}$ as a prefilter. This technique is discussed in the Chap. 22 in more detail and we do not consider $\mathbf{H}^{++}$ furthermore as a receive filter for practical applications.

Finally, some general remarks on receivers with the inverse or the pseudo inverse matrix are given. For systems with $M \leq N$, they enable to exploit the MIMO transmit antenna with the full transmit signal vector and the complete removal of the inter-channel interference at the receiver output. For systems with $M > N$ the filter with pseudo inverse matrix should be positioned at the transmitter as a prefilter and then will completely reduce the inter-channel interference as well. Furthermore, it should be noted that in a real system the inverse or pseudo inverse matrix although analytically elegant can make quite some numerical problems in the signal processor. On account of the channel conditions at a time instant k, the channel matrix $\mathbf{H}$ can be ill-conditioned, which causes increased absolute values of the entries of the inverse matrices of $\mathbf{H}$, $\mathbf{H}^H\mathbf{H}$, and $\mathbf{H}\mathbf{H}^H$ resulting in a noise enhancement at the receiver output.

Example 6

Given a system with $M = 2$ transmit and $N = 1$ receive antennas. The channel with matrix $\mathbf{H} = \left(h_{11} \; h_{21} \right) = \left(1 \; \frac{i}{2} \right)$ shall be time-invariant in the time interval under consideration. We are interested in the receiver with the pseudo inverse and the inter-channel interference reduction.

Solution:

We calculate the pseudo inverse $\mathbf{H}^{++}$ in several steps, $\mathbf{H}\mathbf{H}^H = \frac{5}{4}$, $\left(\mathbf{H}\mathbf{H}^H \right)^{-1} = \frac{4}{5}$ and finally obtain $\mathbf{H}^{++} = \frac{4}{5} \left(1 \; -\frac{i}{2} \right)^T$. If we apply $\mathbf{H}^{++}$ as a receive filter, the inter-channel interference term follows from (18.4) as $\mathbf{G} = \mathbf{H}^{++}\mathbf{H} = \frac{4}{5} \left(1 \; -\frac{i}{2} \right)^T \left(1 \; \frac{i}{2} \right) = \frac{4}{5} \begin{pmatrix} 1 & \frac{i}{2} \\ -\frac{i}{2} & \frac{1}{4} \end{pmatrix}$, which deviates significantly from the identity matrix $\mathbf{I}_2$. Thus, we conclude that the zero-forcing receiver with pseudo inverse $\mathbf{H}^{++}$ is not able to reduce inter-channel interference.

Now we consider a transmit prefilter with the matrix $\mathbf{H}^{++}$ and obtain the corresponding inter-channel interference term $\mathbf{H}\mathbf{H}^{++} = \left(1 \; \frac{i}{2} \right) \frac{4}{5} \left(1 \; -\frac{i}{2} \right)^T = 1$. The receive signal then is $\mathbf{r} = \mathbf{c} + \mathbf{n}$ and this example verifies the general statement before that a prefilter $\mathbf{H}^{++}$ completely removes the inter-channel interference.

18.5   Signal-to-Noise Ratio of Linear Receivers

We consider Fig. 18.1 and are going to calculate the signal-to-noise ratio γ_y of the signal $\mathbf{y}(k)$ at the input of the decision device. As is well known from the theory of digital communications, the higher γ_y the lower the bit error ratio can be at the output of the decision device. The signal part of $\mathbf{y}(k)$ in (18.3) is

$$\mathbf{y}_s(k) = \mathbf{W}(k)\mathbf{H}(k)\mathbf{s}(k) \qquad (18.51)$$

which is superimposed by the noise part

$$\mathbf{y}_n(k) = \mathbf{W}(k)\mathbf{n}(k)$$

In the following, we define the covariance matrix of $\mathbf{s}(k)$

$$\mathbf{R}_{ss} = \mathbf{E}\left[\mathbf{ss}^H\right] \in \mathbb{C}^{M\times M} \qquad (18.52)$$

and of the noise

$$\mathbf{R}_{nn} = \mathbf{E}\left[\mathbf{nn}^H\right] \in \mathbb{C}^{N\times N} \qquad (18.53)$$

Furthermore, we drop k to simplify the notation. As shown in the Appendix B, the mean power of a signal is given by the trace of its covariance matrix. Consequently, for the mean power of $\mathbf{y}_s$ follows

$$\mathbf{E}\left[\|\mathbf{y}_s\|^2\right] = \mathrm{tr}\left(\mathbf{E}\left[\mathbf{y}_s\mathbf{y}_s^H\right]\right) \qquad (18.54)$$

With (18.51), we get $\mathrm{tr}\left(\mathbf{E}\left[\mathbf{y}_s\mathbf{y}_s^H\right]\right) = \mathrm{tr}\left(\mathbf{E}\left[\mathbf{WHss}^H\mathbf{H}^H\mathbf{W}^H\right]\right)$ and using (18.52) provides the result

$$\mathbf{E}\left[\|\mathbf{y}_s\|^2\right] = \mathrm{tr}\left(\mathbf{WHR}_{ss}\mathbf{H}^H\mathbf{W}^H\right) = \mathrm{tr}\left(\mathbf{R}_{ss}\left(\mathbf{WH}\right)^H\mathbf{WH}\right) \qquad (18.55)$$

where we have used in the last step the cyclic permutation rule for the trace of matrices outlined in the Appendix B. Similarly, we obtain for the noise part $\mathbf{y}_n$ the mean power

$$\mathbf{E}\left[\|\mathbf{y}_n\|^2\right] = \mathrm{tr}\left(\mathbf{WR}_{nn}\mathbf{W}^H\right) = \mathrm{tr}\left(\mathbf{R}_{nn}\mathbf{W}^H\mathbf{W}\right) \qquad (18.56)$$

and finally the signal-to-noise ratio

$$\gamma_y = \frac{\mathbf{E}\left[\|\mathbf{y}_s\|^2\right]}{\mathbf{E}\left[\|\mathbf{y}_n\|^2\right]} = \frac{\mathrm{tr}\left(\mathbf{R}_{ss}\mathbf{H}^H\mathbf{W}^H\mathbf{WH}\right)}{\mathrm{tr}\left(\mathbf{R}_{nn}\mathbf{W}^H\mathbf{W}\right)} \qquad (18.57)$$

18.5.1 Signal-to-Noise Ratio with Zero-Forcing Receiver

The receivers are given by the pseudo inverse matrices (18.26) and (18.49). Although we have found that in general the zero-forcing receiver with $\mathbf{H}^{++}$ for $M \geq N$ cannot reduce the inter-channel interference we will give the results for the sake of completeness. We obtain

$$\mathbf{WH} = \begin{cases} \mathbf{H}^+\mathbf{H} = \mathbf{I}_M & ; M \leq N \\ \mathbf{H}^{++}\mathbf{H} = \mathbf{H}^H \left(\mathbf{HH}^H\right)^{-1}\mathbf{H} \; ; & M \geq N \end{cases} \tag{18.58}$$

Consequently, the mean power (18.55) of the signal part at the output of the receive filter is

$$\mathbf{E}\left[\|\mathbf{y}_s\|^2\right] = \begin{cases} \operatorname{tr}\left(\mathbf{R}_{ss}\right) & ; M \leq N \\ \operatorname{tr}\left(\mathbf{R}_{ss}\mathbf{H}^H \left(\mathbf{HH}^H\right)^{-1}\mathbf{H}\right) \; ; & M \geq N \end{cases} \tag{18.59}$$

For the mean power (18.56) of the noise part, we have to determine

$$\mathbf{W}^H\mathbf{W} = \begin{cases} \left(\mathbf{H}^+\right)^H \mathbf{H}^+ = \mathbf{H}\left(\mathbf{H}^H\mathbf{H}\right)^{-1}\left(\mathbf{H}^H\mathbf{H}\right)^{-1}\mathbf{H}^H \; ; \; M \leq N \\ \left(\mathbf{H}^{++}\right)^H \mathbf{H}^{++} = \left(\mathbf{HH}^H\right)^{-1} \qquad\qquad ; \; M \geq N \end{cases} \tag{18.60}$$

where the proof is also straightforward. Then, the mean power of the noise part follows from (18.56)

$$\mathbf{E}\left[\|\mathbf{y}_n\|^2\right] = \begin{cases} \operatorname{tr}\left(\mathbf{R}_{nn}\mathbf{H}\left(\mathbf{H}^H\mathbf{H}\right)^{-1}\left(\mathbf{H}^H\mathbf{H}\right)^{-1}\mathbf{H}^H\right) \; ; \; M \leq N \\ \operatorname{tr}\left(\mathbf{R}_{nn}\left(\mathbf{HH}^H\right)^{-1}\right) \qquad\qquad ; \; M \geq N \end{cases} \tag{18.61}$$

Plugging (18.59) and (18.61) into (18.57) yields the signal-to-noise ratio at the zero-forcing receiver output

$$\gamma_y = \frac{\mathbf{E}\left[\|\mathbf{y}_s\|^2\right]}{\mathbf{E}\left[\|\mathbf{y}_n\|^2\right]} = \begin{cases} \dfrac{\operatorname{tr}(\mathbf{R}_{ss})}{\operatorname{tr}\left(\mathbf{R}_{nn}\mathbf{H}\left(\mathbf{H}^H\mathbf{H}\right)^{-1}\left(\mathbf{H}^H\mathbf{H}\right)^{-1}\mathbf{H}^H\right)} \; ; \; M \leq N \\[4mm] \dfrac{\operatorname{tr}\left(\mathbf{R}_{ss}\mathbf{H}^H\left(\mathbf{HH}^H\right)^{-1}\mathbf{H}\right)}{\operatorname{tr}\left(\mathbf{R}_{nn}\left(\mathbf{HH}^H\right)^{-1}\right)} \; ; \; M \geq N \end{cases} \tag{18.62}$$

Furthermore, the mean noise power $\mathbf{E}\left[\|\mathbf{y}_n\|^2\right]$ in (18.61) is worth to be commentated. If the determinant of $\mathbf{H}^H\mathbf{H}$ and $\mathbf{HH}^H$ exhibits rather small absolute values, then the entries of the corresponding inverse matrices can take on large values, respectively. As a consequence, the noise power at the output of the receive filter can be much larger than $N\sigma_n^2$, thus being enhanced by the receive filter. A matrix with such an unfavorable property is called ill-conditioned. It gets even worse, if the channel matrix $\mathbf{H}$ does not have full rank because the determinant of $\mathbf{H}^H\mathbf{H}$ and $\mathbf{HH}^H$ approaches zero and consequently $\mathbf{E}\left[\|\mathbf{y}_n\|^2\right] \to \infty$. Please note that

$\text{rank}(\mathbf{H}) = \text{rank}\left(\mathbf{H}\mathbf{H}^H\right) = \text{rank}\left(\mathbf{H}^H\mathbf{H}\right) \leq \min\{M, N\}$. Similarities are discussed in Chap. 22 for the output signal of a precoder using an eigenvalue decomposition of $\mathbf{H}^H\mathbf{H}$ and $\mathbf{H}\mathbf{H}^H$, respectively.

The design goal of the mean squared error receiver, which will be discussed in Section 18.6, takes the noise into account and hence can provide a compromise between a reduction of the inter-channel interference and the noise enhancement.

18.5.2   Normalization of the Channel Matrix $\mathbf{H}$

To facilitate a fair comparison of the signal-to-noise ratios of different MIMO schemes a normalization of the channel matrix $\mathbf{H}$ can be done. For that purpose, we define the power gain g_P of the channel as the relation between its mean output power $\mathbf{E}\left[\|\mathbf{Hs}\|^2\right]$ and the mean input power $\mathbf{E}\left[\|\mathbf{s}\|^2\right]$. The input signal shall be uncorrelated, thus $\mathbf{E}\left[\|\mathbf{s}\|^2\right] = M E_S$ and the mean output power then is $\mathbf{E}\left[\|\mathbf{Hs}\|^2\right] = E_S \text{tr}\left(\mathbf{H}^H\mathbf{H}\right)$. Furthermore, we know from Appendix B that $\text{tr}\left(\mathbf{H}^H\mathbf{H}\right) = \text{tr}\left(\mathbf{H}\mathbf{H}^H\right) = \|\mathbf{H}\|_F^2 = \sum_{j=1}^M \sum_{i=1}^N \left|h_{ij}\right|^2$ is the Frobenius norm of a matrix and that the trace operator is commutative. Then, we can define the power gain as

$$g_P = \frac{\mathbf{E}\left[\|\mathbf{Hs}\|^2\right]}{\mathbf{E}\left[\|\mathbf{s}\|^2\right]} = \frac{\text{tr}\left(\mathbf{H}^H\mathbf{H}\right)}{M} = \frac{\|\mathbf{H}\|_F^2}{M} \qquad (18.63)$$

Please remember that the entries of $\mathbf{H}$ are delay spread functions representing the link from the transmitter to the receiver including all building blocks, such as radio channel, filters, modulator, amplifier, etc. Therefore, the overall gain can be adjusted to achieve the normalized channel matrix

$$\mathbf{H}_N = \frac{1}{\sqrt{g_P}}\mathbf{H} = \sqrt{\frac{M}{\|\mathbf{H}\|_F^2}}\mathbf{H} \qquad (18.64)$$

Please note that the entries of $\mathbf{H}$ represent amplitudes rather than powers, which gives rise to apply the square root of g_P. If we replace $\mathbf{H}$ by $\mathbf{H}_N$ and $\mathbf{W}$ by $\mathbf{W}_N$, where $\mathbf{W}_N$ is the receive matrix determined from $\mathbf{H}_N$, then we obtain from (18.57) the normalized signal-to-noise ratio

$$\gamma_{y,N} = \gamma_y\big|_{\mathbf{H}=\mathbf{H}_N, \mathbf{W}=\mathbf{W}_N} \qquad (18.65)$$

Now we introduce the normalized channel matrix $\mathbf{H}_N$ and find with (18.64) $\mathbf{H}^H\mathbf{H} = g_P\mathbf{H}_N^H\mathbf{H}_N$, $\left(\mathbf{H}^H\mathbf{H}\right)^{-1} = \frac{1}{g_P}\left(\mathbf{H}_N^H\mathbf{H}_N\right)^{-1}$, and similarly $\left(\mathbf{H}\mathbf{H}^H\right)^{-1} = \frac{1}{g_P}\left(\mathbf{H}_N\mathbf{H}_N^H\right)^{-1}$. Then follows from (18.62) the normalized signal-to-noise ratio

$$\gamma_{y,N} = \frac{\gamma_y}{g_P} \qquad (18.66)$$

The mean noise power (18.61) turns after normalization into

$$\mathbf{E}\left[\left\|\mathbf{y}_{n,N}\right\|^2\right] = g_P \mathbf{E}\left[\left\|\mathbf{y}_n\right\|^2\right] \tag{18.67}$$

Obviously, for a power gain equal to one $\gamma_{y,N} = \gamma_y$ holds. Assuming a power gain $g_P \neq 1$ we see from (18.66) that the signal-to-noise ratio $\gamma_{y,N}$ after normalization of the channel matrix is by factor $\frac{1}{g_P}$ smaller than the actual one. Hence, there is a need to normalize the channel matrix adequately for a fair comparison.

Example 7

Consider the important practical case that the noise $\mathbf{n}$ and also the signal $\mathbf{s}$ are uncorrelated. Then the covariance matrices are $\mathbf{R}_{nn} = \sigma_n^2 \mathbf{I}_N$ and $\mathbf{R}_{ss} = E_S \mathbf{I}_M$. Determine the signal-to-noise ratio and the mean noise power at the output of the zero-forcing receiver.

Solution:

We consider the first line in (18.62). For the numerator we obtain tr $(\mathbf{R}_{ss}) = M E_S$. With $\mathbf{R}_{nn} = \sigma_n^2 \mathbf{I}_N$ and the cyclic permutation rule for the trace the denominator results in $\sigma_n^2 \mathrm{tr}\left(\left(\mathbf{H}^H\mathbf{H}\right)^{-1}\left(\mathbf{H}^H\mathbf{H}\right)^{-1}\mathbf{H}^H\mathbf{H}\right) = \sigma_n^2 \mathrm{tr}\left(\left(\mathbf{H}^H\mathbf{H}\right)^{-1}\right)$. In the same way we handle the second line in (18.62). Then the signal-to-noise ratio is

$$\gamma_y = \frac{\mathbf{E}\left[\left\|\mathbf{y}_s\right\|^2\right]}{\mathbf{E}\left[\left\|\mathbf{y}_n\right\|^2\right]} = \begin{cases} \frac{M E_S}{\sigma_n^2 \mathrm{tr}\left(\left(\mathbf{H}^H\mathbf{H}\right)^{-1}\right)} & ; \ M \leq N \\[2ex] \frac{M E_S}{\sigma_n^2 \mathrm{tr}\left(\left(\mathbf{H}\mathbf{H}^H\right)^{-1}\right)} & ; \ M \geq N \end{cases} \tag{18.68}$$

Example 8

We continue our Example 7 and consider a system with $M = 1$ transmit and $N = 2$ receive antennas. The channel matrix $\mathbf{H} = \left(h_{11} \ h_{21}\right)^T = \left(1 \ \frac{1}{2}\right)^T$ shall be constant in the time interval under consideration.

The zero-forcing receiver matrix $\mathbf{W} = \mathbf{H}^+ = \left(\mathbf{H}^H\mathbf{H}\right)^{-1}\mathbf{H}^H$ from (18.26) is easily calculated with the following steps: $\mathbf{H}^H\mathbf{H} = \left(1 \ -\frac{1}{2}\right)\left(1 \ \frac{1}{2}\right)^T = \frac{5}{4}$, which is composed of just one element yielding $\left(\mathbf{H}^H\mathbf{H}\right)^{-1} = \frac{4}{5}$. Consequently, we obtain the zero-forcing receiver matrix $\mathbf{H}^+ = \frac{4}{5}\left(1 \ -\frac{1}{2}\right)$, which is just a row vector.

Now we are interested in the mean power of the noise at the receiver output and obtain from the denominator of the first line in (18.68) $\mathbf{E}\left[\left\|\mathbf{y}_n\right\|^2\right] = \sigma_n^2 \mathrm{tr}\left(\left(\mathbf{H}^H\mathbf{H}\right)^{-1}\right) = \frac{4}{5}\sigma_n^2$. Obviously, this receive matrix is even reducing rather than enhancing the mean noise power in this example. However, what counts is the signal-to-noise ratio, which follows from the first line in (18.68) yielding $\gamma_y = \frac{5}{4}\frac{E_S}{\sigma_n^2}$.

Now we consider the figures which result after a normalization of the channel matrix. From (18.64) we get the channel power gain $g_P = \frac{5}{4}$ and consequently from (18.63) $\mathbf{H}_N = \frac{2}{\sqrt{5}}\left(1 \ \frac{1}{2}\right)^T$. With $\mathbf{H}_N^H\mathbf{H}_N = 1$ we obtain from (18.26)

$\mathbf{H}_N^+ = \left(\mathbf{H}_N^H \mathbf{H}_N\right)^{-1} \mathbf{H}_N^H = \frac{2}{\sqrt{5}}\left(1 - \frac{i}{2}\right)$. The mean noise power then is $\mathbf{E}\left[\left\|\mathbf{y}_{n,N}\right\|^2\right] = \sigma_n^2 \mathrm{tr}\left(\left(\mathbf{H}_N^H \mathbf{H}_N\right)^{-1}\right) = \sigma_n^2$. Finally, the normalized SNR is $\gamma_{y,N} = \frac{E_S}{\sigma_n^2}$, which is in line with (18.66).

18.6 Minimum Mean Squared Error receiver

18.6.1 Prerequisites

In contrast to the zero-forcing receiver we now follow a strategy to minimize the quadratic error between the transmit signal $\mathbf{s}$ and the receiver output signal $\mathbf{y}$ and thus include the noise in our design criterion from the beginning. Hence, the target function of the Minimum Mean Squared Error (MMSE) receiver is

$$J = \mathbf{E}\left[\|\Delta\|^2\right] = \min_{\mathbf{W}} \tag{18.69}$$

with the error vector

$$\Delta = \mathbf{s} - \mathbf{y} \tag{18.70}$$

The squared error then is

$$\|\Delta\|^2 = \Delta^H \Delta = \mathrm{tr}\left(\Delta\Delta^H\right) \tag{18.71}$$

From (18.69) follows

$$J = \mathbf{E}\left[\mathrm{tr}\left(\Delta\Delta^H\right)\right] = \mathrm{tr}\left(\mathbf{E}\left[\Delta\Delta^H\right]\right) = \min_{\mathbf{W}} \tag{18.72}$$

where we have used the fact that the trace and the expectation operator are linear and thus can be applied in reverse order. In the course of the computation of the optimal receiver matrix $\mathbf{W}$ we will need the covariance matrix $\mathbf{R}_{ss} = \mathbf{E}\left[\mathbf{s}\mathbf{s}^H\right]$ of the transmit signal $\mathbf{s}$ given by (18.52) and the covariance matrix $\mathbf{R}_{nn} = \mathbf{E}\left[\mathbf{n}\mathbf{n}^H\right]$ of the noise $\mathbf{n}$ defined in (18.53). We state the following prerequisites. In the first place the noise has zero mean, i.e.

$$\mathbf{E}\left[\mathbf{n}\right] = \mathbf{0} \tag{18.73}$$

and secondly, the transmit signal $\mathbf{s}$ is statistically independent of the noise $\mathbf{n}$, i.e. the cross-correlation matrix is

$$\mathbf{R}_{sn} = \mathbf{E}\left[\mathbf{s}\mathbf{n}^H\right] = \mathbf{E}\left[\mathbf{s}\right] \mathbf{E}\left[\mathbf{n}^H\right] \tag{18.74}$$

and with ((18.73)) follows

$$\mathbf{R}_{sn} = \mathbf{0} \tag{18.75}$$

Similarly

$$\mathbf{R}_{ns} = \mathbf{E}\left[\mathbf{ns}^H\right] = \mathbf{R}_{sn}^H = \mathbf{0} \tag{18.76}$$

We further assume that the channel matrix $\mathbf{H}$ and the receiver matrix $\mathbf{W}$ are deterministic but still can be a function of time k.

18.6.2 Receiver Matrix

The partial derivative of J with respect to $\mathbf{W}^*$ set to zero provides the necessary and sufficient condition for the minimum, because the quadratic error is a convex function of $\mathbf{W}$ or $\mathbf{W}^*$. The optimal receiver matrix, which minimizes the squared error, then is

$$\mathbf{W} = \mathbf{R}_{ss}\mathbf{H}^H\left(\mathbf{H}\mathbf{R}_{ss}\mathbf{H}^H + \mathbf{R}_{nn}\right)^{-1} \tag{18.77}$$

provided that the matrix $\mathbf{H}\mathbf{R}_{ss}\mathbf{H}^H + \mathbf{R}_{nn}$ is non-singular. Before we outline the proof a special case is considered.

In many applications, we can assume that the signals emitted by the transmit antenna elements are spatially uncorrelated and each with mean power E_S, thus $\mathbf{R}_{ss} = E_S\mathbf{I}_N$ holds. Also the noise at each receive antenna element can be considered to be spatially uncorrelated each with mean power σ_n^2 resulting in the covariance matrix $\mathbf{R}_{nn} = \sigma_n^2\mathbf{I}_N$. Then we get the optimal MMSE receiver matrix $\mathbf{W} = E_S\mathbf{H}^H\left(\mathbf{H}E_S\mathbf{H}^H + \sigma_n^2\mathbf{I}_N\right)^{-1}$ and finally[1]

$$\mathbf{W} = \mathbf{H}^H\left(\mathbf{H}\mathbf{H}^H + \alpha\mathbf{I}_N\right)^{-1} \tag{18.78}$$

with the ratio of the mean noise power per receive antenna and the mean signal power per transmit antenna

$$\alpha = \frac{\sigma_n^2}{E_S} \tag{18.79}$$

According to communication theory the receive matrix describes a MIMO matched filter. $\mathbf{W}$ is applicable, if the inverse matrix in (18.78) exists.

Now we apply the following matrix inversion lemma (Woodbury identiy), [4, 6, 7]. Let $\mathbf{A}, \mathbf{B}, \mathbf{C}$, and $\mathbf{D}$ be non-singular matrices with compatible dimensions. Then

$$(\mathbf{A} + \mathbf{B}\mathbf{C}\mathbf{D})^{-1} = \mathbf{A}^{-1} - \mathbf{A}^{-1}\mathbf{B}\left(\mathbf{C}^{-1} + \mathbf{D}\mathbf{A}^{-1}\mathbf{B}\right)^{-1}\mathbf{D}\mathbf{A}^{-1} \tag{18.80}$$

holds and we obtain the alternative MMSE receive matrix

$$\mathbf{W} = \left(\mathbf{H}^H\mathbf{H} + \alpha\mathbf{I}_M\right)^{-1}\mathbf{H}^H \; ; \; \alpha \neq 0 \tag{18.81}$$

---

[1]Note $(\alpha\mathbf{A})^{-1} = \frac{1}{\alpha}\mathbf{A}^{-1}$; $\alpha \neq 0$

The proof is given at the end of this Section. We see that both solutions (18.78) and (18.81) respect the noise. For low noise, $\alpha \to 0$, we obtain from (18.78)

$$\mathbf{H}^H \left(\mathbf{H}\mathbf{H}^H\right)^{-1} = \mathbf{H}^{++} \tag{18.82}$$

and from (18.81)

$$\left(\mathbf{H}^H\mathbf{H}\right)^{-1} \mathbf{H}^H = \mathbf{H}^+ \tag{18.83}$$

Obviously, these are the identical matrices, which we achieved also for the zero-forcing receiver. Hence, we conclude that the MMSE receiver degrades to the zero-forcing receiver for high signal-to-noise ratio. With the MMSE receive matrix (18.81) the receiver output signal (18.3) is

$$\mathbf{y} = \mathbf{G}\mathbf{s} + \mathbf{n}' \tag{18.84}$$

with

$$\mathbf{G} = \mathbf{W}\mathbf{H} = \left(\mathbf{H}^H\mathbf{H} + \alpha\mathbf{I}_M\right)^{-1} \mathbf{H}^H\mathbf{H} \tag{18.85}$$

as the remaining inter-channel interference term and the output noise is

$$\mathbf{n}' = \mathbf{W}\mathbf{n} = \left(\mathbf{H}^H\mathbf{H} + \alpha\mathbf{I}_M\right)^{-1} \mathbf{H}^H\mathbf{n} \tag{18.86}$$

Please note that $\mathbf{G}$ approaches $\mathbf{I}_M$ for low noise $\alpha \to 0$. Thus, the lower the noise the better the reduction of the inter-channel interference will be. We are not surprised, because the MMSE solution approaches the zero-forcing matrix. Similar considerations can be done, if the transmitter is furnished with a precoder owing the matrix $\mathbf{W}$ in (18.78).

The signal-to-noise ratio at the MMSE receiver output can be found by inserting the receive matrix (18.81) or its variants into (18.57). In the same way we obtain the mean noise power at the receiver output with (18.56). However, the formulas remain not very transparent and therefore a discussion for specific α is more adequate.

As mentioned before, the MMSE receiver matrix turns into the zero-forcing matrix $\mathbf{W} = \mathbf{H}^+ = \left(\mathbf{H}^H\mathbf{H}\right)^{-1} \mathbf{H}^H$ for a very low receive noise, $\alpha \to 0$, with the same consequences for the signal-to-noise ratio at the receiver output. For $\alpha \neq 0$ the MMSE receiver in (18.81) in comparison owns the term $\alpha\mathbf{I}_M$, which in many cases can enable a proper inverse $\left(\mathbf{H}^H\mathbf{H} + \alpha\mathbf{I}_M\right)^{-1}$ even in the case, where $\mathbf{H}$, and thus $\mathbf{H}^H\mathbf{H}$ do not have full rank or are ill-conditioned. Hence, the MMSE receiver can often provide an adequate solution in the case of a rank deficient channel matrix, where the zero-forcing receiver fails.

Proof of MMSE Receiver Matrix (18.77)

We insert (18.70) into (18.72) and obtain with (18.2)

$$J = \mathrm{tr}\left(\mathbf{E}\left[\Delta\Delta^H\right]\right) = \mathrm{tr}\left(\mathbf{E}\left[(\mathbf{s} - \mathbf{W}\mathbf{r})(\mathbf{s}^H - \mathbf{r}^H\mathbf{W}^H)\right]\right) \tag{18.87}$$

184 18 Principles of Linear MIMO Receivers

and with (18.1)

$$J = \text{tr}\left(\mathbf{E}\left[(\mathbf{s} - \mathbf{WHs} - \mathbf{Wn})(\mathbf{s}^H - \mathbf{s}^H\mathbf{H}^H\mathbf{W}^H - \mathbf{n}^H\mathbf{W}^H)\right]\right) \qquad (18.88)$$

After expanding we get

$$J = \text{tr}\left(\mathbf{R}_{ss} - \mathbf{R}_{ss}\mathbf{H}^H\mathbf{W}^H - \mathbf{R}_{sn}\mathbf{W}^H - \mathbf{WHR}_{ss} + \mathbf{WHR}_{ss}\mathbf{H}^H\mathbf{W}^H\right) +$$

$$+ \text{tr}\left(\mathbf{WHR}_{sn}\mathbf{W}^H - \mathbf{WR}_{ns} + \mathbf{WR}_{ns}\mathbf{H}^H\mathbf{W}^H + \mathbf{WR}_{nn}\mathbf{W}^H\right) \qquad (18.89)$$

Inserting (18.75) and (18.76) and employing the linear property of the trace operator yields

$$J = \text{tr}\left(\mathbf{R}_{ss}\right) - \text{tr}\left(\mathbf{R}_{ss}\mathbf{H}^H\mathbf{W}^H\right) - \text{tr}\left(\mathbf{WHR}_{ss}\right) + \text{tr}\left(\mathbf{WHR}_{ss}\mathbf{H}^H\mathbf{W}^H\right) + \text{tr}\left(\mathbf{WR}_{nn}\mathbf{W}^H\right) \qquad (18.90)$$

We see that J is a quadratic and convex function of $\mathbf{W}$. Therefore, setting the partial derivative with respect to $\mathbf{W}^*$ equal to zero is a necessary and sufficient condition for the minimum. With the cyclic permutation Lemma given in the Appendix B we move $\mathbf{W}$ to the end of the string of matrices in the traces and can directly apply the differentiation Lemmas outlined in the Appendix B. We obtain

$$\frac{\partial\,\text{tr}\left(\mathbf{R}_{ss}\right)}{\partial\,\mathbf{W}^*} = 0 \;;\; \frac{\partial\,\text{tr}\left(\mathbf{R}_{ss}\mathbf{H}^H\mathbf{W}^H\right)}{\partial\,\mathbf{W}^*} = \mathbf{R}_{ss}\mathbf{H}^H \;;\; \frac{\partial\,\text{tr}\left(\mathbf{WHR}_{ss}\right)}{\partial\,\mathbf{W}^*} = 0 \qquad (18.91)$$

$$\frac{\partial\,\text{tr}\left(\mathbf{WHR}_{ss}\mathbf{H}^H\mathbf{W}^H\right)}{\partial\,\mathbf{W}^*} = \mathbf{WHR}_{ss}\mathbf{H}^H \;;\; \frac{\partial\,\text{tr}\left(\mathbf{WR}_{nn}\mathbf{W}^H\right)}{\partial\,\mathbf{W}^*} = \mathbf{WR}_{nn} \qquad (18.92)$$

Setting the partial derivative of J with respect to $\mathbf{W}^*$ equal to zero yields with (18.91) and (18.92)

$$\frac{\partial\,J}{\partial\,\mathbf{W}^*} = -\mathbf{R}_{ss}\mathbf{H}^H + \mathbf{W}\left(\mathbf{HR}_{ss}\mathbf{H}^H + \mathbf{R}_{nn}\right) = 0 \qquad (18.93)$$

from which we conclude the proposition (18.77)

$$\mathbf{W} = \mathbf{R}_{ss}\mathbf{H}^H\left(\mathbf{HR}_{ss}\mathbf{H}^H + \mathbf{R}_{nn}\right)^{-1} \qquad (18.94)$$

and finalize the proof.

Proof of (18.81)

Let $\mathbf{A} = \alpha\mathbf{I}_N, \mathbf{B} = \mathbf{H}, \mathbf{C} = \mathbf{I}_M$, and $\mathbf{D} = \mathbf{H}^H$, then follows from (18.78) with (18.80)
$\mathbf{H}^H\left(\mathbf{HH}^H + \alpha\mathbf{I}_N\right)^{-1} = \mathbf{H}^H\left(\mathbf{A}^{-1} - \mathbf{A}^{-1}\mathbf{H}\left(\mathbf{I}_M + \mathbf{H}^H\mathbf{A}^{-1}\mathbf{H}\right)^{-1}\mathbf{H}^H\mathbf{A}^{-1}\right) =$
$\mathbf{H}^H\mathbf{A}^{-1} - \mathbf{H}^H\mathbf{A}^{-1}\mathbf{H}\left(\mathbf{I}_M + \mathbf{H}^H\mathbf{A}^{-1}\mathbf{H}\right)^{-1}\mathbf{H}^H\mathbf{A}^{-1}$. Now we multiply the first term from the left with the identity matrix $\left(\mathbf{I}_M + \mathbf{H}^H\mathbf{A}^{-1}\mathbf{H}\right)\left(\mathbf{I}_M + \mathbf{H}^H\mathbf{A}^{-1}\mathbf{H}\right)^{-1} = \mathbf{I}_M$ and obtain

$\left(\mathbf{I}_M + \mathbf{H}^H \mathbf{A}^{-1} \mathbf{H}\right) \left(\mathbf{I}_M + \mathbf{H}^H \mathbf{A}^{-1} \mathbf{H}\right)^{-1} \mathbf{H}^H \mathbf{A}^{-1} - \mathbf{H}^H \mathbf{A}^{-1} \mathbf{H} \left(\mathbf{I}_M + \mathbf{H}^H \mathbf{A}^{-1}\right.$
$\left.\mathbf{H}\right)^{-1} \mathbf{H}^H \mathbf{A}^{-1}$. After factoring out $\left(\mathbf{I}_M + \mathbf{H}^H \mathbf{A}^{-1} \mathbf{H}\right)^{-1} \mathbf{H}^H \mathbf{A}^{-1}$ the intermediate
result is $\mathbf{I}_M \left(\mathbf{I}_M + \mathbf{H}^H \mathbf{A}^{-1} \mathbf{H}\right)^{-1} \mathbf{H}^H \mathbf{A}^{-1}$. With $\mathbf{A}^{-1} = \frac{1}{\alpha} \mathbf{I}_N$ follows
$\left(\mathbf{I}_M + \mathbf{H}^H \frac{1}{\alpha} \mathbf{H}\right)^{-1} \mathbf{H}^H \frac{1}{\alpha} = \left(\mathbf{H}^H \mathbf{H} + \alpha \mathbf{I}_M\right)^{-1} \mathbf{H}^H$ and the proof is finished. Please
note that $\alpha \neq 0$ must hold otherwise $\mathbf{A}^{-1}$ does not exist.

18.7 Linear Combiner for Single Input Multiple Output System

18.7.1 Principle of Linear Combining and the Signal-to-Noise Ratio

As depicted in Fig. 18.2, we are now considering a single input multiple output
(SIMO) system with $M = 1$ transmit and $N > 1$ receive antennas. The transmit
signal is denoted as s with the mean power $\mathbf{E}\left[|s|^2\right] = E_S$. The noise shall be spatially
independent with mean power σ_n^2 per antenna and the covariance matrix $\mathbf{R}_{nn} = \sigma_n^2 \mathbf{I}_N$.
The channel matrix of this SIMO system is just a column vector

$$\mathbf{h} = \begin{pmatrix} h_1 \\ h_2 \\ \vdots \\ h_N \end{pmatrix} \qquad (18.95)$$

The linear combiner is described by its receive matrix, which is reduced to a row
vector

$$\mathbf{W} = \left(w_1 \; w_2 \; \cdots \; w_N \right) \qquad (18.96)$$

with complex components w_i ; $i = 1, 2, ..., N$. For the output signal of the linear
combiner we obtain

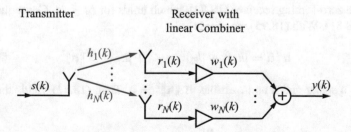

Fig. 18.2 Single input multiple output (SIMO) transmission with linear combiner receiver

$$y = \mathbf{W}\mathbf{h}s + \mathbf{W}\mathbf{n} \tag{18.97}$$

with the signal part

$$y_s = \mathbf{W}\mathbf{h}s \tag{18.98}$$

and the noise part

$$y_n = \mathbf{W}\mathbf{n} \tag{18.99}$$

Please note that the output of the receiver is a scalar rather than a vector. The signal-to-noise ratio is defined as

$$\gamma = \frac{\mathbf{E}\left[|y_s|^2\right]}{\mathbf{E}\left[|y_n|^2\right]} \tag{18.100}$$

For the mean power of y_s we obtain

$$\mathbf{E}\left[|y_s|^2\right] = \mathbf{E}\left[(\mathbf{W}\mathbf{h}s)^H \mathbf{W}\mathbf{h}s\right] = E_S(\mathbf{W}\mathbf{h})^\mathbf{H} \mathbf{W}\mathbf{h} \tag{18.101}$$

and the mean power of the noise part y_n is

$$\mathbf{E}\left[|y_n|^2\right] = \mathbf{E}\left[y_n y_n^H\right] = \mathbf{W}\,\mathbf{E}\left[\mathbf{n}\mathbf{n}^H\right]\mathbf{W}^H = \sigma_n^2 \mathbf{W}\mathbf{W}^H \tag{18.102}$$

As y_n is just a complex scalar, $y_n^H = y_n^*$ holds. Then follows the signal-to-noise ratio from (18.100)

$$\gamma = \frac{\mathbf{E}\left[|y_s|^2\right]}{\mathbf{E}\left[|y_n|^2\right]} = \frac{E_S}{\sigma_n^2} \frac{(\mathbf{W}\mathbf{h})^H \mathbf{W}\mathbf{h}}{\mathbf{W}\mathbf{W}^H} \tag{18.103}$$

18.7.2 MMSE Receiver for SIMO System (Maximum Ratio Combiner)

We would like to find the MMSE receiver matrix, which is a row vector. First we have to clarify, whether (18.78) or (18.81) is applicable. Our solution should hold also for low noise, $\alpha \to 0$, i.e., for the zero-forcing receiver. Because $M = 1$, we have to select the zero-forcing receiver (18.83), which holds for $M \leq N$. Consequently, we apply (18.81). With (18.95) we obtain

$$\mathbf{h}^H\mathbf{h} = |h_1|^2 + |h_2|^2 + \cdots + |h_N|^2 = \|\mathbf{h}\|^2 \tag{18.104}$$

which is a scalar and thus invertible, if $\|\mathbf{h}\|^2 \neq 0$. Then (18.81) yields the receive matrix

$$\mathbf{W} = \left(\|\mathbf{h}\|^2 + \alpha\right)^{-1} \mathbf{h}^H = \frac{\mathbf{h}^H}{\|\mathbf{h}\|^2 + \alpha} = \frac{1}{\|\mathbf{h}\|^2 + \alpha}\left(h_1^*\; h_2^*\; \cdots\; h_N^*\right) \tag{18.105}$$

with the entries

$$w_i = \frac{h_i^*}{\|\mathbf{h}\|^2 + \alpha} \; , \; i = 1, 2, ..., N \qquad (18.106)$$

Obviously, they show up as conjugate complex channel coefficients with some real-valued factor. We see that the optimal receiver is a matched filter known from communications theory. As is well known, a matched filter maximizes the signal-to-noise ratio at its output. This property has given rise to the designation "maximum ratio combiner" for this receiver type. The calculation of the signal-to-noise ratio is straightforward from (18.103). We find for the numerator

$$\mathbf{E}\left[|y_s|^2\right] = \left(\frac{\|\mathbf{h}\|^2}{\|\mathbf{h}\|^2 + \alpha}\right)^2 E_S \qquad (18.107)$$

and for the denominator

$$\mathbf{E}\left[|y_n|^2\right] = \frac{\|\mathbf{h}\|^2}{\left(\|\mathbf{h}\|^2 + \alpha\right)^2} \sigma_n^2 \qquad (18.108)$$

Then the signal-to-noise ratio of the maximum ratio combining scheme follows as

$$\gamma_{MRC} = \frac{\mathbf{E}\left[|y_s|^2\right]}{\mathbf{E}\left[|y_n|^2\right]} = \|\mathbf{h}\|^2 \frac{E_S}{\sigma_n^2} = \frac{1}{\alpha} \left(|h_1|^2 + |h_2|^2 + \cdots + |h_N|^2\right) \qquad (18.109)$$

Finally we obtain the combiner output signal from (18.97)

$$y = \frac{\|\mathbf{h}\|^2}{\|\mathbf{h}\|^2 + \alpha} s + \frac{\mathbf{h}^H}{\|\mathbf{h}\|^2 + \alpha} \mathbf{n} \qquad (18.110)$$

We substantiate that the maximum ratio combiner can completely cancel the inter-channel interference, because the signal part of the combiner output is just composed of the transmit signal s associated with a real-valued factor. For low noise, $\alpha \ll \|\mathbf{h}\|^2$, follows

$$y = s + \frac{\mathbf{h}^H}{\|\mathbf{h}\|^2} \mathbf{n} \qquad (18.111)$$

Figure 18.3 shows the symbol error rate of two arrangements. First, we consider a SIMO system with $N = 2$ receive antennas and a MRC receiver. Second, a SISO system with one antenna on each side serves for a comparison. We consider the symbol error rate of two scenarios, in the presence of white Gaussian noise (AWGN) only and for AWGN together with Rayleigh fading. The modulation scheme is 2-PSK and the receiver employs a threshold detector. A computer simulation was done with the open online platform "webdemo" [8]. To verify the results, the error rates have also been calculated, indicated by "theory" in Fig. 18.3, and the curves coincide perfectly.

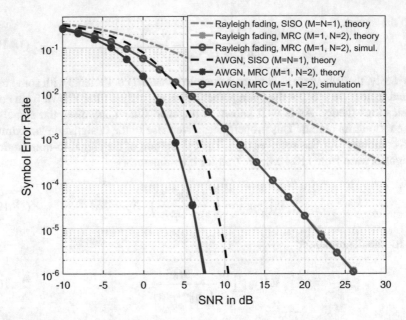

Fig. 18.3 Symbol error rate as a function of signal-to-noise ratio (SNR) for SISO system ($M = N = 1$) compared to SIMO system ($M = 1$, $N = 2$) with maximum ratio combining (MRC), channel with AWGN only and with additional Rayleigh fading, 2-PSK, theoretical and simulated results fit well. *Source* Online platform "webdemo" [8]

As expected, the additional impairment of Rayleigh fading requires significantly more signal-to-noise ratio at the receiver to achieve an adequate symbol error rate in comparison to a channel, which is just corrupted by AWGN. It is widely appreciated that the introduction of the SIMO scheme with maximum ratio combining provides a large improvement compared to the SISO system.

18.7.3 Equal Gain Combiner

The channel vector of the SIMO system is given by (18.95) and by introducing magnitudes and phases follows

$$\mathbf{h} = \left(|h_1| \, e^{j\phi_1} \; |h_2| \, e^{j\phi_2} \; \cdots \; |h_N| \, e^{j\phi_N} \right)^T \tag{18.112}$$

The receive vector of the equal gain combiner is defined by

$$\mathbf{W} = \frac{1}{\|\mathbf{h}\|^2 + \alpha} \left(e^{-j\phi_1} \; e^{-j\phi_2} \; \cdots \; e^{-j\phi_N} \right) \tag{18.113}$$

which means that all the receiver coefficients w_i have the same gain $\left(\|\mathbf{h}\|^2 + \alpha\right)^{-1}$, which has motivated the name "equal gain combiner". However, the receiver coefficients have different phases, which correspond to the negative phases of the respective channel coefficients. In other words, the receiver does not care about the absolute values of the individual channel coefficients. Consequently, $\mathbf{W}$ is technically rather simple to implement. To determine the signal-to-noise ratio at the receiver output we use (18.103). For that purpose we find

$$\mathbf{Wh} = \frac{1}{\|\mathbf{h}\|^2 + \alpha}\left(|h_1| + |h_2| + \cdots + |h_N|\right) \tag{18.114}$$

which is a real scalar. For the denominator of (18.103) we obtain

$$\sigma_n^2 \mathbf{WW}^H = \frac{\sigma_n^2 N}{\left(\|\mathbf{h}\|^2 + \alpha\right)^2} \tag{18.115}$$

Consequently, the signal-to-noise ratio is

$$\gamma_{EGC} = \frac{\mathbf{E}\left[|y_s|^2\right]}{\mathbf{E}\left[|y_n|^2\right]} = \frac{E_S}{N\sigma_n^2}\left(|h_1| + |h_2| + \cdots + |h_N|\right)^2 \tag{18.116}$$

Finally, we obtain for the output signal of the equal gain combiner from (18.97) making use of (18.114) and (18.113)

$$y = \frac{1}{\|\mathbf{h}\|^2 + \alpha}\left(|h_1| + |h_2| + \cdots + |h_N|\right)s +$$

$$+ \frac{1}{\|\mathbf{h}\|^2 + \alpha}\left(e^{-j\alpha_1}\ e^{-j\alpha_2}\ \cdots\ e^{-j\alpha_N}\right)\mathbf{n} \tag{18.117}$$

As we can see, the equal gain combiner output signal contains the transmit signal s associated with a real-valued factor and thus no inter-channel interference is present.

Example 9

Compare the signal-to-noise ratios of the maximum ratio combiner, γ_{MRC}, and the equal gain combiner, γ_{EGC}.

Solution:

We find with (18.109) and (18.116)

$$\frac{\gamma_{MRC}}{\gamma_{EGC}} = N\frac{|h_1|^2 + |h_2|^2 + \cdots + |h_N|^2}{\left(|h_1| + |h_2| + \cdots + |h_N|\right)^2} \tag{18.118}$$

Obviously, the right-hand side is $\leq N$. Consequently, the signal-to-noise ratio of the maximum ratio combiner is at most by factor N larger than that of the equal gain combiner. The equal sign holds for a SISO channel with one coefficient and $N = 1$.

Example 10

Determine the signal-to-noise ratio for a normalized channel vector.

Solution:

We introduce a normalization of the channel matrix using the channel power gain defined in (18.63). With $M = 1$ and $\operatorname{tr}\left(\mathbf{h}^H\mathbf{h}\right) = \|\mathbf{h}\|^2$ from (18.104) we get the power gain for the SIMO channel as $g_P = \|\mathbf{h}\|^2$. To find the normalized results, we have to replace $\mathbf{h}$ by $\frac{\mathbf{h}}{\sqrt{g_P}} = \frac{\mathbf{h}}{\|\mathbf{h}\|}$ in the relevant formulas, which means that all entries of the vector are divided by $\|\mathbf{h}\|$. Then follows from (18.109) the normalized signal-to-noise ratio for a maximum ratio combining receiver

$$\gamma_{MRC,N} = \frac{E_S}{\sigma_n^2} = \frac{1}{\alpha} \tag{18.119}$$

Replacing $|h_i|$ by $\frac{|h_i|}{\|\mathbf{h}\|}$; $i = 1, 2, ..., N$ in (18.116) yields the normalized signal-to-noise ratio for the equal gain combining scheme

$$\gamma_{EGC,N} = \frac{1}{N\alpha} \frac{(|h_1| + |h_2| + \cdots + |h_N|)^2}{\|\mathbf{h}\|^2} \tag{18.120}$$

and finally from (18.118) the ratio

$$\frac{\gamma_{MRC,N}}{\gamma_{EGC,N}} = \frac{\gamma_{MRC}}{\gamma_{EGC}} \tag{18.121}$$

which remains unchanged as in (18.118) because the normalization factor g_P drops out.

18.8 Decision of Receiver Output Signal

As we have seen, a linear receiver with matrix $\mathbf{W}$ tries to minimize the inter-channel interference and to some extend also the noise in the signal components $y_i(k)$ of the output vector (18.2)

$$\mathbf{y}(k) = \left(y_1(k) \; y_2(k) \; \cdots \; y_M(k) \right)^T \tag{18.122}$$

However, the signal components are still corrupted by some noise. As depicted in Fig. 18.1, a decision device following the linear receiver is required to recover the QAM symbols in each component $y_i(k)$. This process is also called signal detection and there are various detection strategies known from communications theory and outlined in Part I. The simplest one is threshold decision, where the complex signal plane can be structured by a rectangular grid with the grid lines as the decision thresholds. The decision device allocates to $y_i(k)$ an estimate $\hat{s}_i(k)$ of the transmit signal alphabet. Depending on the noise and the fading of the channel coefficients, quite some false decisions may occur, which can be quantified by a symbol error probability. Another strategy is maximum likelihood detection of the $y_i(k)$ or of the whole vector $\mathbf{y}(k)$. This method can minimize the error probability either on a "symbol by symbol" basis at every time instant k or by considering finite sequences of symbols, after which a "sequence detection" using the Viterbi algorithm is executed. We know from communications theory that a cascade of a linear receiver and a decision device is sub-optimal for achieving minimal symbol error probability. An optimal solution is the a-posterior, in special cases also the maximum likelihood detector, which are applied directly on the receive signal $\mathbf{r}(k)$ in (18.1). This will lead to a nonlinear receiver and will be presented in the next chapter.

References

1. E.H. Moore, On the reciprocal of the general algebraic matrix. Bull. Am. Math. Soc. **26** (1920)
2. R. Penrose, A generalized inverse for matrices. Proc. Camb. Philos. Soc. **51** (1955)
3. A. Albert, *Regression and the Moore-Penrose Pseudoinverse*, vol. 94 (Elsevier, Academic Press, Amsterdam, 1972)
4. K.B. Petersen, M.S. Pedersen, *The Matrix Cookbook* (Technical University of Denmark, 2012), h. open source https://archive.org/details/imm3274
5. A. Ben-Israel, T.N.E. Greville, *Generalized Inverses* (Springer, Berlin, 2003)
6. A. Hjorungnes, *Complex-valued Matrix Derivatives with Applications in Signal Processing and Communications* (Cambridge University Press, Cambridge, 2011)
7. R.A. Horn, C.R. Johnson, *Matrix Analysis* (Cambridge University Press, Cambridge, 2013)
8. B.B. Chaoudhry, Diversity combining, webdemo, Technical report. Institute of Telecommunications, University of Stuttgart, Germany (2018), http://webdemo.inue.uni-stuttgart.de

Chapter 19
Principles of Nonlinear MIMO Receivers

19.1 Maximum Likelihood MIMO Receiver

Principle

As we have seen in the previous chapter, a linear receiver tries to reduce the impact of inter-channel interference and partially of the noise in the receive signal $\mathbf{y}(k)$ in Fig. 18.1. Next, the signal is subject to a decision also called detection to recover the QAM symbols in each component $y_i(k)$. Various decision strategies are known from communications theory and outlined in Part I. In this section, we will consider a Maximum Likelihood (ML) detector as a receiver. In contrast to the linear receiver, the signal $\hat{\mathbf{s}}(k)$ will be estimated directly from the receive vector

$$\mathbf{r}(k) = \big(r_1(k)\ r_2(k)\ \cdots\ r_N(k) \big)^T \tag{19.1}$$

Hence, a receive matrix $\mathbf{W}$ is not present. In the following, we drop the discrete-time variable k to simplify the notation. The observed receive vector

$$\mathbf{r} = \mathbf{Hs} + \mathbf{n} \tag{19.2}$$

is corrupted by additive noise $\mathbf{n}$, where $\mathbf{Hs}$ is the receive signal in case of a noise-free channel. As for the linear receivers, we assume that the channel matrix $\mathbf{H}$ is precisely known to the receiver. In a practical system, the entries of $\mathbf{H}$ have to be estimated by a separate channel estimator, which is not considered here. The transmit signal vector is given by

$$\mathbf{s} = \big(s_1\ s_2\ \cdots\ s_M \big)^T \tag{19.3}$$

in which each component s_j is taken from a finite QAM symbol alphabet $\mathcal{B}$, e.g., $\mathcal{B}=\{1, j, -1, -j\}$ for 4-ary phase shift keying (4-PSK) or $\mathcal{B} = \{1, -1\}$ for 2-PSK. We assume an additive white Gaussian noise (AWGN) vector $\mathbf{n} = (n_1\ n_2 \ldots n_N)^T$ with the following properties:

© Springer Nature Switzerland AG 2019
J. Speidel, *Introduction to Digital Communications*, Signals and Communication
Technology, https://doi.org/10.1007/978-3-030-00548-1_19

- statistically independent with covariance matrix

$$\mathbf{R}_{nn} = \sigma_n^2 \mathbf{I}_N \qquad (19.4)$$

- all noise components n_i possess the same mean power σ_n^2 and zero mean $\mathbf{E}[n_i] = 0$; $i = 1, 2, \ldots, N$.
- the real part $n_{R,i}$ and the imaginary part $n_{I,i}$ of the noise $n_i = n_{R,i} + j n_{I,i}$ are statistically independent, have the same mean power $\frac{\sigma_n^2}{2}$, and the same Gaussian probability density function

$$p_x(x) = \frac{1}{\sqrt{2\pi}\sigma_x} e^{-\frac{x^2}{2\sigma_x^2}} \; ; \; \sigma_x^2 = \frac{\sigma_n^2}{2} \qquad (19.5)$$

where x stands for $n_{R,i}$ and $n_{I,i}$, $i = 1, 2, \ldots, N$. Consequently, the density function of the noise n_i is given by the product

$$p_{n_i}(n_i) = p_{n_{R,i}}(n_{R,i}) p_{n_{I,i}}(n_{I,i}) = \frac{1}{\pi\sigma_n^2} e^{-\frac{|n_i|^2}{\sigma_n^2}} \; ; \; i = 1, 2, \ldots, N \qquad (19.6)$$

- the multivariate probability density function of the noise vector $\mathbf{n}$ then follows as the product

$$p_{\mathbf{n}}(n_1, n_2, \ldots, n_N) = \left(\frac{1}{\pi\sigma_n^2}\right)^N e^{-\frac{|n_1|^2 + |n_2|^2 + \cdots + |n_N|^2}{\sigma_n^2}} \qquad (19.7)$$

or with shorthand notation

$$p_{\mathbf{n}}(\mathbf{n}) = \left(\frac{1}{\pi\sigma_n^2}\right)^N e^{-\frac{\|\mathbf{n}\|_2^2}{\sigma_n^2}} \qquad (19.8)$$

For the decision process, we first define the following conditional probability density function:

$$p_L(\mathbf{r}\,|\mathbf{Hs}) \qquad (19.9)$$

which is also called likelihood probability density function. It can be interpreted as the density function of $\mathbf{r}$ under the condition that $\mathbf{s}$ was sent, knowing $\mathbf{H}$. Please note that $p_L(\mathbf{r}\,|\mathbf{Hs})$ describes a finite set of probability density functions generated by all possible transmit vectors $\mathbf{s} \in \mathcal{A}$, where $\mathcal{A}$ is the set of all possible transmit vectors. Be it that each of the M components of $\mathbf{s}$ can take on L_Q different QAM symbol values, then $\mathcal{A}$ contains L_Q^M different vectors $\mathbf{s}_m, m = 1, 2, \ldots, L_Q^M$. The maximum likelihood detector selects out of all possible $\mathbf{Hs}$ that estimate $\mathbf{s} = \hat{\mathbf{s}}$, which is maximal likely to the receive vector $\mathbf{r}$, i.e., which has the largest $p_L(\mathbf{r}\,|\mathbf{Hs})$. Hence, the detection criterion is

$$p_L(\mathbf{r}\,|\mathbf{Hs}) = \max_{\mathbf{s} \in \mathcal{A}} \qquad (19.10)$$

from which the optimal estimate

$$\hat{\mathbf{s}} = \arg\max_{\mathbf{s}\in\mathcal{A}} \; p_L\left(\mathbf{r}\,|\mathbf{Hs}\right) \qquad (19.11)$$

results. As is well known from communications theory, if the transmit vectors $\mathbf{s}\in\mathcal{A}$ are equally distributed, then $\hat{\mathbf{s}}$ also maximizes the a-posterior probability and thus minimizes the symbol error probability. With (19.2) and (19.8), we obtain from (19.9)

$$p_L\left(\mathbf{r}\,|\mathbf{Hs}\right) = p_\mathbf{n}\left(\mathbf{r}-\mathbf{Hs}\right) = \left(\frac{1}{\pi\sigma_n^2}\right)^N e^{-\frac{\|\mathbf{r}-\mathbf{Hs}\|^2}{\sigma_n^2}} \qquad (19.12)$$

The argument of the exponential function is always negative. Consequently, the maximal $p_L\left(\mathbf{r}\,|\mathbf{Hs}\right)$ must fulfill the condition

$$\|\mathbf{r}-\mathbf{Hs}\|^2 = \min_{\mathbf{s}\in\mathcal{A}} \qquad (19.13)$$

and the solution formally is

$$\hat{\mathbf{s}} = \arg\min_{\mathbf{s}\in\mathcal{A}} \|\mathbf{r}-\mathbf{Hs}\|^2 \qquad (19.14)$$

Obviously, the statistical detection problem (19.10) translates into the minimization of the Euclidian distance between two vectors, namely, the receive vector $\mathbf{r}$ and the vector $\mathbf{Hs}$, which is the transmit signal $\mathbf{s}$ having passed through the known channel $\mathbf{H}$. Hence, a maximum likelihood detector can be implemented as an algorithm, which calculates a squared error according to (19.13) for all possible transmit signal vectors $\mathbf{s}\in\mathcal{A}$ and selects that $\mathbf{s}=\hat{\mathbf{s}}$, which yields the minimal quadratic error. Of course, the receiver has to know the transmit vector alphabet $\mathcal{A}$, which is quite normal for the design of a digital communications system.

Just a few words about the computational complexity. As already mentioned, if the transmitter is equipped with M antennas and each antenna output signal can take on L_Q different values, then there are L_Q^M different vectors $\mathbf{s}$, for which the detector has to execute (19.13). We conclude that the number of operations in the maximum likelihood detector grows exponentially with the number M of transmit antennas.

Example 1

As a simple example, we take a MIMO transmitter with $M = 2$ antennas. The modulation scheme shall be 2-PSK with the symbol alphabet $\mathcal{B} = \{1, -1\}$. Consequently, $L_Q = 2$ and each component of $\mathbf{s}$ can take on one value out of $\mathcal{B}$ at time instant k. The channel matrix shall be given as

$$\mathbf{H} = \begin{pmatrix} 1 & 0.5 \\ 0 & 1 \\ 1 & 1 \end{pmatrix}$$

Table 19.1 Example, calculation steps for maximum likelihood detection

$\mathbf{s}$	$\begin{pmatrix} 1 \\ 1 \end{pmatrix}$	$\begin{pmatrix} 1 \\ -1 \end{pmatrix}$	$\begin{pmatrix} -1 \\ 1 \end{pmatrix}$	$\begin{pmatrix} -1 \\ -1 \end{pmatrix}$
$\mathbf{Hs}$	$\begin{pmatrix} 1.5 \\ 1.0 \\ 2.0 \end{pmatrix}$	$\begin{pmatrix} 0.5 \\ -1.0 \\ 0 \end{pmatrix}$	$\begin{pmatrix} -0.5 \\ 1.0 \\ 0 \end{pmatrix}$	$\begin{pmatrix} -1.5 \\ -1.0 \\ -2.0 \end{pmatrix}$
$\mathbf{r} - \mathbf{Hs}$	$\begin{pmatrix} -0.4 \\ -2.1 \\ -1.1 \end{pmatrix}$	$\begin{pmatrix} 0.6 \\ -0.1 \\ 0.9 \end{pmatrix}$	$\begin{pmatrix} 1.6 \\ -2.1 \\ 0.9 \end{pmatrix}$	$\begin{pmatrix} 2.6 \\ -0.1 \\ 2.9 \end{pmatrix}$
$\|\mathbf{r} - \mathbf{Hs}\|^2$	5.78	1.18	7.78	15.81

and the noisy receive vector is observed as $\mathbf{r} = \begin{pmatrix} 1.1 & -1.1 & 0.9 \end{pmatrix}^T$. The receiver knows the set $\mathcal{A}$ of all $L_Q^M = 4$ different transmit vectors.

$$\mathcal{A} = \left\{ \begin{pmatrix} 1 \\ 1 \end{pmatrix}, \begin{pmatrix} 1 \\ -1 \end{pmatrix}, \begin{pmatrix} -1 \\ 1 \end{pmatrix}, \begin{pmatrix} -1 \\ -1 \end{pmatrix} \right\} \tag{19.15}$$

Then, the maximum likelihood receiver calculates all vectors $\mathbf{Hs}$ and $\mathbf{r} - \mathbf{Hs}$ as well as the squared error $\|\mathbf{r} - \mathbf{Hs}\|^2$ in Table 19.1. Finally, the minimal $\|\mathbf{r} - \mathbf{Hs}\|^2$, which is 1.18 in our example, is selected and the detector concludes that most likely

$$\hat{\mathbf{s}} = \begin{pmatrix} 1 \\ -1 \end{pmatrix}$$

was sent.

19.2 Receiver with Ordered Successive Interference Cancellation

Prerequisites

We are now coming back to the transmission system depicted in Fig. 18.1 and are going to combine the linear receiver with the decision device. Our target is to successively detect the transmit signal components $s_j(k)$, $j = 1, 2, \ldots, M$ of the transmit vector $\mathbf{s}(k)$. Again to simplify notation, we drop the discrete-time variable k. The starting point of our considerations is the linear receiver. The receive signal is given by

$$\mathbf{r} = \mathbf{Hs} + \mathbf{n} \tag{19.16}$$

with the channel matrix

$$\mathbf{H} = \begin{pmatrix} \mathbf{h}_1 & \mathbf{h}_2 & \cdots & \mathbf{h}_M \end{pmatrix} \tag{19.17}$$

in which $\mathbf{h}_j \in \mathbb{C}^{N \times 1}$, $j = 1, \ldots, M$ are the column vectors. The receiver matrix

$$\mathbf{W} = \begin{pmatrix} \mathbf{w}_1 \\ \mathbf{w}_2 \\ \vdots \\ \mathbf{w}_M \end{pmatrix} \tag{19.18}$$

is structured by its row vectors $\mathbf{w}_i \in \mathbb{C}^{1 \times N}$, $i = 1, \ldots, M$ and can be calculated as the pseudo-inverse or the MMSE receive matrix of the channel matrix. Finally, we get the output of the receiver filter

$$\mathbf{y} = \begin{pmatrix} y_1 \\ y_2 \\ \vdots \\ y_M \end{pmatrix} \tag{19.19}$$

by multiplication

$$\mathbf{y} = \mathbf{Wr} \tag{19.20}$$

Hence, the output signal component y_i is obtained as

$$y_i = \mathbf{w}_i \mathbf{r} \; ; \; i = 1, 2, \ldots, M \tag{19.21}$$

According to Fig. 18.1, a decision device follows and we characterize the input–output relation by the decision function $q(\ldots)$ yielding

$$\hat{s}_i = q(y_i) \; ; \; i = 1, 2, \ldots, M \tag{19.22}$$

The decision device can be a simple threshold detector but also a more sophisticated maximum likelihood detector. If the receiver applies the receive matrix $\mathbf{W}$, the system of equations is solved for all $y_{1,} y_{2,\ldots,} y_M$ in one step and the decided signal components $\hat{s}_1, \hat{s}_2, \ldots, \hat{s}_M$ are obtained in parallel. Now we are going to discuss a method, in which the system of linear equations (19.20) is solved successively in several steps, where in each step the decision operation (19.22) is applied.

Ordered Successive Interference Cancellation

As outlined before, we are looking for an algorithm, with which the $\hat{s}_i = q(y_i)$, $i = 1, 2, \ldots, M$ are calculated one after the other rather than in one step. The algorithm is called ordered successive interference cancellation (OSIC). In principle, the operations (19.20) and (19.22) are merged. With (19.17), we obtain from (19.16)

$$\begin{pmatrix} \mathbf{h}_1 & \cdots & \mathbf{h}_\nu & \cdots & \mathbf{h}_M \end{pmatrix} \begin{pmatrix} s_1 \\ \vdots \\ s_\nu \\ \vdots \\ s_M \end{pmatrix} + \mathbf{n} = \mathbf{r} \qquad (19.23)$$

which is equivalent to

$$\mathbf{h}_1 s_1 + \cdots + \mathbf{h}_{\nu-1} s_{\nu-1} + \mathbf{h}_{\nu+1} s_{\nu+1} + \cdots + \mathbf{h}_M s_M + \mathbf{n} = \mathbf{r} - \mathbf{h}_\nu s_\nu \qquad (19.24)$$

and is the key equation, in which we have moved $\mathbf{h}_\nu s_\nu$ to the right-hand side. The idea is first to find a solution for s_ν, e.g., s_1, and then reduce the dimension of system of linear equations by one. It matters to introduce ν as an index for the iteration step ν.

The algorithm is best explained with $M = 3$ as an example. Let $\mathbf{H} = \begin{pmatrix} \mathbf{h}_1 & \mathbf{h}_2 & \mathbf{h}_3 \end{pmatrix}$, $\mathbf{r}$, and the decision rule $q(\ldots)$ be given. In the course of the iterations, the matrix $\mathbf{H}$ will change and therefore it will be indicated as $\mathbf{H}^{(\nu)}$. In each step, only the first row vector $\mathbf{w}_1^{(\nu)}$ of the receive matrix

$$\mathbf{W}^{(\nu)} = \begin{pmatrix} \mathbf{w}_1^{(\nu)} \\ \mathbf{w}_2^{(\nu)} \\ \mathbf{w}_3^{(\nu)} \end{pmatrix} \qquad (19.25)$$

has to be calculated either from the pseudo-inverse or the MMSE matrix with respect to $\mathbf{H}^{(\nu)}$. The iteration steps are as follows:

step $\nu = 1$:

Let $\qquad \mathbf{H}^{(1)} = \begin{pmatrix} \mathbf{h}_1 & \mathbf{h}_2 & \mathbf{h}_3 \end{pmatrix}$, $\quad \mathbf{r}^{(1)} = \mathbf{r}$

then $\qquad \mathbf{H}^{(1)} \begin{pmatrix} s_1 \\ s_2 \\ s_3 \end{pmatrix} + \mathbf{n} = \mathbf{r}^{(1)}$

calculate $\quad \mathbf{w}_1^{(1)}$ from $\mathbf{H}^{(1)}$ using (19.25)

calculate $\quad y_1 = \mathbf{w}_1^{(1)} \mathbf{r}^{(1)}$ using (19.21)

decide $\qquad \hat{s}_1 = q(y_1)$ using (19.22)

set $\qquad\quad s_1 = \hat{s}_1$ and the new equation system is

$\qquad\qquad \begin{pmatrix} \mathbf{h}_2 & \mathbf{h}_3 \end{pmatrix} \begin{pmatrix} s_2 \\ s_3 \end{pmatrix} + \mathbf{n} = \mathbf{r}^{(1)} - \mathbf{h}_1 \hat{s}_1$

step $\nu = 2$:

Let $\quad \mathbf{H}^{(2)} = \begin{pmatrix} \mathbf{h}_2 & \mathbf{h}_3 \end{pmatrix}, \quad \mathbf{r}^{(2)} = \mathbf{r}^{(1)} - \mathbf{h}_1 \hat{s}_1$

then $\quad \mathbf{H}^{(2)} \begin{pmatrix} s_2 \\ s_3 \end{pmatrix} + \mathbf{n} = \mathbf{r}^{(2)}$

calculate $\quad \mathbf{w}_1^{(2)}$ from $\mathbf{H}^{(2)}$ using (19.25)

calculate $\quad y_2 = \mathbf{w}_1^{(2)} \mathbf{r}^{(2)}$ using (19.21)

decide $\quad \hat{s}_2 = q(y_2)$ using (19.22)

set $\quad s_2 = \hat{s}_2$ and the new equation system is
$\quad \mathbf{h}_3 s_3 + \mathbf{n} = \mathbf{r}^{(2)} - \mathbf{h}_2 \hat{s}_2$

step $\nu = 3$:

Let $\quad \mathbf{H}^{(3)} = \mathbf{h}_3, \quad \mathbf{r}^{(3)} = \mathbf{r}^{(2)} - \mathbf{h}_2 \hat{s}_2$

then $\quad \mathbf{H}^{(3)} s_3 + \mathbf{n} = \mathbf{r}^{(3)}$

calculate $\quad \mathbf{w}_1^{(3)}$ from $\mathbf{H}^{(3)}$ using (19.25)

calculate $\quad y_3 = \mathbf{w}_1^{(3)} \mathbf{r}^{(3)}$ using (19.21)

decide $\quad \hat{s}_3 = q(y_3)$ using (19.22)

set $\quad s_3 = \hat{s}_3$

end $\quad$ In our example with $N = 3$ the algorithm terminates.

The advantage of this algorithm is its low computational complexity and the feature that it reduces inter-channel interference in every decision step. However, decision errors, which may occur at low signal-to-noise ratios, are critical, because they can impact the next decision and thus may cause an error propagation for the following steps. We notice that the algorithm is in principle based on the triangulation of a matrix into a lower or an upper triangular matrix also called L-U decomposition [1], which is continuously applied from one step to the next. This is in principle a linear operation. However, the described OSIC algorithm gets nonlinear owing to the decision made in each step. The algorithm has been practically used in several systems, such as the layered space-time architecture.

19.3 Comparison of Different Receivers

The design criteria for linear and nonlinear receivers have quite some similarities which are now going to be discussed for the zero-forcing (ZF), the minimum mean squared error (MMSE), and the maximum likelihood (ML) receiver. Table 19.2 shows a survey of the different design criteria.

The design of the zero-forcing receiver with and without OSIC does not include the noise at the receiver. Computation of the receiver matrix $\mathbf{W}$ for the MMSE receiver requires the knowledge of the signal-to-noise ratio $\frac{1}{\alpha}$, which is not needed for the maximum likelihood detection. This method operates without any receiver

Table 19.2 Comparison of design criteria for various receivers

Receiver	Target Function	Noise	Result	Output	Method
Zero − Forcing (ZF)	$\|\mathbf{r} - \mathbf{Hs}\|^2$ $= \min_{\mathbf{s} \in \mathbb{C}^{M \times 1}}$	not included	matrix $\mathbf{W}$	$\mathbf{y} = \mathbf{Wr}$	linear
MMSE	$\mathbf{E}\left[\|\mathbf{Wr} - \mathbf{s}\|^2\right]$ $= \min_{\mathbf{s} \in \mathbb{C}^{M \times 1}}$	included	matrix $\mathbf{W}$	$\mathbf{y} = \mathbf{Wr}$	linear
Maximum Likelihood	$\|\mathbf{r} - \mathbf{Hs}\|^2$ $= \min_{\mathbf{s} \in \mathcal{A}}$	included	symbol $\hat{\mathbf{s}}$	$\hat{\mathbf{s}} \in \mathcal{A}$	non− linear
OSIC ZF	as ZF	not included	symbol $\hat{\mathbf{s}}$	$\hat{\mathbf{s}} \in \mathcal{A}$	non− linear
OSIC MMSE	as MMSE	included	symbol $\hat{\mathbf{s}}$	$\hat{\mathbf{s}} \in \mathcal{A}$	non− linear

matrix. Moreover, on the first glance the target functions of the zero-forcing algorithm using the pseudo-inverse matrix and the maximum likelihood receiver look the same. Both receivers minimize the squared error $\|\mathbf{r} - \mathbf{Hs}\|^2$. However, the zero-forcing receiver provides a "soft" output signal $\mathbf{y}$ $\mathbb{C}^{M \times 1}$ with continuous amplitude and phase compared to the output of the maximum likelihood receiver, which is a discrete vector $\hat{\mathbf{s}} \in \mathcal{A}$. Hence, the maximum likelihood scheme minimizes the same target function as the zero-forcing receiver, however, as the result of a discrete minimization problem with the constraint $\hat{\mathbf{s}} \in \mathcal{A}$. This can be formulated as an integer least squares problem for which several mathematical algorithms from the area of integer programming are known [2, 3]. Such methods have been used for lattice decoding and are summarized as sphere decoding algorithm, because they search in a limited hypersphere of the complex vector space rather than performing an overall brute search [4–6] and thus do not always provide the global optimum. In principle, the hypersphere is centered around the receive vector $\mathbf{r}$ and for an efficient search the sphere should cover the lattice points given by the vectors $(\mathbf{Hs} \; ; \; \mathbf{s} \in \mathcal{A})$ located in the vicinity of $\mathbf{r}$. As a result, the complexity of the maximum likelihood algorithm can be significantly reduced and sphere decoding became an important alternative to the much simpler but suboptimal linear receivers.

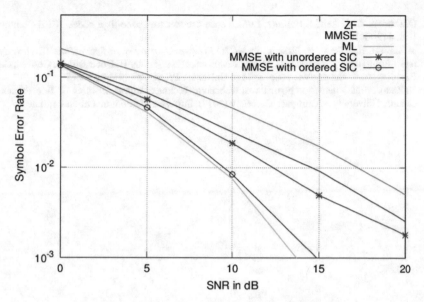

Fig. 19.1 Typical symbol error rate of various receivers for a 2x2 MIMO channel corrupted by white Gaussian noise and Rayleigh fading using 2-PSK, *Source* Online platform "webdemo" [7]

On the other hand, the complexity of the zero-forcing and the minimum mean squared error (MMSE) receiver can also be considerably reduced by introducing successive interference cancellation (OSIC), because only parts of an inverse matrix have to be calculated rather than a full-inverse or pseudo-inverse matrix. However, it should be kept in mind that in case of ill-conditioned matrices the calculation of the inverse matrices may turn out to be numerically not stable.

Figure 19.1 shows a rough comparison of the symbol error rate for various receivers as the result of a computer simulation using the platform "webdemo" [7]. According to our expectations, the maximum likelihood detector (ML) demonstrates the best performance followed by the nonlinear receiver with OSIC. Compared to the zero-forcing receiver (ZF), the minimum mean squared error approach (MMSE) takes the noise into account and thus outperforms the ZF receiver in general.

References

1. R.A. Horn, C.R. Johnson, *Matrix Analysis* (Cambridge University Press, Cambridge, 2013)
2. R. Kannan, Improved algorithms on integer programming and related lattice problems, in *Proceedings of ACM Symposium on Theory of Computation* (1983)
3. U. Finke, M. Pohst, Improved methods for calculating vectors of short length in a lattice, including a complexity analysis. Math. Comput. **44** (1985)
4. E. Viterbo, J. Boutros, A universal lattice code decoder for fading channels. IEEE Trans. Inf. Theory **45** (1999)

5. O. Damen, A. Chkeif, J. Belfiore, Lattice code decoder for space-time codes. IEEE Commun. Lett. **4** (2000)
6. T. Kailath, H. Vikalo, B. Hassibi, in *MIMO Receive Algorithms, in Space-Time Wireless Systems: From Array Processing to MIMO Communications*, ed. by H. Boelcskei, D. Gesbert, C.B. Papadias, A.-J. van der Veen (Cambridge University Press, Cambridge, 2008)
7. N. Zhao, MIMO detection algorithms, webdemo, Technical report, Institute of Telecommunications, University of Stuttgart, Germany (2018), http://webdemo.inue.uni-stuttgart.de

Chapter 20
MIMO System Decomposition into Eigenmodes

In this chapter, we allude to a topic which gives quite some insights into the functionality of an MIMO system. As we have seen, the MIMO channel matrix $\mathbf{H}(k)$ introduces interchannel interference to the received signal

$$\mathbf{r}(k) = \mathbf{H}(k)\mathbf{s}(k) + \mathbf{n}(k) \tag{20.1}$$

We are now interested in the decoupling of the received signal. To achieve this goal $\mathbf{H}(k)$ has to be transformed into a matrix, in which only one diagonal is covered by entries unequal to zero and all remaining elements have to be zero. In the following, we drop the discrete time k to simplify the notation.

20.1 MIMO System Transformation Using Singular Value Decomposition

A first idea to transform the channel matrix $\mathbf{H}$ into diagonal form is the application of the eigenvalue decomposition, as outlined in Appendix B.

$$\mathbf{H} = \mathbf{V}\boldsymbol{\Lambda}\mathbf{V}^{-1} \tag{20.2}$$

where $\boldsymbol{\Lambda}$ is a diagonal matrix containing the eigenvalues of $\mathbf{H}$ in its main diagonal. $\mathbf{V}$ is composed of eigenvectors associated with the respective eigenvalues. However, this approach has various drawbacks both from a mathematical and a technical point of view. First of all, $\mathbf{H}$ must be a square matrix and this fact would restrict the approach to a system with the same number of antennas at transmitter and receiver, $M = N$. Second, not all square matrices can be transformed into diagonal form mathematically. An alternative is the Singular Value Decomposition (SVD), which can be applied to any matrix and which is outlined in detail in Appendix B. In this chapter, we recall the principle steps.

© Springer Nature Switzerland AG 2019
J. Speidel, *Introduction to Digital Communications*, Signals and Communication
Technology, https://doi.org/10.1007/978-3-030-00548-1_20

The singular value decomposition of the channel matrix $\mathbf{H} \in \mathbb{C}^{N \times M}$ is given by

$$\mathbf{H} = \mathbf{U}\mathbf{D}\mathbf{V}^H \qquad (20.3)$$

with

$$\mathbf{D} = \begin{pmatrix} \sqrt{\lambda_1} & 0 & 0 & \cdots & 0 & 0 & \cdots & 0 \\ 0 & \sqrt{\lambda_2} & 0 & \cdots & 0 & 0 & \cdots & 0 \\ & & \ddots & & & & & \\ 0 & 0 & 0 & \cdots & \sqrt{\lambda_P} & 0 & \cdots & 0 \\ 0 & 0 & 0 & \cdots & 0 & 0 & \cdots & 0 \\ & \vdots & & \cdots & & & \ddots & \\ 0 & 0 & 0 & \cdots & 0 & 0 & \cdots & 0 \\ 0 & 0 & 0 & \cdots & 0 & 0 & \cdots & 0 \end{pmatrix} = \begin{pmatrix} \mathbf{\Lambda}_P^{\frac{1}{2}} & 0 & \cdots & 0 \\ 0 & 0 & \cdots & 0 \\ & & \ddots & \\ 0 & 0 & \cdots & 0 \\ 0 & 0 & \cdots & 0 \end{pmatrix} \in \mathbb{R}^{N \times M} \quad (20.4)$$

where $\lambda_1 \geq \lambda_2 \geq \cdots \lambda_P > 0$ and $\lambda_{P+1} = \lambda_{P+2} = \cdots = \lambda_N = 0$ are the N eigenvalues of the matrix

$$\mathbf{Q}_N = \mathbf{H}\mathbf{H}^H \in \mathbb{C}^{N \times N} \qquad (20.5)$$

Please note that $\mathbf{Q}_N$ is a Hermiteian matrix, because $\mathbf{Q}_N^H = \mathbf{Q}_N$, and consequently, all eigenvalues are positive or zero.

$$P = \text{rank}\,(\mathbf{Q}_N) \qquad (20.6)$$

is the rank of the matrix $\mathbf{Q}_N$. In general, the rank of a matrix $\mathbf{H} \in \mathbb{C}^{N \times M}$ is defined as the number of linearly independent rows or columns and thus

$$\text{rank}\,(\mathbf{H}) \leq \min\{M, N\} \qquad (20.7)$$

holds. With (20.5) follows

$$P \leq N \qquad (20.8)$$

$\sqrt{\lambda_i}$; $i = 1, \ldots, P$ are called the singular values of the matrix $\mathbf{H}$.
$\mathbf{U} \in \mathbb{C}^{N \times N}$ and $\mathbf{V} \in \mathbb{C}^{M \times M}$ are unitary matrices, thus

$$\mathbf{U}^{-1} = \mathbf{U}^H \; ; \; \mathbf{V} = \mathbf{V}^H \qquad (20.9)$$

hold. Furthermore, $\mathbf{U}$ is the matrix of the normalized eigenvectors with respect to the eigenvalues $\lambda_1, \lambda_2, \cdots, \lambda_N$. Let

$$\mathbf{\Lambda}_N = \text{diag}\,(\lambda_1, \lambda_2, \ldots, \lambda_P, 0, \ldots, 0) \in \mathbb{R}^{N \times N} \qquad (20.10)$$

be a diagonal matrix composed of the eigenvalues of $\mathbf{Q}_N$. Then the eigenvalue decomposition of $\mathbf{Q}_N$ is

$$\mathbf{U}^H \mathbf{Q}_N \mathbf{U} = \mathbf{\Lambda}_N \tag{20.11}$$

One method to find the matrix $\mathbf{V}$ in (20.3) is the eigenvalue decomposition of the matrix

$$\mathbf{Q}_M = \mathbf{H}^H \mathbf{H} \in \mathbb{C}^{M \times M} \tag{20.12}$$

which is

$$\mathbf{V}^H \mathbf{Q}_M \mathbf{V} = \mathbf{\Lambda}_M \tag{20.13}$$

with the diagonal matrix

$$\mathbf{\Lambda}_M = \operatorname{diag}\,(\lambda_1, \lambda_2, \ldots, \lambda_P, 0, \ldots, 0) \in \mathbb{R}^{M \times M} \tag{20.14}$$

$\mathbf{V}$ is the matrix of eigenvectors with respect to the eigenvalues $\lambda_1, \lambda_2, \cdots, \lambda_M$. Note that the eigenvalues $\lambda_1, \lambda_2, \ldots, \lambda_P$ are the same as for the matrix $\mathbf{Q}_N$. Furthermore

$$\operatorname{rank}\,(\mathbf{Q}_M) = \operatorname{rank}\,(\mathbf{Q}_N) = \operatorname{rank}\,(\mathbf{H}) = P \tag{20.15}$$

holds and the matrices $\mathbf{\Lambda}_M$ and $\mathbf{\Lambda}_N$ contain the same diagonal matrix

$$\mathbf{\Lambda}_P = \operatorname{diag}\,(\lambda_1, \lambda_2, \ldots, \lambda_P) \in \mathbb{R}^{P \times P} \tag{20.16}$$

of the P eigenvalues, which are unequal to zero. Please note in (20.4)

$$\mathbf{\Lambda}_P^{\frac{1}{2}} = \operatorname{diag}\,\left(\sqrt{\lambda_1}, \sqrt{\lambda_2}, \ldots, \sqrt{\lambda_P}\right) \tag{20.17}$$

holds. We obtain from the input–output relation (20.1) with (20.3)

$$\mathbf{r} = \mathbf{U}\mathbf{D}\mathbf{V}^H \mathbf{s} + \mathbf{n} \tag{20.18}$$

and by multiplication of this equation from the left with $\mathbf{U}^H$ follows

$$\mathbf{U}^H \mathbf{r} = \mathbf{U}^H \mathbf{U}\mathbf{D}\mathbf{V}^H \mathbf{s} + \mathbf{U}^H \mathbf{n} \tag{20.19}$$

With $\mathbf{U}^H \mathbf{U} = \mathbf{I}_N$ and the transformed signals

$$\tilde{\mathbf{r}} = \mathbf{U}^H \mathbf{r} \tag{20.20}$$

$$\tilde{\mathbf{s}} = \mathbf{V}^H \mathbf{s} \tag{20.21}$$

and the transformed noise

$$\tilde{\mathbf{n}} = \mathbf{U}^H \mathbf{n} \tag{20.22}$$

a new description of the system referred to as *eigenmode system* based on the eigen-modes can be given as

$$\tilde{\mathbf{r}} = \mathbf{D}\tilde{\mathbf{s}} + \tilde{\mathbf{n}} \tag{20.23}$$

The new *eigenmode channel* is described by the matrix $\mathbf{D}$, which has entries unequal to zero only in a diagonal. All other entries are zero. Equation (20.23) clearly reveals that the eigenmode branches and thus the signal and noise components are decoupled, which is the basic idea of the approach. This can be even better seen when we write down the ith equation of (20.23)

$$\tilde{r}_i = \sqrt{\lambda_i}\tilde{s}_i + \tilde{n}_i \; ; \; i = 1, 2, \dots, P \tag{20.24}$$

and

$$\tilde{r}_i = \tilde{n}_i \; ; \; i = P + 1, 2, \dots, N \tag{20.25}$$

20.2   Implementation of the MIMO Eigenmode Decomposition

Both Eqs. (20.24) and (20.25) can be implemented with the block diagram depicted in Fig. 20.1a.

In essence, the branches with indices $i = 1, 2, \dots, P$ are called the eigenmodes of the MIMO system, because they carry the information from the transmitter to the receiver. The remaining branches $i = P + 1, 2, \dots, N$ contain only noise rather than information and are therefore often not included in the definition of the MIMO

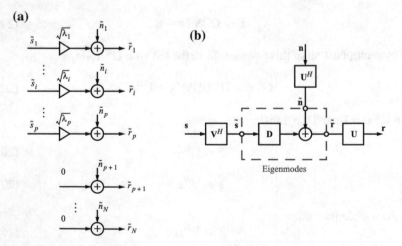

Fig. 20.1 a Block diagram of MIMO system decomposed into parallel eigenmodes; **b** block diagram of the MIMO eigenmodes using vector and matrix notation

Fig. 20.2 Block diagram of
the MIMO system with
eigenmodes $\tilde{\mathbf{s}}$, $\tilde{\mathbf{n}}$, and $\tilde{\mathbf{r}}$ as
input and output,
respectively

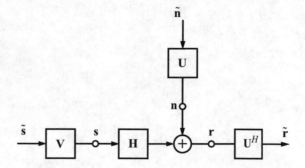

eigenmodes in the narrow sense. Hence, these branches do not contribute to the
MIMO channel capacity. Figure 20.1b represents the implementation of (20.18) as a
block diagram.

An alternative implementation of (20.23) is shown in Fig. 20.2, where the kernel
is given by the original MIMO system. From (20.20) – (20.22) follows

$$\mathbf{r} = \mathbf{U}\tilde{\mathbf{r}} \tag{20.26}$$

$$\mathbf{s} = \mathbf{V}\tilde{\mathbf{s}} \tag{20.27}$$

$$\mathbf{n} = \mathbf{U}\tilde{\mathbf{n}} \tag{20.28}$$

Chapter 21
Channel Capacity of Single-User Transmission Systems

In this chapter, we allude to a topic which is important for the design of a communications system. We will answer the question how many bit/s can be transmitted per symbol or equivalently per channel use. For a certain bandwidth of the channel, the interesting point is how many bit/s per Hz bandwidth can be achieved as a maximum. The channel capacity was introduced by Ch. Shannon in his pioneering work [1] in the year 1948 for single input single output (SISO) channels. The extension to MIMO channels was given by I. Telatar, [2]. The capacity of various models for stochastic MIMO channels have been intensively studied, e.g., in [3]. In the following, we start with the SISO channel capacity for a real-valued signal, extend it to complex signals, and finally derive the channel capacity of a MIMO scheme.

21.1 Channel Capacity of SISO System

21.1.1 AWGN Channel with Real Signals and Noise

We shortly review the basics of the channel capacity C given by Shannon [1]. In general, C is formulated as the maximal mutual information $I(X, Y)$ between an input X and an output Y in Fig. 21.1a.

$$C = \max\, I(X, Y) \tag{21.1}$$

where X and Y are random variables or stochastic processes. Maximization is done for overall degrees of freedom of the system. The mutual information is a measure, which quantifies the information we can get about X by the observation of Y. This is actually the situation in information transmission, as we can just observe the signal Y at the receiver and try to retrieve as much as possible information about X. The capacity C is measured in bit/channel use. In Fig. 21.1b, we consider the simple arrangement, which is the additive white Gaussian noise (AWGN) channel. First, we assume that the transmit signal $s(t) = s_R(t)$, the noise $n(t) = n_R(t)$, and the

© Springer Nature Switzerland AG 2019
J. Speidel, *Introduction to Digital Communications*, Signals and Communication
Technology, https://doi.org/10.1007/978-3-030-00548-1_21

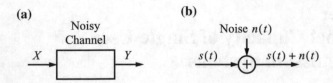

Fig. 21.1 **a** Noisy channel with input X and output Y, **b** Additive white Gaussian noise (AWGN) channel

receive signal $r(t) = r_R(t)$ are real-valued. Obviously, $r(t) = s(t) + n(t)$ holds. For the AWGN channel, the maximization operation in (21.1) can be done only over the probability density function of the input signal $s(t)$, because it is the only degree of freedom here. Often the maximization is dropped, and just $I(X, Y)$ is called channel capacity, which is not precise enough, and should be better denoted just as a system capacity. However, we will still use the term "channel capacity" and the actual meaning will be understood from the context. The signal and the noise shall have the following mean power

$$\mathbf{E}[s_R^2] = \frac{E_S}{2} \tag{21.2}$$

$$\mathbf{E}[n_R^2] = \frac{\sigma_n^2}{2} \tag{21.3}$$

The transmit signal contains discrete symbols with a spacing T on the time axis. Hence, the symbol rate is

$$v_S = \frac{1}{T} \tag{21.4}$$

Consequently, we can define alternatively the channel capacity C' in bit/s as

$$C' = v_S C = \frac{1}{T} C \tag{21.5}$$

We persuade ourselves that C' is measured in $\frac{\text{bit}}{\text{s}}$ simply by checking the dimension on the right-hand side of (21.5), $\frac{\text{symbol}}{\text{s}} \frac{\text{bit}}{\text{symbol}} = \frac{\text{bit}}{\text{s}}$.

Now, we introduce a strict bandwidth limitation with cut-off frequency f_c to the channel and assume that the receive signal $s_R(t)$ and also the noise $n_R(t)$ are strictly band-limited with f_c. We know from the first Nyquist condition that we can transmit without intersymbol interference over an ideal lowpass channel with the frequency response

$$G(f) = \begin{cases} 1 \; ; \; f \le f_c \\ 0 \; ; \; f > f_c \end{cases} \tag{21.6}$$

with cut-off frequency f_c at the maximal symbol rate

$$\frac{1}{T} = 2f_c \tag{21.7}$$

Obviously the same condition results, if we limit the spectrum of the signal and the noise to f_c with the ideal lowpass filter (21.6) and use a sampling frequency $\frac{1}{T}$, which just satisfies the sampling theorem. Then, we get from (21.5)

$$C' = 2f_cC \tag{21.8}$$

If we would sample above the lower limit given by the sampling theorem, i.e., $\frac{1}{T} > 2f_c$, the discrete-time signal would contain additional samples. However, they are redundant, as they contain no additional information. Shannon has shown [1] that C is given by the logarithm of the signal-to-noise ratio as

$$C = \frac{1}{2} \log_2 \left(1 + \frac{\mathrm{E}[s_R^2]}{\mathrm{E}[n_R^2]} \right) \tag{21.9}$$

from which we conclude with (21.2) and (21.3)

$$C = \frac{1}{2} \log_2 \left(1 + \frac{E_S}{\sigma_n^2} \right) \tag{21.10}$$

and (21.8) yields

$$C' = f_c \log_2 \left(1 + \frac{E_S}{\sigma_n^2} \right) \tag{21.11}$$

Without proof the maximum of C occurs, if the signal $s_R(t)$ has a Gaussian probability density function. We clearly see from (21.11) that C' increases linearly with the cut-off frequency f_C of the signal, however only logarithmic with respect to the signal-to-noise ratio $\frac{E_S}{\sigma_n^2}$.

21.1.2 AWGN Channel with Complex Signals and Noise

Now, we are going to extend the previous considerations to the case of a complex modulation scheme known as quadrature amplitude modulation QAM with a complex carrier $e^{j2\pi f_0 t}$ and complex signals. We know from the modulation theory that we can describe this scheme by means of complex baseband signals and noise after demodulation. We define the signal and the noise by its real parts (index R), its imaginary parts (index I), and with the following properties

Signal

$$s = s_R + js_I \tag{21.12}$$

- s_R and s_I are statistically independent and both with zero mean

$$\mathbf{E}[s_R] = \mathbf{E}[s_I] = 0 \tag{21.13}$$

- Mean power

$$\mathbf{E}\left[s_R^2\right] = \mathbf{E}\left[s_I^2\right] = \frac{E_S}{2} \tag{21.14}$$

Then it follows

$$\mathbf{E}\left[|s|^2\right] = \mathbf{E}\left[ss^*\right] = \mathbf{E}\left[s_R^2 + s_I^2\right] = \mathbf{E}\left[s_R^2\right] + \mathbf{E}\left[s_I^2\right] = E_S \tag{21.15}$$

A complex stochastic process with this property is called cyclic symmetrical.

Noise

$$n = n_R + jn_I \tag{21.16}$$

- n_R and n_I are statistically independent AWGN and both with zero mean

$$\mathbf{E}[n_R] = \mathbf{E}[n_I] = 0 \tag{21.17}$$

- Mean power

$$\mathbf{E}\left[n_R^2\right] = \mathbf{E}\left[n_I^2\right] = \frac{\sigma_n^2}{2} \tag{21.18}$$

Similar to (21.15) follows

$$\mathbf{E}\left[|n|^2\right] = \mathbf{E}\left[nn^*\right] = \mathbf{E}\left[n_R^2 + n_I^2\right] = \mathbf{E}\left[n_R^2\right] + \mathbf{E}\left[n_I^2\right] = \sigma_n^2 \tag{21.19}$$

We also see that n is a cyclic symmetrical stochastic process.

Channel Capacity

The real and imaginary part carry independent information. Therefore, we get twice the capacity as in (21.9)

$$C = \log_2\left(1 + \frac{\mathbf{E}\left[|s|^2\right]}{\mathbf{E}\left[|n|^2\right]}\right) = \log_2\left(1 + \frac{E_S}{\sigma_n^2}\right) \tag{21.20}$$

and with (21.7) follows

$$C' = \frac{1}{T}C = 2f_c \log_2\left(1 + \frac{E_S}{\sigma_n^2}\right) \qquad (21.21)$$

Please note that $2f_c$ is the bandwidth of the transmit signal spectrum after modulation measured in the frequency range $|f| \geq 0$ and f_c is the cut-off frequency of the equivalent baseband system.

21.2 Channel Capacity of MIMO Systems with Statistically Independent Transmit Signals and Noise

21.2.1 Prerequisites

As depicted in Fig. 21.2, we are now considering a MIMO system with the input signal vector $\mathbf{s}(k) \in \mathbb{C}^{M \times 1}$, a frequency flat fading channel with matrix $\mathbf{H}(k) \in \mathbb{C}^{N \times M}$, an additive noise vector $\mathbf{n}(k) \in \mathbb{C}^{N \times 1}$, and the receive vector $\mathbf{r}(k) \in \mathbb{C}^{N \times 1}$. In the course of our investigation, we will introduce a prefilter with matrix $\mathbf{A} \in \mathbb{C}^{M \times M}$. Then, the transmit signal changes from $\mathbf{s}(k)$ to $\underline{\mathbf{s}}(k)$. However, if not otherwise stated, we first drop the prefilter, thus in Fig. 21.2

$$\mathbf{A} = \mathbf{I}_M \qquad (21.22)$$

holds and we will come back to the prefilter in Sects. 21.3 and 21.5. To simplify the notation we drop k in the following.

The signal and the noise shall comply with the following prerequisites,

Original Transmit Signal

$$\mathbf{s} = \left(s_1 \; s_2 \; \cdots \; s_M\right)^T \; ; \; s_j = s_{R,j} + \mathrm{j}s_{I,j} \; ; \; j = 1, 2, \ldots, M \qquad (21.23)$$

$$s_{R,j} \; ; \; s_{I,j} \; ; \; j = 1, 2, \ldots, M \text{ statistically independent, zero mean} \qquad (21.24)$$

$$\mathbf{E}\left[s_{R,j}^2\right] = \mathbf{E}\left[s_{I,j}^2\right] = \frac{E_S}{2} \;\Rightarrow\; \mathbf{E}\left[|s_j|^2\right] = E_S \; ; \; j = 1, 2, \ldots, M \qquad (21.25)$$

Fig. 21.2 MIMO system with statistically independent input signal **s**, noise **n**, prefilter **A**, and channel **H**

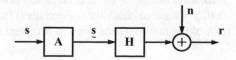

$$s_j \; ; \; s_m \; ; \; j, m = 1, 2, \ldots, M \; ; \; j \neq m \; ; \text{ statistically independent} \tag{21.26}$$

From (21.26) follows the spatial covariance matrix of $\mathbf{s}$

$$\mathbf{R}_{ss} = \mathbf{E}\left[\mathbf{s}\mathbf{s}^H\right] = E_S \mathbf{I}_M \tag{21.27}$$

Consequently, the total mean transmit power is

$$\mathbf{E}\left[\|\mathbf{s}\|^2\right] = \text{tr}\ (\mathbf{R}_{ss}) = M E_S \tag{21.28}$$

Noise

$$\mathbf{n} = \left(n_1 \; n_2 \; \cdots \; n_N\right)^T \; ; \; n_i = n_{R,i} + jn_{I,i} \; ; \; i = 1, 2, \ldots, N \tag{21.29}$$

$$n_{R,i} \; ; \; n_{I,i} \; ; \; i = 1, 2, \ldots, N \text{ statistically independent, zero mean} \tag{21.30}$$

$$\mathbf{E}\left[n_{R,i}^2\right] = \mathbf{E}\left[n_{I,i}^2\right] = \frac{\sigma_n^2}{2} \; \Rightarrow \; \mathbf{E}\left[|n_i|^2\right] = \sigma_n^2 \; ; \; i = 1, 2, \ldots, N \tag{21.31}$$

$$n_i \; ; \; n_m \; ; \; i, m = 1, 2, \ldots, N \; ; \; i \neq m \; ; \text{ statistically independent} \tag{21.32}$$

From (21.32), it follows the spatial covariance matrix of $\mathbf{n}$

$$\mathbf{R}_{nn} = \mathbf{E}\left[\mathbf{n}\mathbf{n}^H\right] = \sigma_n^2 \mathbf{I}_N \tag{21.33}$$

and the total mean power of the noise at the receiver is

$$\mathbf{E}\left[\|\mathbf{n}\|^2\right] = \text{tr}\ (\mathbf{R}_{nn}) = N \sigma_n^2 \tag{21.34}$$

Let us give some remarks on these prerequisites. Statistical independence of real and imaginary part of the transmit signal is mostly in compliance with reality, because the user allocates to the QAM symbols independent data streams to fully exploit channel capacity. The statistical independence of the output signals of the different antenna elements strongly depends on the spatial correlation condition at the transmitter of a wireless link, such as a rich scattering environment. To allocate the same mean signal power E_S to every transmit antenna output signal is a good starting point. However, we will see in Sect. 21.5 that a power allocation prefilter with matrix $\mathbf{A}$ at the transmitter can maximize the channel capacity. The assumption of the noise with zero mean and equal mean noise power at each receive antenna element fits well with reality. This also holds for the statistical independence of the real and the imaginary part of the noise and the noise at different antennas.

21.2.2 Instantaneous MIMO Channel Capacity

We have seen in the previous section that the output signals of the transmit antennas can be assumed as statistically independent. The same holds for the noise at the receiver. If the receive signals of the N branches would also be statistically independent, then the total capacity could be calculated easily as the sum of the individual capacities of the branches. However, the transmit signals are passing through the channel, undergo inter-channel interference, and thus the receive signals are spatially correlated. Nevertheless, the idea of independent receive signals is attractive and we follow this approach by applying the decomposition of the MIMO scheme into independent eigenmode branches.

Capacity of the MIMO Eigenmodes

We briefly review the main results of the eigenmode decomposition outlined in Chap. 20. The singular value decomposition (SVD) of the channel matrix $\mathbf{H} \in \mathbb{C}^{N \times M}$ is given by

$$\mathbf{H} = \mathbf{U}\mathbf{D}\mathbf{V}^H \tag{21.35}$$

with

$$\mathbf{D} = \begin{pmatrix} \sqrt{\lambda_1} & 0 & 0 \cdots & 0 & 0 \cdots 0 \\ 0 & \sqrt{\lambda_2} & 0 \cdots & 0 & 0 \cdots 0 \\ & & \ddots & & \\ 0 & 0 & 0 \cdots \sqrt{\lambda_P} & 0 \cdots 0 \\ 0 & 0 & 0 \cdots & 0 & 0 \cdots 0 \\ \vdots & \cdots & & \ddots \\ 0 & 0 & 0 \cdots & 0 & 0 \cdots 0 \\ 0 & 0 & 0 \cdots & 0 & 0 \cdots 0 \end{pmatrix} \in \mathbb{R}^{N \times M} \tag{21.36}$$

$\lambda_1 \geq \lambda_2 \geq \cdots \lambda_P > 0$, and $\lambda_{P+1} = \lambda_{P+2} = \cdots = \lambda_N = 0$ are the N eigenvalues of the Hermiteian matrix $\mathbf{Q}_N = \mathbf{H}\mathbf{H}^H \in \mathbb{C}^{N \times N}$ and $P = \text{rank}(\mathbf{Q}_N) \leq N$ is the rank of $\mathbf{Q}_N$.

$\sqrt{\lambda_i}$, $i = 1, \ldots, P$ are called the singular values of $\mathbf{H}$. The unitary matrix $\mathbf{U} \in \mathbb{C}^{N \times N}$ is the matrix of the normalized eigenvectors with respect to the eigenvalues $\lambda_1, \lambda_2, \cdots, \lambda_N$ of $\mathbf{Q}_N$. The matrix $\mathbf{Q}_N$ can be decomposed into the diagonal matrix

$$\mathbf{\Lambda}_N = \text{diag}(\lambda_1, \lambda_2, \ldots, \lambda_P, 0, \ldots, 0) \in \mathbb{R}^{N \times N} \tag{21.37}$$

The unitary matrix $\mathbf{V} \in \mathbb{C}^{M \times M}$ in (21.35) can be found by the eigenvalue decomposition of the matrix $\mathbf{Q}_M = \mathbf{H}^H \mathbf{H} \in \mathbb{C}^{M \times M}$ using the diagonal matrix

$$\mathbf{\Lambda}_M = \text{diag}(\lambda_1, \lambda_2, \ldots, \lambda_P, 0, \ldots, 0) \in \mathbb{R}^{M \times M} \tag{21.38}$$

with the eigenvalues $\lambda_1, \lambda_2, \cdots, \lambda_M$ of $\mathbf{Q}_M$. Then $\mathbf{V}$ represents the matrix of eigenvectors with respect to the eigenvalues of $\mathbf{Q}_M$. Note that rank $(\mathbf{Q}_M)$ = rank $(\mathbf{Q}_N)$ = rank $(\mathbf{H})$ = P holds and that the matrices $\mathbf{\Lambda}_M$ and $\mathbf{\Lambda}_N$ contain the same diagonal matrix

$$\mathbf{\Lambda}_P = \text{diag } (\lambda_1, \lambda_2, \ldots, \lambda_P) \, \epsilon \, \mathbb{R}^{P \times P} \tag{21.39}$$

of the P eigenvalues, which are unequal to zero. Then, we obtain from the input–output relation of the MIMO system

$$\mathbf{r} = \mathbf{Hs} + \mathbf{n} \tag{21.40}$$

the input–output relation of the eigenmode system

$$\tilde{\mathbf{r}} = \mathbf{D}\tilde{\mathbf{s}} + \tilde{\mathbf{n}} \tag{21.41}$$

with the eigenmode signals

$$\tilde{\mathbf{r}} = \mathbf{U}^H \mathbf{r} \tag{21.42}$$

$$\tilde{\mathbf{s}} = \mathbf{V}^H \mathbf{s} \tag{21.43}$$

and the eigenmode noise

$$\tilde{\mathbf{n}} = \mathbf{U}^H \mathbf{n} \tag{21.44}$$

Equation (21.41) represents the linear system of equations

$$\tilde{r}_i = \sqrt{\lambda_i}\tilde{s}_i + \tilde{n}_i \; ; \; i = 1, 2, \ldots, P \tag{21.45}$$

$$\tilde{r}_i = \tilde{n}_i \; ; \; i = P + 1, 2, \ldots, N \tag{21.46}$$

where (21.45) determines the eigenmodes in the narrow sense. The remaining eigenmodes in (21.46) do not carry information and thus provide no contribution to the channel capacity. Figure 21.3 illustrates the eigenmode decomposition of the MIMO system as a block diagram.

Before we calculate the total capacity, let us first check the statistical properties of $\tilde{\mathbf{s}}$ and $\tilde{\mathbf{n}}$. We obtain with (21.43) and by assuming that $\mathbf{V}$ is deterministic

$$\mathbf{R}_{\tilde{s}\tilde{s}} = \mathbf{E}\left[\tilde{\mathbf{s}}\tilde{\mathbf{s}}^H\right] = \mathbf{E}\left[\mathbf{V}^H \mathbf{s}\mathbf{s}^H \mathbf{V}\right] = \mathbf{V}^H \mathbf{R}_{ss}\mathbf{V} \tag{21.47}$$

As we would like to have statistically independent eigenmode branches to follow our goal to calculate the total channel capacity from the sum of the P individual eigenmodes, we have to make sure that the components $\tilde{s}_j$ of $\tilde{\mathbf{s}}$ are independent and have the same mean power.

$$\mathbf{E}\left[\left|\tilde{s}_j\right|^2\right] = E_S \; ; \; j = 1, 2, \ldots, M \tag{21.48}$$

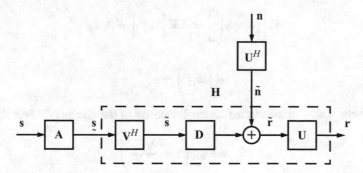

Fig. 21.3 MIMO system decomposition into eigenmodes with statistically independent input signal **s**, noise **n**, prefilter **A**, and channel **H** (for **n** = **0**)

Hence the covariance matrix must be

$$\mathbf{R}_{\tilde{s}\tilde{s}} = E_S \mathbf{I}_M \tag{21.49}$$

This is fulfilled if $\mathbf{R}_{ss} = E_S \mathbf{I}_M$, which is the prerequisite given by (21.27). Furthermore, we derive the mean value from (21.43) with (21.24)

$$\mathbf{E}\left[\tilde{\mathbf{s}}\right] = \mathbf{V}^H \mathbf{E}\left[\mathbf{s}\right] = \mathbf{0} \tag{21.50}$$

which indicates that all components $\tilde{s}_j$ have zero mean in compliance with the requirement (21.24). In a similar way, we get for the eigenmode of the noise assuming that **U** is deterministic

$$\mathbf{R}_{\tilde{n}\tilde{n}} = \mathbf{U}^H \mathbf{R}_{nn} \mathbf{U} \tag{21.51}$$

and with (21.33)

$$\mathbf{R}_{\tilde{n}\tilde{n}} = \sigma_n^2 \mathbf{I}_N \tag{21.52}$$

from which we see that the components $\tilde{n}_i$ of $\tilde{\mathbf{n}}$ are independent, and have the same mean power σ_n^2. Also the mean value is zero, because we get from (21.44) with (21.30)

$$\mathbf{E}\left[\tilde{\mathbf{n}}\right] = \mathbf{U}^H \mathbf{E}\left[\mathbf{n}\right] = \mathbf{0} \tag{21.53}$$

Furthermore, if **n** is AWGN with zero mean, we can substantiate that the same holds for $\tilde{\mathbf{n}}$, because a linear operation with a matrix **U** in (21.44) keeps the noise Gaussian.

As a conclusion, the individual eigenmodes are statistically independent and fulfill the same requirements as **s** and **n** given in Sect. 21.1.2. Consequently, they are adequate for the calculation of the total channel capacity as the sum of the individual and independent eigenmode branches. For branch i, we can apply (21.20), which requires the mean power of the signal and noise part of $\tilde{r}_i$ in (21.45), which are

$$\mathbf{E}\left[\left|\sqrt{\lambda_i}\tilde{s}_i\right|^2\right] = \lambda_i \mathbf{E}\left[\left|\tilde{s}_i\right|^2\right] = \lambda_i E_S \tag{21.54}$$

and

$$\mathbf{E}\left[\left|\tilde{n}_i\right|^2\right] = \sigma_n^2, \tag{21.55}$$

respectively. Then, we obtain the channel capacity of the eigenmode branch i from (21.20)

$$C_i = \log_2\left(1 + \frac{\lambda_i E_S}{\sigma_n^2}\right) \tag{21.56}$$

and the total MIMO system capacity becomes

$$C = \sum_{i=1}^{P} C_i = \sum_{i=1}^{P} \log_2\left(1 + \frac{\lambda_i E_S}{\sigma_n^2}\right) \tag{21.57}$$

measured in bit/symbol or in bit/channel use. Applying (21.21) yields

$$C' = 2 f_c \sum_{i=1}^{P} \log_2\left(1 + \frac{\lambda_i E_S}{\sigma_n^2}\right) \tag{21.58}$$

and is measured in bit/s.

Discussion of MIMO Channel Capacity(21.57)

- We see that only P eigenmode paths contribute to the channel capacity, although there are M transmit and N receive antennas. As $P = \text{rank}\left(\mathbf{HH}^H\right) = \text{rank}\left(\mathbf{H}^H\mathbf{H}\right) = \text{rank}\left(\mathbf{H}\right)$, the number of linearly independent lines or columns in the channel matrix determine the number of contributing eigenmodes rather than the number of transmit and receive antennas.
- Let $\lambda_1 = \lambda_2 = \cdots = \lambda_P = \lambda$ and $\mathbf{H}$ with full rank. Then $C = P \log_2\left(1 + \frac{\lambda E_S}{\sigma_n^2}\right)$ is proportional to $P = \min\{M, N\}$. In case of $M = N$ and a full rank channel matrix with $P = M$ the capacity is $C = M \log_2\left(1 + \frac{\lambda E_S}{\sigma_n^2}\right)$ and hence proportional to the number of transmit (and receive) antennas.
- Let λ_{max} be the largest eigenvalue of $\mathbf{HH}^H$ and assume that the remaining eigenvalues are very small compared to λ_{max}. Then, the eigenmode corresponding to λ_{max} dominates and the channel capacity approximately is $C = \log_2\left(1 + \frac{\lambda_{max} E_S}{\sigma_n^2}\right)$.
- As $H(k)$ is time varying, also all eigenvalues and thus the capacity $C(k)$ depend on time. This is the reason why $C(k)$ is called instantaneous capacity.

21.2.3 Alternative Formulas for the MIMO Channel Capacity

Now, we are going to derive some useful alternative formulas for the MIMO channel capacity.

Channel Capacity as a Function of Eigenvalues

We start with (21.57) and use the basic relation of the logarithm,
$$\log(x_1 x_2) = \log(x_1) + \log(x_2) \text{ yielding}$$

$$C = \log_2 \left[\prod_{i=1}^{P} \left(1 + \frac{\lambda_i E_S}{\sigma_n^2} \right) \right] \tag{21.59}$$

Moreover, the argument of the logarithm can be interpreted as the determinant of the diagonal matrix $\mathbf{I}_P + \frac{E_S}{\sigma_n^2} \mathbf{\Lambda}_P$, where $\mathbf{\Lambda}_P$ is given in (21.39). Accordingly we find

$$C = \log_2 \left[\det \left(\mathbf{I}_P + \frac{E_S}{\sigma_n^2} \mathbf{\Lambda}_P \right) \right] \tag{21.60}$$

Now we introduce $\mathbf{\Lambda}_M$ or $\mathbf{\Lambda}_N$, which have additional zeros in their main diagonal compared to $\mathbf{\Lambda}_P$. However, the argument of the logarithm in (21.59) will just get some additional factors 1, which do not change the result. Hence (21.60) is identical to

$$C = \log_2 \left[\det \left(\mathbf{I}_M + \frac{E_S}{\sigma_n^2} \mathbf{\Lambda}_M \right) \right] \tag{21.61}$$

and

$$C = \log_2 \left[\det \left(\mathbf{I}_N + \frac{E_S}{\sigma_n^2} \mathbf{\Lambda}_N \right) \right] \tag{21.62}$$

Channel Capacity as a Function of the Channel Matrix

We are now looking for a relation between the channel capacity and the channel matrix $\mathbf{H}$. Such a formula would be useful, because no singular values of the channel matrix would have to be calculated. We first give the result and then the proof:

$$C = \log_2 \left[\det \left(\mathbf{I}_N + \frac{E_S}{\sigma_n^2} \mathbf{H} \mathbf{H}^H \right) \right] \tag{21.63}$$

and

$$C = \log_2 \left[\det \left(\mathbf{I}_M + \frac{E_S}{\sigma_n^2} \mathbf{H}^H \mathbf{H} \right) \right] \tag{21.64}$$

Proof of (21.63)

To this end we consider (21.61). We know that the eigenvalue decomposition of $\mathbf{Q}_M = \mathbf{H}^H \mathbf{H}$ is given by $\mathbf{V}^H \mathbf{H}^H \mathbf{H} \mathbf{V} = \mathbf{\Lambda}_M$. Then we obtain from (21.61)

$C = \log_2\left[\det\left(\mathbf{I}_M + \frac{E_S}{\sigma_n^2}\mathbf{V}^H\mathbf{H}^H\mathbf{H}\mathbf{V}\right)\right]$. Next we use the cyclic permutation rule for determinants from Appendix B.

Given $\mathbf{A} \in \mathbb{C}^{M \times N}$ and $\mathbf{B} \in \mathbb{C}^{N \times M}$, then $\det(\mathbf{I}_M + \mathbf{AB}) = \det(\mathbf{I}_N + \mathbf{BA})$ is true.

Thus, we can rewrite the capacity as $C = \log_2\left[\det\left(\mathbf{I}_N + \frac{E_S}{\sigma_n^2}\mathbf{H}\mathbf{V}\mathbf{V}^H\mathbf{H}^H\right)\right]$ and use the fact that $\mathbf{V}$ is a unitary matrix to get the final result $C = \log_2\left[\det\left(\mathbf{I}_N + \frac{E_S}{\sigma_n^2}\mathbf{H}\mathbf{H}^H\right)\right]$. In a similar way (21.64) can be proven.

21.3 MIMO Channel Capacity for Correlated Transmit Signals

We are now going to abandon the requirement that the input signal of the channel is spatially uncorrelated. According to Fig. 21.2 we assume an original signal $\mathbf{s}$ as in Sect. 21.2 with $\mathbf{R}_{ss} = E_S\mathbf{I}_M$, however we introduce the prefilter matrix $\mathbf{A} \in \mathbb{C}^{M \times M}$ that generates a channel input signal $\underset{\sim}{\mathbf{s}} \in \mathbb{C}^{M \times 1}$ with covariance matrix

$$\mathbf{R}_{\underset{\sim}{s}\underset{\sim}{s}} = \mathbf{E}\left[\underset{\sim}{\mathbf{s}}\,\underset{\sim}{\mathbf{s}}^H\right] \tag{21.65}$$

Please note that $\mathbf{R}_{\underset{\sim}{s}\underset{\sim}{s}}$ is a Hermiteian matrix, else arbitrary. The input signal $\mathbf{s}$ complies with the prerequisites for the calculation of the total channel capacity using singular value decomposition, as outlined in the Sect. 21.2. The output signal is given by

$$\mathbf{r} = \mathbf{H}\mathbf{A}\mathbf{s} + \mathbf{n} \tag{21.66}$$

With $\underset{\sim}{\mathbf{s}} = \mathbf{A}\mathbf{s}$ the covariance matrix $\mathbf{R}_{\underset{\sim}{s}\underset{\sim}{s}}$ is given by

$$\mathbf{R}_{\underset{\sim}{s}\underset{\sim}{s}} = \mathbf{A}\mathbf{R}_{ss}\mathbf{A}^H = E_S\mathbf{A}\mathbf{A}^H \tag{21.67}$$

where the last step holds for $\mathbf{R}_{ss} = E_S\mathbf{I}_M$. If we partition

$$\mathbf{R}_{\underset{\sim}{s}\underset{\sim}{s}} = \mathbf{R}_{\underset{\sim}{s}\underset{\sim}{s}}^{\frac{1}{2}}\left(\mathbf{R}_{\underset{\sim}{s}\underset{\sim}{s}}^{\frac{1}{2}}\right)^H \tag{21.68}$$

a comparison with (21.67) yields the precoding matrix

$$\mathbf{A} = \frac{1}{\sqrt{E_S}}\mathbf{R}_{\underset{\sim}{s}\underset{\sim}{s}}^{\frac{1}{2}} \tag{21.69}$$

$\mathbf{R}_{\underset{\sim}{SS}}^{\frac{1}{2}}$ is the square root matrix of $\mathbf{R}_{\underset{\sim}{SS}}$.

As we recognize from (21.66), the system capacity with prefilter can be easily determined from (21.63) and (21.64), if $\mathbf{H}$ is replaced there by $\mathbf{HA}$ yielding

$$C = \log_2 \left[\det \left(\mathbf{I}_N + \frac{E_S}{\sigma_n^2} \mathbf{HAA}^H \mathbf{H}^H \right) \right] \qquad (21.70)$$

and

$$C = \log_2 \left[\det \left(\mathbf{I}_M + \frac{E_S}{\sigma_n^2} \mathbf{AA}^H \mathbf{H}^H \mathbf{H} \right) \right] \qquad (21.71)$$

where we have used the cyclic permutation rule for determinants in (21.71). Hence, with the prefilter matrix $\mathbf{A}$, the covariance matrix $\mathbf{R}_{SS}$ of the channel input signal $\underset{\sim}{s}$ and the system capacity can be adjusted. Then, we obtain with (21.67) from (21.70) and (21.71) two equivalent forms for the channel capacity

$$C = \log_2 \left[\det \left(\mathbf{I}_N + \frac{1}{\sigma_n^2} \mathbf{H} \mathbf{R}_{\underset{\sim}{SS}} \mathbf{H}^H \right) \right] \qquad (21.72)$$

$$C = \log_2 \left[\det \left(\mathbf{I}_M + \frac{1}{\sigma_n^2} \mathbf{R}_{\underset{\sim}{SS}} \mathbf{H}^H \mathbf{H} \right) \right] \qquad (21.73)$$

21.4 Channel Capacity for Correlated MIMO Channel

We are going to determine the channel capacity of a MIMO channel with transmit and receive correlation. To this end, we can use the Kronecker model from Sect. 17.4.4 with the channel matrix given by (17.41) resulting in $\mathbf{H} = \left(\mathbf{R}_{rx}^{\frac{1}{2}} \right)^H \mathbf{H}_w \mathbf{R}_{tx}^{\frac{1}{2}}$. Then we obtain from (21.63)

$$C = \log_2 \left[\det \left(\mathbf{I}_N + \frac{E_S}{\sigma_n^2} \mathbf{H}_w \mathbf{R}_{tx}^{\frac{1}{2}} \left(\mathbf{R}_{tx}^{\frac{1}{2}} \right)^H \mathbf{H}_w^H \mathbf{R}_{rx}^{\frac{1}{2}} \left(\mathbf{R}_{rx}^{\frac{1}{2}} \right)^H \right) \right] \qquad (21.74)$$

where we have used the cyclic permutation rule. In general, correlation of the delay spread functions of the channel results in a loss of the channel capacity, as detailed investigations with short term and long term statistical parameters have shown, [3]. The reader can convince oneself by experimenting some scenarios using the platform "web demo" provided by [4]. An example is depicted in Fig. 21.4 for a MIMO system with two transmit and two receive antennas.

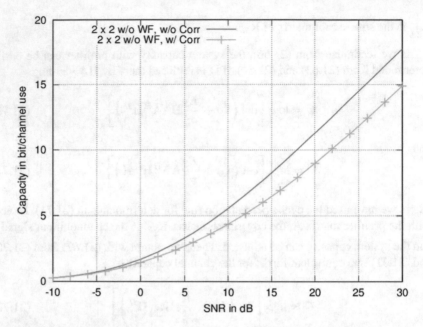

Fig. 21.4 Capacity of a MIMO system with $M = 2$ transmit and $N = 2$ receive antennas under Gaussian noise, transmit and receive correlation $\rho = 0.8$. Upper curve without (w/o) and lower curve with (w/) transmit and receive correlation, equal transmit power (without water filling WF). Source: Online platform "webdemo" [4]

The MIMO channel is corrupted by Gaussian noise. Transmit and receive correlation is present with the correlation matrices $\mathbf{R}_{tx} = \mathbf{R}_{rx} = \begin{pmatrix} 1 & \rho \\ \rho & 1 \end{pmatrix}$ and $\rho = 0.8$. The degradation of the channel capacity under channel correlation (lower curve) is clearly visible.

21.5 Maximizing MIMO System Capacity Using the Water Filling Algorithm

21.5.1 Prefilter for Transmit Power Allocation

In the previous sections, we have not maximized the MIMO system capacity using the degrees of freedom. Among others, we have assumed that the mean output power E_S is the same for all M transmit antenna elements. Now, we are going to distribute the total mean transmit power ME_S individually over the antenna elements with the goal to maximize the capacity of the MIMO system. In Fig. 21.2, we have already taken precaution by introducing a filter with the matrix $\mathbf{A}$, which we will call power allocation filter. The block diagram in Fig. 21.3 shows also the decomposition of the MIMO channel matrix $\mathbf{H}$ into eigenmodes according to (21.35). We assume that the input signal $\mathbf{s}$ is still spatially uncorrelated with mean power ME_S and covariance matrix $\mathbf{R}_{ss} = E_S \mathbf{I}_M$. The matrix $\mathbf{A}$ together with $\mathbf{V}^H$ shall generate an input signal $\tilde{\mathbf{s}}$ of the eigenmode channel $\mathbf{D}$ with a covariance matrix

$$\mathbf{R}_{\tilde{s}\tilde{s}} = E_S \, \text{diag} \, (a_1, a_2, \ldots, a_M) \tag{21.75}$$

where

$$\sum_{j=1}^{M} a_j = M \, , \text{ and } a_j \geq 0 \, ; \quad j = 1, 2, \ldots, M \tag{21.76}$$

So, we are going to weight the input power of each eigenmode with a dedicated positive coefficient a_j , $j = 1, 2, \ldots, M$. We substantiate from (21.75) that the vector $\tilde{\mathbf{s}}$ remains spatially uncorrelated, as required for the calculation of the total capacity as the sum of the individual and independent capacities of the eigenmodes. It will be shown that a precoding matrix $\mathbf{A}$, which is generating the covariance matrix (21.75) is given by

$$\mathbf{A} = \frac{1}{\sqrt{E_S}} \mathbf{V} \mathbf{R}_{\tilde{s}\tilde{s}}^{\frac{1}{2}} = \mathbf{V} \text{diag} \left(\sqrt{a_1}, \sqrt{a_2}, \ldots, \sqrt{a_M} \right) \tag{21.77}$$

From Fig. 21.3 we see that

$$\tilde{\mathbf{s}} = \mathbf{V}^H \mathbf{A} \mathbf{s} \tag{21.78}$$

and with (21.77) follows

$$\tilde{\mathbf{s}} = \text{diag} \left(\sqrt{a_1}, \sqrt{a_2}, \ldots, \sqrt{a_M} \right) \mathbf{s} \tag{21.79}$$

Please note that the power allocation matrix $\mathbf{A}$ shows up as a full rather than a diagonal matrix due to $\mathbf{V}$. We can write (21.79) in component form as

$$\tilde{s}_i = \sqrt{a_i} s_i \, ; \quad i = 1, 2, \ldots, M \tag{21.80}$$

and from the input–output relation (21.45) of the eigenmode channel $\mathbf{D}$, we obtain for the ith branch of the eigenmode decomposition

$$\tilde{r}_i = \sqrt{\lambda_i} \sqrt{a_i} s_i + \tilde{n}_i \, ; \quad i = 1, 2, \ldots, P \tag{21.81}$$

This is portrait in Fig. 21.5. The resulting total capacity is given by

$$C = \log_2 \left[\det \left(\mathbf{I}_M + \frac{E_S}{\sigma_n^2} \text{diag} \, (a_1, a_2, \ldots, a_M) \, \mathbf{\Lambda}_M \right) \right] \tag{21.82}$$

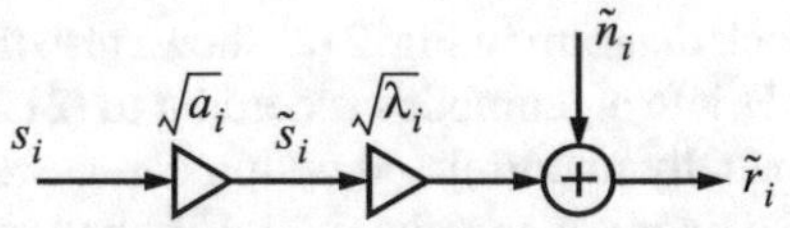

Fig. 21.5 ith branch of the eigenmode decomposition of the MIMO system with individual transmit power allocation coefficient a_i

The proof is given at the end of this section.

In the next section, we are going to calculate the optimal coefficients a_i.

Proof of (21.77)

With (21.78) and (21.27) we obtain $\mathbf{R}_{\tilde{s}\tilde{s}} = \mathbf{E}\left[\tilde{s}\tilde{s}^H\right] = \mathbf{V}^H \mathbf{A}\mathbf{A}^H \mathbf{V} E_S$. From the decomposition $\mathbf{R}_{\tilde{s}\tilde{s}} = \mathbf{R}_{\tilde{s}\tilde{s}}^{\frac{1}{2}} \left(\mathbf{R}_{\tilde{s}\tilde{s}}^{\frac{1}{2}}\right)^H = \sqrt{E_S}\mathbf{V}^H \mathbf{A} \left(\mathbf{V}^H \mathbf{A}\right)^H \sqrt{E_S}$ we conclude $\mathbf{R}_{\tilde{s}\tilde{s}}^{\frac{1}{2}} = \sqrt{E_S}\mathbf{V}^H \mathbf{A}$ yielding with (21.75) the result $\mathbf{A} = \frac{1}{\sqrt{E_S}}\mathbf{V}\mathbf{R}_{\tilde{s}\tilde{s}}^{\frac{1}{2}} = \mathbf{V}\mathrm{diag}\left(\sqrt{a_1}, \sqrt{a_2}, \dots, \sqrt{a_M}\right)$ and the proof is finished.

Proof of (21.82)

We apply (21.71) and insert $\mathbf{A}$ from (21.77) yielding with $\mathbf{A}\mathbf{A}^H = \mathbf{V}\mathrm{diag}(a_1, a_2 \dots, a_M)\mathbf{V}^H$ and with the cyclic permutation rule for determinants

$$C = \log_2 \left[\det \left(\mathbf{I}_M + \frac{E_S}{\sigma_n^2}\mathrm{diag}(a_1, a_2 \dots, a_M)\mathbf{V}^H \mathbf{H}^H \mathbf{H}\mathbf{V}\right)\right].$$ As $\mathbf{V}^H \mathbf{H}^H \mathbf{H}\mathbf{V} = \mathbf{\Lambda}_M$ the proof of (21.82) ends.

21.5.2 Computation of the Optimal Power Allocation Coefficients a_i

Maximization with Constraints

From Fig. 21.5 and the prerequisites, we see that (21.57) can be used for determining the total capacity, if we replace λ_i by $\lambda_i a_i$. Then, the capacity with transmit power loading prefilter $\mathbf{A}$ in (21.77) is obtained as

$$C = \sum_{i=1}^{P} \log_2 \left(1 + \frac{a_i \lambda_i E_S}{\sigma_n^2}\right) \tag{21.83}$$

which has to be maximized with respect to $a_1, \dots, a_P$ and under the two constraints

$$\sum_{j=1}^{P} a_j = M \iff g\left(a_1, a_{2,} \dots, a_P\right) = \sum_{j=1}^{P} a_j - M = 0 \tag{21.84}$$

and

$$a_j \geq 0 \; ; \; j = 1, 2, \ldots, P \tag{21.85}$$

Please note that we have restricted (21.76) to $P \leq M$, because there are only P eigenmodes and hence only P free coefficients a_i are left for the maximization. The remaining coefficients $a_i \; ; \; i = P + 1, \ldots, M$ are set equal to zero.

General Solution with Lagrange Method

The Lagrange method defines a new target function composed as a linear combination of the previous function C and the constraint $g(a_1, a_2, \ldots, a_P)$

$$J = C + L g = \max_{\{a_1, \ldots, a_P, L\}} \tag{21.86}$$

L is the Lagrange multiplier and is considered as a new free parameter. The constraint (21.85), which is an inequality, has to be treated separately in a second step. We recognize J as a concave function of the free parameters $a_1, \ldots, a_P, L$. Consequently, setting the partial derivatives of J with respect to the parameters equal to zero is a necessary and sufficient condition for the maximum

$$\frac{\partial J}{\partial a_m} = \frac{\partial C}{\partial a_m} + L \frac{\partial g}{\partial a_m} = 0 \; ; \; m = 1, 2, \ldots, P \tag{21.87}$$

We express $\log_2(x) = \eta \ln(x)$ with $\eta = \frac{1}{\ln(2)}$ and knowing that $\frac{\partial \ln(bx)}{\partial x} = b \frac{1}{x}$ we obtain from (21.87)

$$\eta \frac{1}{1 + \frac{a_m \lambda_m E_S}{\sigma_n^2}} \frac{\lambda_m E_S}{\sigma_n^2} + L = 0 \tag{21.88}$$

from which follows

$$a_m = -\frac{\eta}{L} - \frac{\sigma_n^2}{\lambda_m E_S} \; ; \; m = 1, 2, \ldots, P \tag{21.89}$$

Next we calculate

$$\frac{\partial J}{\partial L} = g = 0 \tag{21.90}$$

which yields the constraint (21.84)

$$\sum_{j=1}^{P} a_j = M \tag{21.91}$$

as is typically for the Lagrange method. Because L is a free parameter, we can redefine the Lagrange multiplier as $K = -\frac{\eta}{L}$ and consider K as a new parameter. Then (21.89) yields

$$a_m = K - \frac{\sigma_n^2}{\lambda_m E_S} \; ; \; m = 1, 2, \ldots, P \qquad (21.92)$$

Equations (21.91) and (21.92) are $P + 1$ equations for the $P + 1$ unknowns $a_1, \ldots,$ a_P, and K. However, the second constraint (21.85) still has to be met. Therefore, we conclude from (21.92) the optimal solution in a first approach

$$a_m^{opt} = \left(K^{opt} - \frac{\sigma_n^2}{\lambda_m E_S} \right)_+ \; ; \; m = 1, 2, \ldots, P \qquad (21.93)$$

with the function

$$(x)_+ = \begin{cases} x \; ; \; x > 0 \\ 0 \; ; \; x \le 0 \end{cases} \qquad (21.94)$$

and from (21.91) follows

$$\sum_{j=1}^{P} a_j^{opt} = M \qquad (21.95)$$

Let

$$a_m^{opt} \begin{cases} > 0 \; ; \quad m = 1, 2, \ldots, P_0 \\ = 0 \; ; \; m = P_0 + 1, \ldots, P \end{cases} \qquad (21.96)$$

Then we conclude from (21.93)

$$a_m^{opt} = \begin{cases} K^{opt} - \frac{\sigma_n^2}{\lambda_m E_S} \; ; \quad m = 1, 2, \ldots, P_0 \\ \quad = 0 \qquad \quad ; \; m = P_0 + 1, \ldots, P \end{cases} \qquad (21.97)$$

Inserting (21.97) into (21.95) yields

$$\sum_{j=1}^{P_0} \left(K^{opt} - \frac{\sigma_n^2}{\lambda_j E_S} \right) = M \qquad (21.98)$$

from which we obtain

$$K^{opt} = \frac{M}{P_0} + \frac{1}{P_0} \sum_{j=1}^{P_0} \frac{\sigma_n^2}{\lambda_j E_S} \qquad (21.99)$$

In summary, (21.96), (21.97), and (21.99) represent the final conditions for the solution.

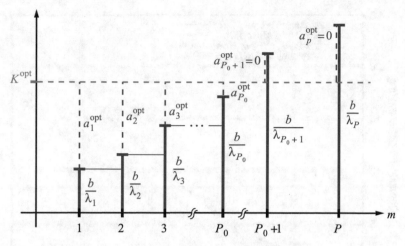

Fig. 21.6 Graphical interpretation of the water filling algorithm to maximize capacity. Note $b = \frac{\sigma_n^2}{E_S}$

21.5.3 Graphical Interpretation of the Water Filling Solution

$\lambda_1 \geq \lambda_2 \geq \cdots \geq \lambda_P > 0$ shall be given. As depicted in Fig. 21.6, the terms $\frac{\sigma_n^2}{\lambda_m E_S}$; $m = 1, 2, \ldots, P$ are modeled to built up the ground of a vessel. If we are pouring water with a volume of $\sum_{j=1}^{P} a_j$ into that vessel, the level will rise up to K^{opt} given in (21.99).

However, for the given shape of the ground, the total "water" M distributes in such a way that $\sum_{j=1}^{P_0} a_j^{opt} = M$ holds, because no more water is available. Consequently, $a_m^{opt} = 0$ for $m = P_0 + 1, \ldots, P$. From (21.97) we also conclude

$$a_m^{opt} + \frac{\sigma_n^2}{\lambda_m E_S} = K^{opt} \; ; \; m = 1, 2, \ldots, P_0 \tag{21.100}$$

and K^{opt} is determining the final water level. From Fig. 21.6 we see for $m > P_0$ the ground of the vessel $\frac{\sigma_n^2}{\lambda_m E_S}$ exceeds the final water level, because no more water is available and the remaining a_m are zero. In general, the solution is iterative. With the optimal coefficients, we obtain the maximal capacity from (21.83) as

$$C^{opt} = \sum_{i=1}^{P_0} \log_2 \left(1 + \frac{a_i^{opt} \lambda_i E_S}{\sigma_n^2} \right) \tag{21.101}$$

For a small b in Fig. 21.6, i.e., a high signal-to-noise ratio $\frac{E_S}{\sigma_n^2}$ all a_j^{opt} tend to have the same size and thus the water filling algorithm can not change the capacity significantly. A similar result is obtained, if all singular values $\sqrt{\lambda_i}$ are approximately

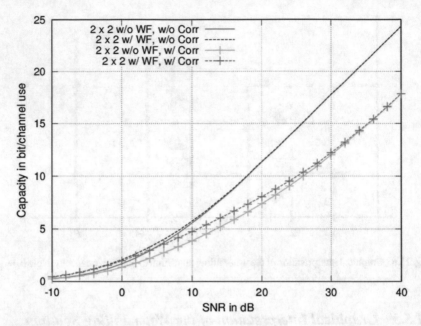

Fig. 21.7 Capacity of a MIMO system with $M = 2$ transmit and $N = 2$ receive antennas under Gaussian noise, transmit and receive correlation $\rho = 0.95$. Upper two curves without (w/o) and lower two curves with (w/) transmit and receive correlation, with and without water filling (WF). Source: Online platform "webdemo" [4]

equal. This is reflected in Fig. 21.7 as the result of a computer simulation using the "webdemo" [4]. The lower two curves represent a 2x2 MIMO channel with strong transmit and receive correlation ($\mathbf{R}_{tx} = \mathbf{R}_{rx} = \begin{pmatrix} 1 & \rho \\ \rho & 1 \end{pmatrix}$, $\rho = 0.95$). Apparently, the capacity gains vanish for increasing SNR $\frac{E_S}{\sigma_n^2}$.

21.5.4   Iterative Solution and Example

In the following, the principle algorithm is given as a pseudo code.

Begin

$\quad$ Set $a_{P+1} < 0$

$\quad$ For $m = P,\ P-1, ..., 1$ do

$\qquad$ if $a_{m+1} > 0$ set $K_m = K^{opt}$ else $K_m = \frac{1}{m} \left(M + \frac{\sigma_n^2}{E_S} \sum_{j=1}^{m} \frac{1}{\lambda_j} \right)$

$\qquad$ $a_m = K_m - \frac{\sigma_n^2}{\lambda_m E_S}$

$\qquad$ if $a_m > 0$ $\quad$ set $a_m^{opt} = a_m$ and $K_m = K^{opt}$

$\qquad$ if $a_m \leq 0$ $\quad$ set $a_m^{opt} = 0$

        End

End

Example 1

Given a MIMO system with $M = 8$ transmit antennas and $\frac{\sigma_n^2}{E_S} = 19$. The channel matrix $\mathbf{H}$ has the rank $P = 4$. A singular value decomposition has provided the eigenvalues $\lambda_1 = 19$, $\lambda_2 = \frac{19}{2}$, $\lambda_3 = \frac{19}{10}$, $\lambda_4 = 1$, $\lambda_5 = \lambda_6 = \lambda_7 = \lambda_8 = 0$. Find the optimal power allocation coefficients and the maximal capacity.

Solution: The iteration steps according to the pseudo code are as follows.

$m = 4$

$$K_4 = 10$$
$$a_4 = -9, \quad a_4^{opt} = 0$$

$m = 3$

$$K_3 = 7$$
$$a_3 = -3, \quad a_3^{opt} = 0$$

$m = 2$

$$K_2 = \frac{11}{2}$$
$$a_2 = \frac{7}{2} \quad a_2^{opt} = \frac{7}{2} \quad K_2 = K^{opt} = \frac{11}{2}$$

$m = 1$

$$a_1 = \frac{9}{2} \quad a_1^{opt} = \frac{9}{2}$$

The remaining coefficients are $a_5 = a_6 = a_7 = a_8 = 0$. We also see that $P_0 = 2$.

Now, we are going to calculate the resulting capacity in bit/channel use taking (21.83) with the result

$$C^{opt} \approx \log_2\left(\frac{11}{2}\right) + \log_2\left(\frac{11}{4}\right) \approx 2.46 + 1.46 = 3.92$$

Without optimal power loading the capacity (bit/channel use) is

$$C = \log_2(2) + \log_2\left(\frac{3}{2}\right) + \log_2\left(\frac{11}{10}\right) + \log_2\left(\frac{20}{19}\right) \approx 1.79$$

according to (21.57). In this example, the channel capacity has more than doubled by employing the optimal transmit power loading.

Example 2

Given the results from Example 1, $E_S = 1$ and the transform matrix $\mathbf{V}$. Find the matrix $\mathbf{A}$ of the power allocation filter and the covariance matrix $\mathbf{R}_{\tilde{s}\tilde{s}}$ of the input signal $\tilde{s}$ of the eigenmode channel $\mathbf{D}$. Calculate the covariance matrix $\mathbf{R}_{\underset{\sim}{s}\underset{\sim}{s}}$ of the input signal $\underset{\sim}{s}$ of the channel $\mathbf{H}$.

Solution: $\mathbf{A}$ can be calculated with (21.77) as the 8x8 matrix
$\mathbf{A} = \mathbf{V}\,\mathrm{diag}\left(\sqrt{\frac{9}{2}}, \sqrt{\frac{7}{2}}, 0, 0, 0, 0, 0, 0\right)$ and $\mathbf{R}_{\tilde{s}\tilde{s}}$ follows from (21.75) as
$\mathbf{R}_{\tilde{s}\tilde{s}} = \mathrm{diag}\left(\frac{9}{2}, \frac{7}{2}, 0, 0, 0, 0, 0, 0\right)$.
The covariance matrix $\mathbf{R}_{\underset{\sim}{s}\underset{\sim}{s}}$ is given by (21.67) as $\mathbf{R}_{\underset{\sim}{s}\underset{\sim}{s}} = E_S \mathbf{A}\mathbf{A}^H$ from which follows

$\mathbf{R}_{\underset{\sim}{s}\underset{\sim}{s}} = \mathbf{V}\mathrm{diag}\left(\frac{9}{2}, \frac{7}{2}, 0, 0, 0, 0, 0, 0\right)\mathbf{V}^H$, which is in general not a diagonal matrix.

Thus, the output signal $\underset{\sim}{s}$ of the prefilter is spatially correlated, as expected.

As a conclusion, the power allocation filter $\mathbf{A}$ is condensing the available total mean power $8E_S$ covered by the output signal vector $\mathbf{s}$ of the eight transmit antenna elements to just two eigenmodes. All other eigenmodes do not carry power. However, this does not mean that we can switch off any transmit antenna element.

Example 3

Given the matrices $\mathbf{V}$ and the diagonal matrix $\mathbf{F}$. Show with the following example that $\mathbf{G} = \mathbf{V}\mathbf{F}\mathbf{V}^H$ is a diagonal matrix only if $\mathbf{F} = \mathbf{I}_2$ is the identity matrix.

$$\mathbf{V} = \begin{pmatrix} 1 & j \\ j & 1 \end{pmatrix}, \quad \mathbf{F} = \begin{pmatrix} 1 & 0 \\ 0 & f \end{pmatrix}$$

Solution:
We obtain $\mathbf{G} = \mathbf{V}\mathbf{F}\mathbf{V}^H = \begin{pmatrix} 1+f & j(f-1) \\ -j(f-1) & 1+f \end{pmatrix}$ and recognize that $\mathbf{G}$ is a diagonal matrix only for $f = 1$ resulting in $\mathbf{F} = \mathbf{I}_2$.

21.6 Capacity of a Stochastic MIMO Channel

We have already pointed out that the capacity $C(k)$ considered so far is the instantaneous capacity, as it depends on time. Capacity is an important parameter for the quality of service and therefore the instantaneous capacity is of limited value for guaranteeing a satisfying operation of a system over a long time period. To this end, two new measures have been introduced for characterizing quality of service of a MIMO system namely the "ergodic capacity" and the "outage capacity", which will be addressed briefly in the following.

21.6.1 Ergodic Channel Capacity

We consider $\mathbf{H}(k)$ as a stochastic process and define the ergodic capacity as the expectation of $C(k)$

$$C_{erg} = \mathbf{E}[C] = \int_0^\infty C p_C(C) dC \qquad (21.102)$$

where $p_C(C)$ is the probability density function of the capacity C. Please note that $C \geq 0$ and hence the integration starts at zero. $p_C(C)$ can be found from adequate channel models or measurements of the MIMO system, which is not considered here in more detail.

21.6.2 Outage Capacity

Knowing the probability density function $p_C(C)$ of the capacity C, we can define the outage capacity C_{out} to be considered as the capacity threshold where the system starts to leave the guaranteed quality of service. This event will occur with the outage probability

$$P_{out} = \Pr[0 \leq C \leq C_{out}] - \int_0^{C_{out}} p_C(C) dC \qquad (21.103)$$

As an example, if the system is required to operate at full quality with a probability of $\geq 99.99\,\%$, the required outage probability has to be $P_{out} \leq 0.01\%$.

References

1. C. Shannon, A mathematical theory of communication. Bell Syst. Tech. J. **27**, 379 (1948)
2. I. Telatar, Capacity of multi-antenna Gaussian channels. Eur. Trans. Telecommun. **10**, 585 (1999)
3. M. Kiessling, Statistical analysis and transmit prefiltering for MIMO wireless systems in correlated fading environments, Ph.D. dissertation, University of Stuttgart, Institute of Telecommunications, Shaker (2004). ISBN 3-8322-3444-6
4. X. Meng, X. Wang, Capacity limits of MIMO channels, webdemo, Technical report, Institute of Telecommunications, University of Stuttgart, Germany (2018), http://webdemo.inue.uni-stuttgart.de

Chapter 22
MIMO Systems with Precoding

22.1 Principle of MIMO Precoding

In Chap. 18, we have investigated the zero-forcing and the minimum mean squared error (MMSE) receiver, which are able to remove or at least minimize the inter-channel interference to the expense of a potential increase of the mean noise power at the receiver output. To maximize the channel capacity, we have already investigated a prefilter in Chap. 21, which acts as a power allocation filter at the transmitter. Now, we are going to consider prefilters also denoted as precoders to reduce inter-channel interference and thus move the receive filter in principle to the transmitter. As for single input single output wire-line and wireless systems, one motivation is to relocate the hardware complexity of the receiver to some extend to the transmitter [1–3]. This strategy is advantageous in the downlink scenario from the base station to the user, where the receivers are the individual user terminals, which then could be less complex. In most cases, the resulting hardware increase of the base station transmitter can be afforded, because its cost is shared among the large number of the users. However, there is a significant burden, because precoding requires knowledge about the channel parameters at the transmitter side to be able to adjust the precoder. Consequently, the channel estimator, which is located at the receiver, has to sent appropriate channel parameters, e.g., the full channel matrix $\mathbf{H}$ to the transmitter via a feedback channel. This arrangement is thus denoted as closed- loop scheme. Precoding is also used in the downlink of multi-user scenarios and described in Chap. 24. Precoders, which do not require channel knowledge, are called space-time encoders, are open loop schemes, and discussed in Chap. 23.

In Fig. 22.1, the principle block diagram of a MIMO downlink transmission with a prefilter matrix $\mathbf{A}(k)$ is shown. $\mathbf{c}(k)$ is the input signal vector containing the symbols from a QAM mapper.

The output of the precoder emits the signal vector $\mathbf{s}(k) \in \mathbb{C}^{M \times 1}$ using a MIMO antenna with M elements. The fading channel is modeled by the channel matrix $\mathbf{H}(k) \in \mathbb{C}^{N \times M}$, which owns a frequency-flat spectrum. $\mathbf{r}(k) \in \mathbb{C}^{N \times 1}$ is the receive vector and $\mathbf{n}(k) \in \mathbb{C}^{N \times 1}$ the additive noise with zero mean. While the dimensions of $\mathbf{s}$, $\mathbf{H}$,

© Springer Nature Switzerland AG 2019
J. Speidel, *Introduction to Digital Communications*, Signals and Communication Technology, https://doi.org/10.1007/978-3-030-00548-1_22

Fig. 22.1 MIMO
transmission with precoding
matrix **A** at the transmitter

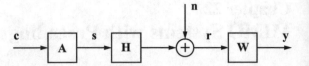

n, and **r** are the same as in the previous chapters, the selection of the dimension of
A and **c** deserve some discussion. If we would define **A** as a $M \times M$ matrix such as
in Chap. 21 on MIMO capacity, it would be prohibitive to apply the pseudo inverse
$\mathbf{H}^+$ or $\mathbf{H}^{++}$ (both $M \times N$) of **H**, which could play the role as a prefilter in our further
investigations. Therefore, we define $\mathbf{A} \in \mathbb{C}^{M \times N}$ and as a consequence $\mathbf{c} \in \mathbb{C}^{N \times 1}$.

In the next section, we will discuss a scheme, which just requires the precoding
matrix, and the receive matrix **W** in Fig. 22.1 is not present. Alternatively, we investigate in Sect. 22.3 also some precoding schemes, which require an additional receive
matrix. In the following, we first consider the general block diagram in Fig. 22.1 and
specialize later. We drop the discrete time variable k to simplify the notation and
obtain for the transmit signal

$$\mathbf{s} = \mathbf{Ac} \tag{22.1}$$

and for the receive vector $\mathbf{r} \in \mathbb{C}^{N \times 1}$

$$\mathbf{r} = \mathbf{HAc} + \mathbf{n} \tag{22.2}$$

We recognize that the signal part of the receive signal is

$$\mathbf{r}_s = \mathbf{HAc} \tag{22.3}$$

where **HA** characterizes the inter-channel interference. Please recall that the inter-
channel interference is completely removed if **HA** is a diagonal matrix in particular
$\mathbf{HA} = \mathbf{I}_N$ or is a scalar. The noise part in the receive signal is just

$$\mathbf{r}_n = \mathbf{n} \tag{22.4}$$

and obviously untouched by the precoder. For the mean power of $\mathbf{r}_s$ we obtain

$$\mathbf{E}\left[\|\mathbf{r}_s\|^2\right] = \operatorname{tr}\left(\mathbf{R}_{cc}\,(\mathbf{HA})^H\,\mathbf{HA}\right) \tag{22.5}$$

with the covariance matrix of **c**

$$\mathbf{R}_{cc} = \mathbf{E}\left[\mathbf{cc}^H\right] \in \mathbb{C}^{N \times N} \tag{22.6}$$

(22.5) is easily shown with (22.3) as follows.

$\mathbf{E}\left[\|\mathbf{r}_s\|^2\right] = \mathbf{E}\left[\text{tr}\left(\mathbf{r}_s\mathbf{r}_s^H\right)\right] = \text{tr}\left(\mathbf{H}\mathbf{A}\,\mathbf{E}\left[\mathbf{c}\mathbf{c}^H\right](\mathbf{H}\mathbf{A})^H\right) = \text{tr}\left(\mathbf{H}\mathbf{A}\mathbf{R}_{cc}(\mathbf{H}\mathbf{A})^H\right) =$ $\text{tr}\left(\mathbf{R}_{cc}(\mathbf{H}\mathbf{A})^H\mathbf{H}\mathbf{A}\right)$, where we have used for the last term the cyclic permutation rule for the trace of a matrix product, as outlined in the Appendix B.

With the mean noise power

$$\mathbf{E}\left[\|\mathbf{r}_n\|^2\right] = \mathbf{E}\left[\|\mathbf{n}\|^2\right] \tag{22.7}$$

we finally obtain the signal-to-noise ratio at the receiver

$$\gamma_r = \frac{\mathbf{E}\left[\|\mathbf{r}_s\|^2\right]}{\mathbf{E}\left[\|\mathbf{r}_n\|^2\right]} = \frac{\text{tr}\left(\mathbf{R}_{cc}(\mathbf{H}\mathbf{A})^H\mathbf{H}\mathbf{A}\right)}{\mathbf{E}\left[\|\mathbf{n}\|^2\right]} \tag{22.8}$$

For a system with receive filter $\mathbf{W}$ we get from Fig. 22.1 and with (22.2)

$$\mathbf{y} = \mathbf{W}\mathbf{r} = \mathbf{W}\mathbf{H}\mathbf{A}\mathbf{c} + \mathbf{W}\mathbf{n} \tag{22.9}$$

with the signal part

$$\mathbf{y}_s = \mathbf{W}\mathbf{H}\mathbf{A}\mathbf{c} \tag{22.10}$$

and the noise part

$$\mathbf{y}_n = \mathbf{W}\mathbf{n} \tag{22.11}$$

The mean power of $\mathbf{y}_s$ is easily obtained from (22.5) by just replacing $\mathbf{H}$ by $\mathbf{W}\mathbf{H}$

$$\mathbf{E}\left[\|\mathbf{y}_s\|^2\right] = \text{tr}\left(\mathbf{R}_{cc}(\mathbf{W}\mathbf{H}\mathbf{A})^H\mathbf{W}\mathbf{H}\mathbf{A}\right) \tag{22.12}$$

The mean noise power is

$$\mathbf{E}\left[\|\mathbf{y}_n\|^2\right] = \text{tr}\left(\mathbf{R}_{nn}\mathbf{W}^H\mathbf{W}\right) \tag{22.13}$$

where we have used again the cyclic permutation rule. Consequently, for the signal-to-noise ratio at the receiver output follows

$$\gamma_y = \frac{\mathbf{E}\left[\|\mathbf{y}_s\|^2\right]}{\mathbf{E}\left[\|\mathbf{y}_n\|^2\right]} = \frac{\text{tr}\left(\mathbf{R}_{cc}(\mathbf{W}\mathbf{H}\mathbf{A})^H\mathbf{W}\mathbf{H}\mathbf{A}\right)}{\text{tr}\left(\mathbf{R}_{nn}\mathbf{W}^H\mathbf{W}\right)} \tag{22.14}$$

The mean power of the transmit signal $\mathbf{s}$ is also of interest and we obtain the result just by replacing in (22.12) $\mathbf{W}\mathbf{H}\mathbf{A}$ by $\mathbf{A}$ yielding

$$\mathbf{E}\left[\|\mathbf{s}\|^2\right] = \text{tr}\left(\mathbf{R}_{cc}\mathbf{A}^H\mathbf{A}\right) \tag{22.15}$$

We observe that $\mathbf{E}\left[\|\mathbf{s}\|^2\right]$ could be enhanced by the precoder $\mathbf{A}$ compared to the mean power

$$\mathbf{E}\left[\|\mathbf{c}\|^2\right] = \text{tr}\left(\mathbf{R}_{cc}\right) \tag{22.16}$$

of **c**. This may cause an overload of the channel input and even a violation of power limits for wireless transmitters. In this case the mean power of **c** must be reduced, which impacts the signal-to-noise ratio at the receiver accordingly. Another remedy to overcome this drawback is the use of the Tomlinson–Harashima scheme, which can limit the amplitudes of the components of the transmit signal vector $\mathbf{s} = \begin{pmatrix} s_1 \cdots s_M \end{pmatrix}^T$ by applying modulo operations on the input symbols, [4–6].

22.2   Zero-Forcing and MMSE Precoding

A single prefilter matrix **A** shall be designed to reduce the inter-channel interference without the need for a receive filter. Thus, the matrix **W** in Fig. 22.1 is dropped and we focus on the receive signal **r**.

22.2.1   Zero-Forcing Precoder

In the following, we check whether the pseudo inverse of the channel matrix **H** from Chap. 18 depending on the number of transmit and receive antennas, M and N, can remove inter-channel interference.

$$\mathbf{A} = \begin{cases} \mathbf{H}^+ = \left(\mathbf{H}^H\mathbf{H}\right)^{-1}\mathbf{H}^H \in \mathbb{C}^{M\times N} & ; \ M \leq N \\ \mathbf{H}^{++} = \mathbf{H}^H\left(\mathbf{H}\mathbf{H}^H\right)^{-1} \in \mathbb{C}^{M\times N} & ; \ M \geq N \end{cases} \tag{22.17}$$

The prerequisites for this investigation are the full rank of $\mathbf{H}^H\mathbf{H}$ and $\mathbf{H}\mathbf{H}^H$ i.e., rank $\left(\mathbf{H}^H\mathbf{H}\right) = M$, if $M \leq N$ and rank $\left(\mathbf{H}\mathbf{H}^H\right) = N$ if $M \geq N$, respectively. Otherwise the inverse matrices in (22.17) do not exist. Please note that rank $\left(\mathbf{H}^H\mathbf{H}\right) = $ rank $(\mathbf{H})$ and rank $\left(\mathbf{H}\mathbf{H}^H\right) = $ rank $(\mathbf{H})$ hold in each of the two cases.

Then the inter-channel interference in the receive signal (22.2) is given by

$$\mathbf{HA} = \begin{cases} \mathbf{H}\mathbf{H}^+ = \mathbf{H}\left(\mathbf{H}^H\mathbf{H}\right)^{-1}\mathbf{H}^H \neq \mathbf{I}_N & ; \ M \leq N \\ \mathbf{H}\mathbf{H}^{++} = \mathbf{I}_N & ; \ M \geq N \end{cases} \tag{22.18}$$

and we substantiate that only the pseudo-inverse $\mathbf{H}^{++}$ as a prefilter owns the ability to completely remove inter-channel interference in general and we get from (22.2) the receive signal

$$\mathbf{r} = \mathbf{c} + \mathbf{n} \ ; \ M \geq N \tag{22.19}$$

which is composed of the transmit vector **c** just corrupted by the noise **n**. Therefore the prefilter with $\mathbf{H}^{++}$ is the most interesting one. However, it is restricted to systems with $M \geq N$.

Apparently, the precoder matrix $\mathbf{A} = \mathbf{H}^{++}$ is the same as for the zero-forcing receiver investigated in Chap. 18. However, the receive signal (22.19) is significantly different, because the noise $\mathbf{n}$ is untouched by the precoder. Remember, the use of a zero-forcing receive matrix $\mathbf{W}$ without a precoder results in $\mathbf{y} = \mathbf{s} + \mathbf{Wn}$ with a potential enhancement of the noise by the receive filter.

The signal-to-noise ratio at the receiver of the precoding scheme with $\mathbf{H}^{++}$ directly follows from (22.19)

$$\gamma_r = \frac{\mathbf{E}\left[\|\mathbf{c}\|^2\right]}{\mathbf{E}\left[\|\mathbf{n}\|^2\right]} \; ; \; M \geq N \tag{22.20}$$

which can also be deduced from (22.8) with (22.16) and (22.18).

Another interesting observation, albeit critical, is that the mean power $\mathbf{E}\left[\|\mathbf{s}\|^2\right]$ of the transmit signal $\mathbf{s}$ could be increased compared to the mean power of $\mathbf{c}$. This is discussed shortly in the following. To this end we refer to (22.15) and first determine $\mathbf{A}^H\mathbf{A}$ with (22.17). For the sake of completeness, we do not drop the case with the prefilter $\mathbf{H}^+$ in the following. It is straightforward to show that

$$\mathbf{A}^H\mathbf{A} = \begin{cases} \left(\mathbf{H}^+\right)^H \mathbf{H}^+ = \mathbf{H}\left(\mathbf{H}^H\mathbf{H}\right)^{-1}\left(\mathbf{H}^H\mathbf{H}\right)^{-1}\mathbf{H}^H \; ; \; M \leq N \\ \left(\mathbf{H}^{++}\right)^H \mathbf{H}^{++} = \left(\mathbf{H}\mathbf{H}^H\right)^{-1} \qquad\qquad ; \; M \geq N \end{cases} \tag{22.21}$$

holds yielding from (22.15)

$$\mathbf{E}\left[\|\mathbf{s}\|^2\right] = \begin{cases} \mathrm{tr}\left(\mathbf{R}_{cc}\mathbf{H}\left(\mathbf{H}^H\mathbf{H}\right)^{-1}\left(\mathbf{H}^H\mathbf{H}\right)^{-1}\mathbf{H}^H\right) \; ; \; M \leq N \\ \mathrm{tr}\left(\mathbf{R}_{cc}\left(\mathbf{H}\mathbf{H}^H\right)^{-1}\right) \qquad\qquad ; \; M \geq N \end{cases} \tag{22.22}$$

For illustration we consider the following example.

Example 1

We are going to discuss the mean power of the transmit signal $\mathbf{s}$. In many applications $\mathbf{c}$ is uncorrelated with covariance matrix

$$\mathbf{R}_{cc} = E_S\mathbf{I}_N \tag{22.23}$$

and consequently its mean signal power is obtained from (22.16) as

$$\mathbf{E}\left[\|\mathbf{c}\|^2\right] = NE_S \tag{22.24}$$

From (22.22) then follows

$$\mathbf{E}\left[\|\mathbf{s}\|^2\right] = \begin{cases} E_S\mathrm{tr}\left(\left(\mathbf{H}^H\mathbf{H}\right)^{-1}\right) \; ; \; M \leq N \\ E_S\mathrm{tr}\left(\left(\mathbf{H}\mathbf{H}^H\right)^{-1}\right) \; ; \; M \geq N \end{cases} \tag{22.25}$$

where the cyclic permutation rule was applied for the case $M \leq N$. We assume that the matrices $\mathbf{H}^H \mathbf{H}$ and $\mathbf{H}\mathbf{H}^H$ have the full rank, say $P = \min\{M, N\}$. As the matrices are Hermiteian, their eigenvalues $\lambda_1, \lambda_2, \ldots, \lambda_P$ are positive and unequal to zero. Then from (22.25) follows with (22.24)[1]

$$
\mathbf{E}\left[\|\mathbf{s}\|^2\right] = \frac{\mathbf{E}\left[\|\mathbf{c}\|^2\right]}{N} \sum_{i=1}^{P} \lambda_i^{-1} \tag{22.26}
$$

We clearly recognize that the mean power of $\mathbf{s}$ can be significantly larger than that of $\mathbf{c}$, if some eigenvalues λ_i exhibit small values. This is equivalent to say that the determinants of the two matrices $\mathbf{H}^H \mathbf{H}$ and $\mathbf{H}\mathbf{H}^H$ are small. Matrices with this property are called ill conditioned. The mean signal power of $\mathbf{s}$ will be enhanced in this case and may even cause an overload of the channel input. To make matters worse, if the matrices do not have full rank, $P < \min\{M, N\}$, then some eigenvalues are zero and the mean power of the transmit signal $\mathbf{s}$ even approaches infinity, theoretically. Finally, please note that not only the mean power but also the magnitude of the signal $\mathbf{s}$ at the output of the zero-forcing precoder can be enhanced in case of ill-conditioned matrices.

22.2.2 MMSE Precoder

In this section, we are looking for a precoder which minimizes the difference Δ between the original transmit signal $\mathbf{c}$ and the receive signal $\mathbf{r}$ in Fig. 22.1

$$
\Delta = \mathbf{c} - \mathbf{r} \tag{22.27}
$$

in the mean squared error (MMSE) sense similar to an MMSE receiver discussed in Chap. 18. Thus, the minimization problem can be formulated as

$$
J = \mathbf{E}\left[\|\Delta\|^2\right] = \mathrm{tr}\left(\mathbf{E}\left[\Delta \Delta^H\right]\right) = \min_{\mathbf{A}} \tag{22.28}
$$

Below we prove that the squared error is

$$
\begin{aligned}
J = \mathrm{tr}&\left\{\mathbf{R}_{cc} + \left(\mathbf{R}_{cn}^H - \mathbf{R}_{cc}\right)(\mathbf{HA})^H + \mathbf{HA}\left(\mathbf{R}_{cn} - \mathbf{R}_{cc}\right)\right\} \\
&+ \mathrm{tr}\left\{\mathbf{HAR}_{cc}(\mathbf{HA})^H - \mathbf{R}_{cn} - \mathbf{R}_{cn}^H + \mathbf{R}_{nn}\right\}
\end{aligned} \tag{22.29}
$$

---

[1]Note from Appendix B: For an $M \times M$ matrix $\mathbf{Q}$ with eigenvalues $\lambda_1, \lambda_2, \ldots, \lambda_M$ holds: $\mathrm{tr}(\mathbf{Q}) = \sum_{i=1}^{M} \lambda_i$, $\mathrm{tr}(\mathbf{Q}^{-1}) = \sum_{i=1}^{M} \lambda_i^{-1}$ (if non zero eigenvalues), and $\det(\mathbf{Q}) = \lambda_1 \lambda_2 \cdots \lambda_M$.

with the covariance matrices $\mathbf{R}_{cc}$ in (22.6) of the signal and $\mathbf{R}_{nn} = \mathbf{E}\left[\mathbf{nn}^H\right]$ of the noise as well as the cross-correlation matrices $\mathbf{R}_{cn} = \mathbf{E}\left[\mathbf{cn}^H\right]$ and $\mathbf{R}_{nc} = \mathbf{E}\left[\mathbf{nc}^H\right] = \mathbf{R}_{cn}^H$. The precoder matrix minimizing J turns out to be

$$\mathbf{A} = \left(\mathbf{H}^H\mathbf{H}\right)^{-1}\mathbf{H}^H\left(\mathbf{R}_{cc} - \mathbf{R}_{cn}^H\right)\mathbf{R}_{cc}^{-1} \; ; \; M \leq N \tag{22.30}$$

Now, we assume that $\mathbf{c}$ and $\mathbf{n}$ are uncorrelated yielding $\mathbf{R}_{cn} = \mathbf{E}\left[\mathbf{c}\right]\mathbf{E}\left[\mathbf{n}^H\right] = 0$, where the last term follows from the zero mean of the noise. Then the MMSE precoding matrix finally is

$$\mathbf{A} = \left(\mathbf{H}^H\mathbf{H}\right)^{-1}\mathbf{H}^H = \mathbf{H}^+ \; ; \; M \leq N \tag{22.31}$$

and identical with the zero-forcing precoding matrix for $M \leq N$. Please note that the condition $M \leq N$ in (22.30) and (22.31) guarantees that $\left(\mathbf{H}^H\mathbf{H}\right)^{-1}$ exists, if $\mathbf{H}$ has full rank M.

However, from (22.18), we see that the inter-channel interference cannot be removed by an MMSE precoding matrix $\mathbf{A} = \mathbf{H}^+$. Below, we also show that this solution yields from (22.29) under the realistic assumptions $\mathbf{R}_{cn} = \mathbf{E}\left[\mathbf{cn}^H\right] = 0$ and $\mathbf{R}_{cc} = E_S\mathbf{I}_N$ the MMSE

$$J(\mathbf{H}^+) = J_{min} = \mathbf{E}\left[\|\mathbf{n}\|^2\right] \tag{22.32}$$

The approach (22.28) has provided us a precoder matrix only for the case $M \leq N$. One may speculate that

$$\mathbf{A} = \mathbf{H}^{++} = \mathbf{H}^H\left(\mathbf{HH}^H\right)^{-1} \; ; \; M \geq N \tag{22.33}$$

could be the solution for $M \geq N$. This is really true. The proof is given below, where we also show that the achieved MMSE J_{min} is the same as in (22.32). This result is not surprising, because the prefilter $\mathbf{H}^{++}$ is completely removing the inter-channel interference according to (22.18), second line, resulting in the receive signal $\mathbf{r} = \mathbf{c} + \mathbf{n}$ yielding $\Delta = \mathbf{c} - \mathbf{r} = -\mathbf{n}$ and thus

$$J(\mathbf{H}^{++}) = \mathbf{E}\left[\|\mathbf{n}\|^2\right] = J_{min} \tag{22.34}$$

In summary, we conclude that the precoding matrices for the zero-forcing and the MMSE precoder are identical. Consequently, the same holds for the signal-to-noise ratios at the receiver. Both MMSE precoders provide the same MMSE under the condition of an uncorrelated transmit signal $\mathbf{c}$, an uncorrelated noise $\mathbf{n}$, and $\mathbf{R}_{cn} = 0$. Only the precoder matrix $\mathbf{H}^{++}$ can completely remove the inter-channel interference.

Proof of (22.29) and (22.30)

With (22.27) and (22.2) we obtain $\Delta\Delta^H = (\mathbf{c} - \mathbf{HAc} - \mathbf{n})\left(\mathbf{c}^H - \mathbf{c}^H\mathbf{A}^H\mathbf{H}^H - \mathbf{n}^H\right)$. After multiplication, the expected value follows as $\mathbf{E}\left[\Delta\Delta^H\right] = \mathbf{R}_{cc} + \left(\mathbf{R}_{cn}^H - \mathbf{R}_{cc}\right)\mathbf{A}^H\mathbf{H}^H + \mathbf{HA}\left(\mathbf{R}_{cn} - \mathbf{R}_{cc}\right) + \mathbf{HAR}_{cc}\mathbf{A}^H\mathbf{H}^H - \mathbf{R}_{cn} - \mathbf{R}_{cn}^H + \mathbf{R}_{nn}$. Next, we apply the

trace operator yielding (22.29) and differentiate with respect to $\mathbf{A}^*$ using the differentiation rules summarized in Appendix B. Then follows $\frac{\partial J}{\partial \mathbf{A}^*} = \mathbf{H}^H \left(\mathbf{R}_{cn}^H - \mathbf{R}_{cc} \right) + \mathbf{H}^H \mathbf{H} \mathbf{A} \mathbf{R}_{cc}$. Setting this derivative equal to zero yields
$\mathbf{A} = \left(\mathbf{H}^H \mathbf{H} \right)^{-1} \mathbf{H}^H \left(\mathbf{R}_{cc} - \mathbf{R}_{cn}^H \right) \mathbf{R}_{cc}^{-1}$ and the proof is finished.

Proof of (22.32)

From the first line of (22.18) follows $\mathbf{H}\mathbf{A} = \mathbf{H}\mathbf{H}^+ = (\mathbf{H}\mathbf{A})^H$. Then we obtain from (22.29) with the prerequisite $\mathbf{R}_{cn} = 0$
$J = \mathrm{tr}\,(\mathbf{R}_{cc}) + \mathrm{tr}\,(\mathbf{R}_{nn}) - 2\mathrm{tr}\left(\mathbf{R}_{cc}\mathbf{H}\mathbf{H}^+\right) + \mathrm{tr}\left(\mathbf{R}_{cc}\mathbf{H}\mathbf{H}^+\mathbf{H}\mathbf{H}^+\right)$. With the prerequisite $\mathbf{R}_{cc} = E_S \mathbf{I}_N$ and the cyclic permutation rule follows
$\mathrm{tr}\left(\mathbf{R}_{cc}\mathbf{H}\mathbf{H}^+\right) = E_S \mathrm{tr}\left(\mathbf{H}\left(\mathbf{H}^H\mathbf{H}\right)^{-1}\mathbf{H}^H\right) = E_S \mathrm{tr}\left(\left(\mathbf{H}^H\mathbf{H}\right)^{-1}\mathbf{H}^H\mathbf{H}\right) = NE_S$.
In a similar way we get
$\mathrm{tr}\left(\mathbf{R}_{cc}\mathbf{H}\mathbf{H}^+\mathbf{H}\mathbf{H}^+\right) = E_S \mathrm{tr}\left(\mathbf{H}\left(\mathbf{H}^H\mathbf{H}\right)^{-1}\mathbf{H}^H\mathbf{H}\left(\mathbf{H}^H\mathbf{H}\right)^{-1}\mathbf{H}^H\right) = NE_S$.
Finally, we obtain $J = \mathrm{tr}\,(\mathbf{R}_{nn}) = \mathbf{E}\left[\|\mathbf{n}\|^2\right]$ and the proof ends.

Proof of (22.34)

We use the second line of (22.18). Plugging $\mathbf{H}\mathbf{A} = \mathbf{I}_N = (\mathbf{H}\mathbf{A})^H$ into (22.29) directly yields the result $J = \mathrm{tr}\,(\mathbf{R}_{nn}) = \mathbf{E}\left[\|\mathbf{n}\|^2\right]$ and the proof is finished.

22.3   Precoding Based on Singular Value Decomposition

22.3.1   SVD-Based Precoder and Receiver

Precoder and Receiver Matrix

Using the theory of the eigenmode decomposition we now determine the precoder matrix as

$$\mathbf{A} = \mathbf{V} \tag{22.35}$$

where the unitary matrix $\mathbf{V} \in \mathbb{C}^{M \times M}$ stems from the right-hand side of the singular value decomposition of $\mathbf{H} \in \mathbb{C}^{N \times M}$

$$\mathbf{H} = \mathbf{U}\mathbf{D}\mathbf{V}^H \tag{22.36}$$

Then the receive signal (22.2) is

$$\mathbf{r} = \mathbf{H}\mathbf{V}\mathbf{c} + \mathbf{n} \tag{22.37}$$

Though the inter-channel interference $\mathbf{H}\mathbf{V}$ is not completely reduced, we require a receive filter with matrix $\mathbf{W}$, which we select according to the left-hand side matrix of (22.36)

$$\mathbf{W} = \mathbf{U}^H \in \mathbb{C}^{N \times N} \tag{22.38}$$

The output signal of the receiver is $\mathbf{y} = \mathbf{Wr}$ and with (22.37) and (22.38) follows

$$\mathbf{y} = \mathbf{U}^H \mathbf{HVc} + \mathbf{n}' \tag{22.39}$$

with the filtered noise

$$\mathbf{n}' = \mathbf{U}^H \mathbf{n} \tag{22.40}$$

Please note that the dimensions of the matrices and vectors partly differ from Sect. 22.1, namely $\mathbf{c} \in \mathbb{C}^{M \times 1}$, $\mathbf{A} = \mathbf{V} \in \mathbb{C}^{M \times M}$, and $\mathbf{y} \in \mathbb{C}^{N \times 1}$. On the first glance it seems not plausible that the input and output vectors, $\mathbf{c}$ and $\mathbf{y}$, of the system have different dimensions, M and N, respectively. We will see later that $|N - P|$ components of $\mathbf{y}$ will be zero, where P is the rank of $\mathbf{H}$.

Replacing in (22.39) $\mathbf{H}$ by (22.36) and knowing that $\mathbf{U}$ and $\mathbf{V}$ are unitary matrices we get the final result

$$\mathbf{y} = \mathbf{Dc} + \mathbf{n}' \tag{22.41}$$

with

$$\mathbf{D} = \left(\begin{array}{cccc|c} \sqrt{\lambda_1} & 0 & 0 & 0 & \\ 0 & \sqrt{\lambda_2} & 0 & 0 & \\ & & \ddots & & \mathbf{0} \\ 0 & 0 & 0 & \sqrt{\lambda_P} & \\ \hline & \mathbf{0} & & & \mathbf{0} \end{array} \right) \in \mathbb{R}^{N \times M} \tag{22.42}$$

$\sqrt{\lambda_i}$; $i = 1, 2, \ldots, P$ are the singular values of $\mathbf{H}$ and $P = \text{rank}(\mathbf{H}) \leq \min\{M, N\}$ holds. The matrix $\mathbf{D}$ contains at the bottom $N - P$ lines with zeros and $M - P$ columns with zeros on the right-hand side. Therefore (22.41) boils down to

$$y_i = \sqrt{\lambda_i} c_i + n_i' \; ; \; i = 1, 2, \ldots, P \tag{22.43}$$

Obviously, to recover $c_1, c_2, \ldots, c_M$ from the replicas $y_1, y_2, \ldots, y_M$ the channel matrix must have the rank $P = M$. The remaining receiver output signals

$$y_i = n_i' \; ; \; i = M + 1, \ldots, N \tag{22.44}$$

just contain noise and have to be discarded by the receiver. We substantiate that the multi-user interference is completely removed, because the transmit signal c_i is just multiplied by a factor $\sqrt{\lambda_i}$.

Example 2

We consider a MIMO system with $M = 3$ transmit and $N = 4$ receive antennas. $\mathbf{H}$ shall have rank P. Then we can write (22.41) as

$$\begin{pmatrix} y_1 \\ y_2 \\ y_3 \\ y_4 \end{pmatrix} = \begin{pmatrix} \sqrt{\lambda_1} & 0 & 0 \\ 0 & \sqrt{\lambda_2} & 0 \\ 0 & 0 & \sqrt{\lambda_3} \\ 0 & 0 & 0 \end{pmatrix} \begin{pmatrix} c_1 \\ c_2 \\ c_3 \end{pmatrix} + \begin{pmatrix} n'_1 \\ n'_2 \\ n'_3 \\ n'_4 \end{pmatrix} \tag{22.45}$$

The last row will be discarded, because y_4 is just noise. If the channel matrix has full rank $P = 3$, all symbols c_1, c_2, c_3 can be recovered. In case of $P = 2$ the singular value $\sqrt{\lambda_3} = 0$ and $y_3 = n'_3$ follows. Consequently, c_3 cannot be recovered. We also see that a weak eigenmode λ_i provides a low signal part $\sqrt{\lambda_i} c_i$.

Signal-to-Noise Ratio

To determine the signal-to-noise ratio of the SVD-based precoding scheme we consider in (22.39) the signal part

$$\mathbf{y}_s = \mathbf{U}^H \mathbf{H} \mathbf{V} \mathbf{c} \tag{22.46}$$

and the noise part

$$\mathbf{y}_n = \mathbf{n}' = \mathbf{U}^H \mathbf{n} \tag{22.47}$$

Then we obtain the mean signal power at the receiver

$$\mathbf{E}\left[\|\mathbf{y}_s\|^2\right] = \mathbf{E}\left[\operatorname{tr}\left(\mathbf{y}_s \mathbf{y}_s^H\right)\right] = \operatorname{tr}\left(\mathbf{R}_{cc} \mathbf{V}^H \mathbf{H}^H \mathbf{H} \mathbf{V}\right) \tag{22.48}$$

Using (22.36) yields $\mathbf{E}\left[\|\mathbf{y}_s\|^2\right] = \operatorname{tr}\left(\mathbf{R}_{cc} \mathbf{D}^T \mathbf{D}\right)$. With

$$\mathbf{D}^T \mathbf{D} = \Lambda_M = \operatorname{diag}\left(\lambda_1, \lambda_2, \ldots, \lambda_P, 0, \ldots, 0\right) \in \mathbb{R}^{M \times M} \tag{22.49}$$

follows the result

$$\mathbf{E}\left[\|\mathbf{y}_s\|^2\right] = \operatorname{tr}\left(\mathbf{R}_{cc} \Lambda_M\right) \tag{22.50}$$

In case of a channel matrix $\mathbf{H}$ with $M \le N$ and full rank, $P = M$ holds.

The mean power of the noise part $\mathbf{y}_n$ does not change, because in the Appendix B it is shown that a unitary matrix maintains the mean power. Consequently

$$\mathbf{E}\left[\|\mathbf{y}_n\|^2\right] = \mathbf{E}\left[\|\mathbf{n}\|^2\right] \tag{22.51}$$

holds and for the signal-to-noise ratio at the receiver output follows with (22.48) and (22.50)

$$\gamma_y = \frac{\mathbf{E}\left[\|\mathbf{y}_s\|^2\right]}{\mathbf{E}\left[\|\mathbf{y}_n\|^2\right]} = \frac{\operatorname{tr}\left(\mathbf{R}_{cc} \mathbf{V}^H \mathbf{H}^H \mathbf{H} \mathbf{V}\right)}{\mathbf{E}\left[\|\mathbf{n}\|^2\right]} = \frac{\operatorname{tr}\left(\mathbf{R}_{cc} \Lambda_M\right)}{\mathbf{E}\left[\|\mathbf{n}\|^2\right]} \tag{22.52}$$

With the same argument as for (22.51), we substantiate that the SVD-based precoder does not change the mean power of the transmit signal $\mathbf{s}$, hence

$$\mathbf{E}\left[\|\mathbf{s}\|^2\right] = \mathbf{E}\left[\|\mathbf{c}\|^2\right] = \operatorname{tr}\left(\mathbf{R}_{cc}\right) \tag{22.53}$$

holds.

Example 3

In many applications, the symbol vector $\mathbf{c} \in \mathbb{C}^{M \times 1}$ is uncorrelated with covariance matrix $\mathbf{R}_{cc} = E_S \mathbf{I}_M$. Then we obtain for the signal-to-noise ratio from (22.52) with the cyclic permutation rule

$$\gamma_y = \frac{\mathbf{E}\left[\|\mathbf{y}_s\|^2\right]}{\mathbf{E}\left[\|\mathbf{y}_n\|^2\right]} = \frac{E_S \text{tr}\left(\mathbf{H}^H \mathbf{H}\right)}{\mathbf{E}\left[\|\mathbf{n}\|^2\right]} = \frac{E_S \text{tr}\left(\Lambda_M\right)}{\mathbf{E}\left[\|\mathbf{n}\|^2\right]} = \frac{E_S \sum_{i=1}^{P} \lambda_i}{\mathbf{E}\left[\|\mathbf{n}\|^2\right]} \qquad (22.54)$$

22.3.2   Comparison of Zero-Forcing and SVD-Based Precoding

In Table 22.1, the main features of the zero-forcing and the SVD-based precoding for an uncorrelated transmit signal $\mathbf{c}$ with covariance matrix $\mathbf{R}_{cc} = E_S \mathbf{I}_M$ are compared. As seen before, the zero-forcing and the MMSE precoder own the same matrices. Therefore we just mention the first one in the following. The zero-forcing precoder with $\mathbf{A} = \mathbf{H}^{++}$ can completely cancel the inter-channel interference, which is also true for the SVD-based precoding. Moreover, both schemes are not enhancing the noise at the receiver. The SVD-based method leaves the transmit mean power unchanged, whereas the zero-forcing precoder can cause an unfavorable enhancement of the transmit power. The zero-forcing precoding scheme does not require a receive filter.

Table 22.1 Comparison: SVD versus zero-forcing/MMSE-based precoder for uncorrelated transmit signal $\mathbf{c}$ with $\mathbf{R}_{cc} = E_S \mathbf{I}_M$

Feature	SVD-based precoding	Zero-forcing (MMSE) precoding
Precoder matrix	$\mathbf{A} = \mathbf{V}$	$\mathbf{A} = \mathbf{H}^{++} = \mathbf{H}^H \left(\mathbf{H}\mathbf{H}^H\right)^{-1}$ $M \geq N$
Receive matrix	$\mathbf{W} = \mathbf{U}^H$	not required
Receive signal	$\mathbf{y} = \mathbf{D}\mathbf{c} + \mathbf{U}^H \mathbf{n}$	$\mathbf{r} = \mathbf{c} + \mathbf{n}$
Inter-channel interference	completely removed	completely removed
Signal-to-noise ratio	$\gamma_y = \frac{E_S \text{tr}(\mathbf{H}^H \mathbf{H})}{\mathbf{E}[\|\mathbf{n}\|^2]}$	$\gamma_r = \frac{N E_S}{\mathbf{E}[\|\mathbf{n}\|^2]}$; $M \geq N$
Mean power of transmit signal	$M E_S$	$E_S \text{tr}\left(\left(\mathbf{H}\mathbf{H}^H\right)^{-1}\right)$; $M \geq N$
Mean power of receive noise	$\mathbf{E}\left[\|\mathbf{n}\|^2\right]$	$\mathbf{E}\left[\|\mathbf{n}\|^2\right]$

References

1. R. Fischer, *Precoding and Signal Shaping for Digital Transmission* (Wiley, New York, 2002)
2. M. Joham, W. Utschick, J. Nossek, Linear transmit processing in MIMO communications systems. IEEE Trans. Signal Process. **53**, 2700–2712 (2005)
3. M. Vu, A. Paulraj, MIMO wireless linear precoding. IEEE Signal Process. Mag. **24**, 86–105 (2007)
4. H. Harashima, H. Miyakawa, Matched-transmission technique for channels with intersymbol interference. IEEE Trans. Commun. **20**, 774–780 (1972)
5. M. Tomlinson, New automatic equalizer employing modulo arithmetic. Electron. Lett. **7**, 138–139 (1971)
6. R. Fischer, C. Windpassinger, A. Lampe, J. Huber, Space-time transmission using Tomlinson-Harashima precoding, in *4th ITG- Conference on Source and Channel Coding* (2002)

Chapter 23
Principles of Space-Time Coding

23.1 Space-Time Block Coding

Figure 23.1 shows the principle block diagram of a MIMO transmitter with space-time encoding. The incoming bit sequence $b(n)$ is fed into the QAM mapper, which periodically maps κ consecutive bits to a QAM symbol $c(k')$, constituting a 2^κ-ary QAM. $b(n)$ may contain redundancy bits from a forward error correction encoder [1–3].

We focus on the description of the space-time encoder, which allocates to the input QAM symbols $c(k')$ dedicated redundancy symbols in the space-time domain to improve transmission quality. The operation of the encoder is block-wise. The sequence $c(k')$ of symbols from a QAM mapper shall occur at time instances $t = k'T'$; $k' \in \mathbb{Z}$ with symbol rate $v_S = \frac{1}{T'}$. The space-time encoder outputs a sequence of space-time symbol vectors, which appear at the time instances $t = kT$; $k \in \mathbb{Z}$ and with the rate $v_{ST} = \frac{1}{T}$. In the time frame $N_c T'$, which we also call the block length, a QAM symbol vector

$$\mathbf{c}(k') = \big(c(k') \; c(k'+1) \; \cdots \; c(k'+N_c-1) \big)^T \qquad (23.1)$$

is mapped to the output matrix

$$\mathbf{S}(k) = \big(\mathbf{s}(k) \; \mathbf{s}(k+1) \; \cdots \; \mathbf{s}(k+L-1) \big)$$

$$= \begin{pmatrix} s_1(k) & s_1(k+1) & \cdots & s_1(k+L-1) \\ \vdots & \vdots & \ddots & \vdots \\ s_j(k) & s_j(k+1) & \cdots & s_j(k+L-1) \\ \vdots & \vdots & \ddots & \vdots \\ s_M(k) & s_M(k+1) & \cdots & s_M(k+L-1) \end{pmatrix} \qquad (23.2)$$

The original version of this chapter was revised: a few typographical errors were corrected. The correction to this chapter can be found at https://doi.org/10.1007/978-3-030-00548-1_25

© Springer Nature Switzerland AG 2019
J. Speidel, *Introduction to Digital Communications*, Signals and Communication Technology, https://doi.org/10.1007/978-3-030-00548-1_23

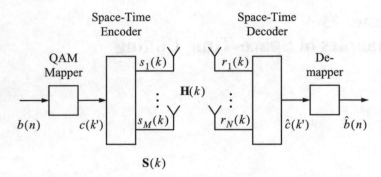

Fig. 23.1 Principle block diagram of a MIMO system with space-time encoding and decoding

in the time frame LT. In other words, N_c QAM symbols are mapped to L space-time symbol vectors

$$\mathbf{s}(k+l) = \big(s_1(k+l)\ s_2(k+l) \cdots s_M(k+l) \big)^T\ ;\ l = 0, 1, \ldots, L-1 \quad (23.3)$$

which are the column vectors of $\mathbf{S}(k)$. There are various denominations for $\mathbf{S}(k)$ either transmission matrix, space-time coding matrix, space-time codeword matrix, or space-time code block. The mapping of the vector $\mathbf{c}(k')$ to the matrix $\mathbf{S}(k)$ can be done in various ways, which is the subject of the space-time coding theory and will be addressed in the following sections in more detail. For example, a linear space-time block encoder employs linear combining on the input symbols, such as addition and subtraction of the original symbols, their real or imaginary parts, or their conjugate complex values. For a (nonlinear) trellis space-time encoder, the input–output relation is defined by a trellis diagram.

We recognize that in general there are different time bases at the input and at the output of the space-time encoder indicated by the discrete-time variables k' and k, respectively. For synchronous block-wise operation between the input and the output,

$$N_c T' = LT \qquad\qquad (23.4)$$

must hold. Please note that the definition of $\mathbf{S}(k)$ in (23.2) does not respect causality. Consequently from an implementation point of view, the space-time symbol vectors $\mathbf{s}(k), \mathbf{s}(k+1), \ldots$ have to be delayed by $N_c T'$ time intervals, because in general the space-time encoder can start the mapping operation not before all N_c QAM symbols have been entered.

From (23.2), we see that row j of $\mathbf{S}(k)$ represents a sequence of L samples sent out by the transmit antenna j. On the other hand, column $k+l$ is the space-time symbol vector emitted by all M transmit antenna elements at the same time instant $k+l$. An important parameter of any encoder is the code rate, which is defined as the ratio between the number of symbols at the input and the number of symbols at the output, both considered in the same time frame LT. In contrast to an encoder for temporal forward error correction on bit level, the output of the space-time encoder is a vector, which is counted as one element despite of the fact that it is composed of M components. Hence, the spatial code rate of the space-time encoder is defined as

$$r_s = \frac{\text{number of input symbols in time frame } LT}{\text{number of output symbol vectors in time frame } LT} \qquad (23.5)$$

For synchronous block-wise operation specified by (23.4), the spatial code rate is

$$r_s = \frac{N_c}{L} \qquad (23.6)$$

As already mentioned, the time basis at the input and the output of the space-time encoder are in general different. Given the QAM symbol rate at the input as

$$v_S = \frac{1}{T'} \qquad (23.7)$$

we conclude from (23.4) and (23.6) for the space-time symbol vector rate

$$v_{ST} = \frac{1}{T} = \frac{L}{N_c} v_S = \frac{v_S}{r_s} \qquad (23.8)$$

We can determine the redundant symbols allocated by the space-time encoder in the time frame (23.4) recognizing that we have in total LM symbols at the output and N_c QAM symbols at the input yielding $LM - N_c = N_c \left(\frac{M}{r_s} - 1 \right)$ redundant output symbols. From this consideration follows that the space-time encoder allocates redundancy, if $LM > N_c$ or with (23.6)

$$r_s < M \qquad (23.9)$$

There is no redundancy if $LM = N_c$. From $L \in \mathbb{N}$ follows for this special case that N_c has to be an integer multiple of M and we obtain

$$r_s = M \qquad (23.10)$$

This is exactly the code rate of a serial-to-parallel converter, as will be outlined in Sect. 23.2. From the viewpoint of classical forward error correction for bit streams, where the maximal code rate is one, a code rate larger than one of the space-time encoders is surprising. Obviously, systems with more than one transmit antenna can demonstrate that property.

We conclude from (23.8), if $r_S = 1$ then the clock rates of the output and the input sequences are identical, $v_{ST} = v_S$. As a consequence, the bandwidth of the baseband of the space-time symbol sequence $\mathbf{s}(k)$ is not increased compared to the bandwidth of QAM symbol sequence $c(k')$ although $N_c(M - 1)$ redundant symbols are contained in each output block of length LT. For $r_s < 1$, the speed of the output sequence and accordingly the bandwidth increase by factor $\frac{1}{r_s} = \frac{v_{ST}}{v_S}$.

For the sake of completeness, we calculate the code rates of the building blocks of the transmitter in Fig. 23.1. Assume that in the time interval (23.4), the data source feeds κN_b information bits into the forward error correction encoder (not depicted in Fig. 23.1) and that we have at its output κN_c bits including $\kappa(N_c - N_b)$ redundancy bits. Furthermore, the QAM mapper shall assign one QAM symbol to a sequence of κ consecutive input bits, which results in a 2^κ-ary QAM scheme with κ bit/symbol. Then, the code rates are as follows:

- code rate of (temporal) forward error correction including temporal interleaving, $r_t = \frac{N_b}{N_c}$
- "code rate" of QAM mapper, $r_m = \kappa$
- overall code rate at the output of the space-time encoder, $r_{total} = \frac{\kappa N_b}{L} = \kappa \frac{N_b}{N_c} \frac{N_c}{L} = r_m r_t r_s$

As depicted in Fig. 23.1, at the receiving end a linear space-time decoder with N receive antennas can be employed, which performs in principle the reverse operation of the encoder. Alternatively, a nonlinear receiver with N receive antennas and a maximum likelihood detector is feasible. Then for ideal conditions the original bit sequence is recovered, $\hat{b}(n) = b(n)$.

The receive signal can be determined as follows. The transmit signal vector at time instant $k + l$ is $\mathbf{s}(k + l)$ according to (23.3). Then the receive vector is

$$\mathbf{r}(k + l) = \mathbf{H}(k + l)\mathbf{s}(k + l) + \mathbf{n}(k + l) \; ; \; l = 0, 1, \ldots, L - 1 \qquad (23.11)$$

where $\mathbf{H}(k + l)$ is the channel matrix and $\mathbf{n}(k + l)$ is the noise vector. Now we can assemble all receive vectors in a receive matrix, which is the response to the transmit signal matrix $\mathbf{S}(k)$ in (23.2)

$$\left(\mathbf{r}(k) \cdots \mathbf{r}(k + L - 1) \right) =$$

$$= \left(\mathbf{H}(k)\mathbf{s}(k) \cdots \mathbf{H}(k + L - 1)\mathbf{s}(k + L - 1) \right) + \left(\mathbf{n}(k) \cdots \mathbf{n}(k + L - 1) \right)$$
$$(23.12)$$

If the channel exhibits block fading, its channel matrix is approximately unchanged in the time interval of the block, say from time instant k to $k + L - 1$,

$$\mathbf{H}(k + l) \approx \mathbf{H}(k) \; ; \; l = 1, \ldots, L - 1 \qquad (23.13)$$

Then follows from (23.12) with (23.2), the matrix of the receive vectors approximately

$$\left(\mathbf{r}(k) \cdots \mathbf{r}(k + L - 1) \right) = \mathbf{H}(k)\mathbf{S}(k) + \left(\mathbf{n}(k) \cdots \mathbf{n}(k + L - 1) \right) \qquad (23.14)$$

In summary, we see that multiple transmit antennas enable the concatenation of symbols of the input symbol sequence $c(k')$ in the temporal and the spatial domain, which justifies the name space-time coding. This method is a powerful alternative to binary forward error correction. As an example, it can be seen from (23.9) that a spatial code rate $r_s = 1$ adds redundancy to the output sequence for $M > 1$ without increasing the speed (23.8) of the output signal. The space-time encoder can be combined with a multi-antenna receiver to exploit additional receive diversity. In principle, the scheme can even be equipped with forward error correction as a layered space-time coding technique, as discussed in Sect. 23.5 or as a concatenated encoding scheme allowing iterative (Turbo) decoding at the receiver. Obviously, space-time coding is an open-loop procedure, as it does not require channel information at the transmitter through feedback from the receiver. A comprehensive survey on space-time coding is given by [4]. In the next sections, we will outline some examples of space-time coding schemes in more detail.

23.2   Spatial Multiplexing

Spatial multiplexing can be considered as a very simple form of linear space-time encoding. As no redundancy is allocated to the information symbols $c(k')$, this scheme is often not ranked as a space-time encoder. The principle is shown in Fig. 23.2. At the transmitter, a spatial demultiplexer is employed, which is just a serial-to-parallel converter.

The input QAM symbol sequence

$$c(k'), c(k' + 1), \ldots, c(k' + M - 1), \ldots \tag{23.15}$$

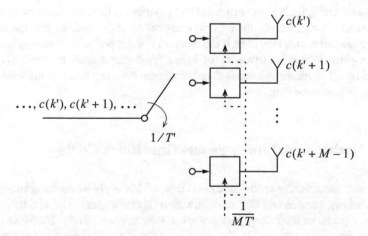

Fig. 23.2   Principle of a transmitter with spatial demultiplexer

with symbol rate $v_S = \frac{1}{T'}$ is entering a shift register. After MT' clock cycles (block length), the space-time symbol vector

$$\mathbf{s}(k) = \big(c(k')\ c(k'+1)\ \cdots\ c(k'+M-1)\big)^T \qquad (23.16)$$

is available at the parallel outputs of the shift register and each component is forwarded to the respective antenna. In the following MT' clock cycles, the next space-time symbol vector

$$\mathbf{s}(k+1) = \big(c(k'+M)\ c(k'+M+1)\ \cdots\ c(k'+2M-1)\big)^T \qquad (23.17)$$

is generated and so forth. Please note that in contrast to the general scheme in Fig. 23.1, no processing other than storage and serial-to-parallel conversion is performed. Furthermore, the clock rate of the output space-time symbol vectors is $v_{ST} = \frac{1}{MT'} = \frac{v_S}{M}$ and thus by factor $\frac{1}{M}$ lower than the QAM symbol rate v_S at the input. We easily check that for $M=1$ the clock rates are identical, as expected. Similarly as depicted in Fig. 23.1, at the receiving end a zero-forcing, MMSE, or maximum likelihood receiver with N receive antennas can be employed providing M output signals to the space-time decoder, which in this case is just a spatial multiplexer. It operates as a simple parallel-to-serial converter and outputs a serial symbol sequence, which is identical to (23.15), if no transmission errors are present. We can easily calculate the spatial code rate as follows: M QAM symbols $c(k')$ are entering in the time interval $MT' = LT$ (with $L=1$) and one symbol vector is going out. Thus

$$r_s = M \qquad (23.18)$$

With spatial multiplexing, one original QAM symbol is transmitted by no more than one antenna. Consequently, there is no transmit diversity and we say the transmit diversity has order one. However, if the receiver is equipped with N receive antennas, they can provide receive diversity of order N, because each transmit symbol is received by all N receive antennas. The maximum diversity order of the total spatial multiplexing scheme then is N.

23.3   Orthogonal, Linear Space-Time Block Coding

We are now considering space-time encoders, which apply linear combining on the input symbols, such as addition or subtraction of the original symbols, their real or imaginary parts, or their conjugate complex values, respectively. Motivated by the Alamouti encoding scheme [5] described in the next section, the rows of the space-time coding matrix $\mathbf{S}$ turn out to be pairwise orthogonal. The same holds for the columns.

23.3.1 The Alamouti Encoder for MISO System with Two Transmit Antennas

The Transmission Scheme

The Alamouti space-time encoder was presented in [5] and is a simple but effective technique to achieve transmit diversity. It can be shown that this scheme is unique, as it employs the only 2×2 space-time block coding matrix with complex entries to achieve full spacial code rate $r_s = 1$ and full transmit diversity of the order two. In this section, we consider a multiple input single output (MISO) transmission scheme with $M = 2$ transmit and $N = 1$ receive antennas. In the next section, this approach is extended to a MIMO system with two receive antennas. The principal block diagram is depicted in Fig. 23.3. The multiple input single output (MISO) channel is described by the channel matrix, which is actually a row vector

$$\mathbf{H}(k) = \mathbf{h}^T(k) = \left(h_{11}(k) \; h_{12}(k) \right) \tag{23.19}$$

The Alamouti space-time encoding scheme operates under the assumption of block fading. This means that the channel coefficients do not change during a block time, here from time instants k to $k + 1$. Consequently

$$h_{11}(k) \approx h_{11}(k+1) \; ; \; h_{12}(k) \approx h_{12}(k+1) \; \Rightarrow \mathbf{h}^T(k) \approx \mathbf{h}^T(k+1) \tag{23.20}$$

Furthermore, we assume that the channel coefficients are exactly known at the receiver. In reality, they can only be estimated with some deviation from the actual values. However, the channel estimation scheme is not considered here and not shown in Fig. 23.3.

We will see that the time basis of the input and the output of the encoder are the same, thus $k' = k$. The space-time encoder maps the input QAM symbol vector

$$\mathbf{c}(k) = \left(c(k) \; c(k+1) \right)^T \tag{23.21}$$

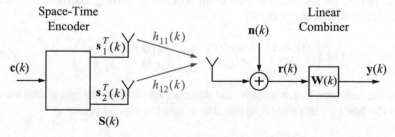

Fig. 23.3 Block diagram of a 2×1 multiple input single output (MISO) transmission scheme with Alamouti space-time encoder

to the space-time coding matrix with $L = 2$

$$\mathbf{S}(k) = \begin{pmatrix} s_1(k) & s_1(k+1) \\ s_2(k) & s_2(k+1) \end{pmatrix} = \big(\mathbf{s}(k) \;\; \mathbf{s}(k+1)\big) \tag{23.22}$$

where we have introduced the space-time coding vectors at the output

$$\mathbf{s}(k) = \big(s_1(k) \; s_2(k)\big)^T \;\; ; \;\; \mathbf{s}(k+1) = \big(s_1(k+1) \; s_2(k+1)\big)^T \tag{23.23}$$

as column vectors of $\mathbf{S}(k)$. In Fig. 23.3, the row vectors $\mathbf{s}_j^T(k) = \big(s_j(k) \; s_j(k+1)\big)$ of $\mathbf{S}(k)$ are shown, which indicate the output symbols of the antenna j for the time instants k and $k+1$, where $j = 1, 2$. The encoder responds to the input symbol $c(k)$ by the vector $\mathbf{s}(k)$, and the receive signal is

$$r(k) = \mathbf{h}^T(k)\mathbf{s}(k) + n(k) \tag{23.24}$$

The next input symbol $c(k+1)$ generates the output vector $\mathbf{s}(k+1)$ and the receive signal

$$r(k+1) = \mathbf{h}^T(k+1)\mathbf{s}(k+1) + n(k+1) \tag{23.25}$$

The block-wise processing can be simply described by introducing the following vector notation with (23.24) and (23.25):

$$\begin{pmatrix} r(k) \\ r(k+1) \end{pmatrix} = \begin{pmatrix} \mathbf{h}^T(k) \\ \mathbf{h}^T(k+1) \end{pmatrix} \big(\mathbf{s}(k) \;\; \mathbf{s}(k+1)\big) + \begin{pmatrix} n(k) \\ n(k+1) \end{pmatrix} \tag{23.26}$$

With (23.22) follows

$$\begin{pmatrix} r(k) \\ r(k+1) \end{pmatrix} = \begin{pmatrix} \mathbf{h}^T(k) \\ \mathbf{h}^T(k+1) \end{pmatrix} \mathbf{S}(k) + \begin{pmatrix} n(k) \\ n(k+1) \end{pmatrix} \tag{23.27}$$

The space-time coding matrix is designed such that the signal vector $\mathbf{s}(k)$ can be recovered at the receiver in the noise-free case by multiplying (23.27) from the left by an appropriate matrix. The Alamouti space-time coding matrix is defined for block fading channels (23.20) as

$$\mathbf{S}(k) = \begin{pmatrix} c(k+1) & -c^*(k) \\ c(k) & c^*(k+1) \end{pmatrix} = \begin{pmatrix} s_1(k) & -s_2^*(k) \\ s_2(k) & s_1^*(k) \end{pmatrix} \tag{23.28}$$

where the last term is just a shorthand notation. Obviously, the space-time encoder keeps the first output vector $\mathbf{s}(k)$ and determines the second vector as

$$\mathbf{s}(k+1) = \big(-s_2^*(k) \; s_1^*(k)\big)^T \tag{23.29}$$

Now, we introduce (23.19), (23.20), and (23.28) in (23.27) and obtain after some simple manipulations

$$\underbrace{\begin{pmatrix} r(k) \\ r^*(k+1) \end{pmatrix}}_{\mathbf{r}(k)} = \underbrace{\begin{pmatrix} h_{11}(k) & h_{12}(k) \\ h_{12}^*(k) & -h_{11}^*(k) \end{pmatrix}}_{\mathbf{U}(k)} \underbrace{\begin{pmatrix} s_1(k) \\ s_2(k) \end{pmatrix}}_{\mathbf{s}(k)} + \underbrace{\begin{pmatrix} n(k) \\ n^*(k+1) \end{pmatrix}}_{\mathbf{n}(k)} \qquad (23.30)$$

Before we are going to investigate the receive matrix $\mathbf{W}(k)$ we look at some properties of $\mathbf{S}(k)$.

Properties of the Alamouti Space-Time Coding Matrix $\mathbf{S}(k)$

In the following, we drop k to simplify the notation.

- $\mathbf{S}$ has the property

$$\mathbf{S}\mathbf{S}^H = \alpha \mathbf{I}_2 \qquad (23.31)$$

  with

$$\alpha = |s_1|^2 + |s_2|^2 \qquad (23.32)$$

Consequently, $\frac{1}{\sqrt{\alpha}}\mathbf{S}$ is a unitary matrix and the rows of $\mathbf{S}$ are orthogonal. The proof is straightforward. Similarly,

$$\mathbf{S}^H \mathbf{S} = \alpha \mathbf{I}_2 \qquad (23.33)$$

holds and the columns of $\mathbf{S}$ are orthogonal as well.
- $\mathbf{S}$ has full rank 2 and thus the Alamouti precoder can provide full transmit diversity of order two.

Determination of Receive Signal and Receiver Matrix

Obviously, $\mathbf{U}(k)$ in (23.30) has the property

$$\mathbf{U}^H \mathbf{U} = \beta \mathbf{I}_2 \qquad (23.34)$$

where

$$\beta(k) = |h_{11}(k)|^2 + |h_{12}k)|^2 \geq 0 \qquad (23.35)$$

Similar to (23.33), we conclude that $\frac{1}{\sqrt{\beta}}\mathbf{U}$ is a unitary matrix. To recover the signal vector $\mathbf{s}(k)$, the linear receiver in Fig. 23.3 is equipped with a matrix

$$\mathbf{W}(k) = \mathbf{U}^H(k) = \begin{pmatrix} h_{11}^*(k) & h_{12}(k) \\ h_{12}^*(k) & -h_{11}(k) \end{pmatrix} \qquad (23.36)$$

which yields the output signal vector

$$\mathbf{y}(k) = \begin{pmatrix} y(k) \\ y(k+1) \end{pmatrix} = \mathbf{U}^H(k) \begin{pmatrix} r(k) \\ r^*(k+1) \end{pmatrix}$$

and with (23.30) and (23.34) follows

$$\mathbf{y}(k) = \beta(k)\mathbf{s}(k) + \mathbf{U}^H(k)\mathbf{n}(k) \tag{23.37}$$

Obviously, the output of the receive filter is composed of the signal part $\beta(k)\mathbf{s}(k)$ and the noise part $\mathbf{n}'(k) = \mathbf{U}^H(k)\mathbf{n}(k)$.

The coefficient $\beta(k)$ demonstrates the expected transmit diversity. As we can see from (23.35), the signal part only fades out, if $h_{11}(k) = 0$ and $h_{12}(k) = 0$ at the same time. However, this event occurs with small probability. Consequently, the space-time coding scheme with two transmit antennas is very effective to achieve low symbol error rate in particular for situations, where the (temporal) forward error correction cannot help, because one link is in deep fade. The removal of the inter-channel interference in (23.37) by the receive matrix is also appreciated.

Unitary matrices preserve the mean power of signals, as is shown in Appendix B. If the channel coefficients in (23.35) are normalized such that $\beta = 1$, then the mean powers of the signal and the noise part in the receive signal (23.37) are not enhanced by the receive matrix.

The signal $\mathbf{y}(k)$ in Fig. 23.3 is subject to a final decision, e.g., using the maximum likelihood detection algorithm. Please note that there is no need to feedback the channel coefficients to the transmitter, because the space-time encoding matrix is independent of the channel coefficients. This is different from closed-loop systems, which use a transmit filter for allocating individual power to the antennas or a zero-forcing precoder.

We also recognize that in the time interval $[k, k+1]$ two QAM symbols $c(k)$ and $c(k+1)$ are input and two space-time coding vectors are output resulting in the spatial code rate $r_s = 1$. Therefore, also the clock rates at the input and at the output are identical, $v_{ST} = v_S$. The space-time encoder adds redundancy, because the two input symbols generate four output samples contained in the coding matrix $\mathbf{S}(k)$, which results in a redundancy of two symbols. As already pointed out, to calculate the spatial code rate for space-time coding, we consider all transmit antennas as a single unit emitting a vector rather than counting the individual samples. In contrast, a code rate 1 indicates for conventional forward error correction that no redundancy is allocated to the information bits.

Please note that the definition of $\mathbf{S}(k)$ does not respect the causality of a real system. Causality can be simply obtained by introducing an adequate delay to ensure that all QAM symbols have entered the encoder before the output is generated. The Alamouti scheme has found frequent applications both for wireless and wire-line digital communications [6]. In [7, 8], an application for digital transmission over more than two copper wires is described, which became part of the final standard to establish MIMO in-home power line communications.

23.3.2 The Alamouti Space-Time Encoder for a 2×2 MIMO System

The principle block diagram of the system is depicted in Fig. 23.4. The transmitter is unchanged and operates with two antennas. As the receiver employs also two antennas, the 2×2 MIMO channel matrix is

$$\mathbf{H}(k) = \begin{pmatrix} h_{11}(k) & h_{12}(k) \\ h_{21}(k) & h_{22}(k) \end{pmatrix} \qquad (23.38)$$

Similar to (23.27), the signal at the receiver i is given by

$$\begin{pmatrix} r_i(k) \\ r_i(k+1) \end{pmatrix} = \begin{pmatrix} \mathbf{h}_i^T(k) \\ \mathbf{h}_i^T(k+1) \end{pmatrix} \mathbf{S}(k) + \begin{pmatrix} n_i(k) \\ n_i(k+1) \end{pmatrix} \; ; \; i = 1, 2 \qquad (23.39)$$

with the channel row vectors $\mathbf{h}_i^T(k) = \begin{pmatrix} h_{i1}(k) & h_{i2}(k) \end{pmatrix} \; ; \; i = 1, 2$ and the same space-time coding matrix $\mathbf{S}(k)$ as given in (23.28). Obviously, the upper branch of the receiver was already investigated in Sect. 23.3.1 and the lower branch is quite similar. Therefore, we allocate an index i to $\mathbf{r}, \mathbf{n}, \mathbf{U}, \mathbf{W}, \mathbf{y}$ and replace h_{11} by h_{i1}, and h_{12} by h_{i2}. Then follows from (23.30)

$$\underbrace{\begin{pmatrix} r_i(k) \\ r_i^*(k+1) \end{pmatrix}}_{\mathbf{r}_i(k)} = \underbrace{\begin{pmatrix} h_{i1}(k) & h_{i2}(k) \\ h_{i2}^*(k) & -h_{i1}^*(k) \end{pmatrix}}_{\mathbf{U}_i(k)} \underbrace{\begin{pmatrix} s_1(k) \\ s_2(k) \end{pmatrix}}_{\mathbf{s}(k)} + \underbrace{\begin{pmatrix} n_i(k) \\ n_i^*(k+1) \end{pmatrix}}_{\mathbf{n}_i(k)} \; ; \; i = 1, 2$$

$$(23.40)$$

where $i = 1$ and $i = 2$ indicate the upper and the lower branch in Fig. 23.4, respectively. Furthermore

$$\mathbf{U}_i^H \mathbf{U}_i = \beta_i \mathbf{I}_2 \; ; \; i = 1, 2 \qquad (23.41)$$

and

$$\beta_i(k) = |h_{i1}(k)|^2 + |h_{i2}(k)|^2 \geq 0 \; ; \; i = 1, 2 \qquad (23.42)$$

hold. Hence, $\frac{1}{\sqrt{\beta_i}} \mathbf{U}_i$ is a unitary matrix. The receiver matrices follow from (23.36) as

$$\mathbf{W}_i(k) = \mathbf{U}_i^H(k) = \begin{pmatrix} h_{i1}^*(k) & h_{i2}(k) \\ h_{i2}^*(k) & -h_{i1}(k) \end{pmatrix} \; ; \; i = 1, 2 \qquad (23.43)$$

The output of the receiver is

$$\mathbf{y}_i(k) = \begin{pmatrix} y_i(k) \\ y_i(k+1) \end{pmatrix} = \mathbf{U}_i^H(k) \begin{pmatrix} r_i(k) \\ r_i^*(k+1) \end{pmatrix} \; ; \; i = 1, 2 \qquad (23.44)$$

and with (23.40) it can be written as

$$\mathbf{y}_i(k) = \beta_i(k)\mathbf{s}(k) + \mathbf{U}_i^H(k)\mathbf{n}_i(k) \; ; \; i = 1, 2 \qquad (23.45)$$

The receive signal of both branches exhibits a signal part $\beta_i(k)\mathbf{s}(k)$ and a noise part $\mathbf{n}_i'(k) = \mathbf{U}_i^H(k)\mathbf{n}_i(k) \; ; \; i = 1, 2.$

According to Fig. 23.4, we calculate the total receiver output signal block-wise as

$$\mathbf{y}_3(k) = \begin{pmatrix} y(k) \\ y(k+1) \end{pmatrix} = \mathbf{y}_1(k) + \mathbf{y}_2(k) = \begin{pmatrix} y_1(k) + y_2(k) \\ y_1(k+1) + y_2(k+1) \end{pmatrix} \qquad (23.46)$$

With (23.45) and the Frobenius norm of the channel matrix $\mathbf{H}(k)$

$$\|\mathbf{H}(k)\|_F^2 = \sum_{i=1}^{2} \sum_{j=1}^{2} |h_{ij}(k)|^2 = \beta_1(k) + \beta_2(k) \qquad (23.47)$$

the output vector eventually is

$$\mathbf{y}_3(k) = \|\mathbf{H}(k)\|_F^2 \, \mathbf{s}(k) + \mathbf{U}_1^H(k)\mathbf{n}_1(k) + \mathbf{U}_2^H(k)\mathbf{n}_2(k) \qquad (23.48)$$

Consequently, the signal part in $\mathbf{y}_3(k)$ only vanishes, if all four channel coefficient are zero at the same time and the resilience against fading is significantly increased using a second receive antenna. If the channel coefficients in (23.42) are normalized such that $\beta_1 = \beta_2 = 1$, then the noise part in the receive signal (23.48) is not enhanced by the receiver matrices.

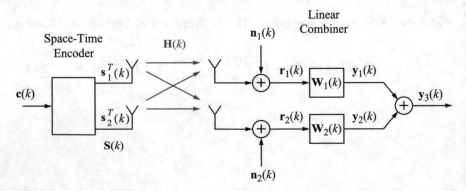

Fig. 23.4 Block diagram of 2×2 MIMO transmission scheme with Alamouti space-time encoder

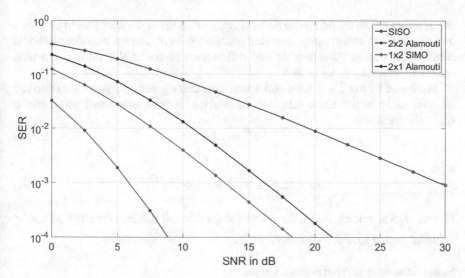

Fig. 23.5 Symbol error rate (SER) as a function of the signal-to-noise ratio (SNR) for the Alamouti space-time coding scheme with one and two receive antennas. A frequency flat channel with Rayleigh fading and 4-PSK are used. For comparison, a 1×1 (SISO) and a 1×2 SIMO system without precoding are also shown

Figure 23.5 shows the symbol error rate (SER) as a function of the signal-to-noise ratio (SNR). An i.i.d. channel model with Rayleigh fading, frequency flat spectrum, and 4-PSK are used. Apparently, the 2×1 Alamouti precoding with one receive antenna (2nd curve from top) outperforms an SISO system (1st curve from top). As expected, a further improvement is achieved using the Alamouti precoding with two receive antennas (2×2 Alamouti, 4th curve from top). For comparison, an 1×2 SIMO scheme (3rd curve fom top) without precoding and two receive antennas is also given providing respectable performance as well.

23.3.3   *Orthogonal Space-Time Block Codes for More Than Two Transmit Antennas*

This mathematically demanding subject goes back to [9]. We introduce the shorthand notation for the input QAM symbols in (23.1) as

$$c(k' + l) = c_{l+1} \; ; \; l = 0, 1, \ldots, N_c - 1 \qquad (23.49)$$

For linear space-time block codes, the entries of the $M \times L$ space-time coding matrix **S** are composed of $\pm c_l$ and $\pm c_l^*$ as with the Alamouti encoder and $\pm \mathrm{Re}\,[c_l]$, $\pm \mathrm{Im}\,[c_l]$ as well as linear combinations thereof, $l = 0, 1, \ldots, N_c - 1$. Major design targets are as follows: The decoding should be simple, i.e., a linear decoder should employ a unitary matrix to avoid matrix inversion processing. Furthermore, the scheme should achieve full transmit diversity M and full receive diversity N resulting in a total maximum diversity order of MN.

Motivated by the 2×2 Alamouti space-time coding matrix, a generalization can be done using $M \times L$ space-time coding matrices $\mathbf{S}$ with orthogonal rows, which fulfill the condition

$$\mathbf{S}\mathbf{S}^H = \alpha \mathbf{I}_M \tag{23.50}$$

with

$$\alpha = |c_1|^2 + |c_2|^2 + \cdots + |c_{N_c}|^2 \tag{23.51}$$

Hence, $\frac{1}{\sqrt{\alpha}}\mathbf{S}$ is unitary. In [9], the authors show that full transmit diversity M can be achieved with a spatial code rate $r_s = \frac{N_c}{L}$.

Some Examples of Space-Time Codes

Without proof, we review some important space-time coding matrices with complex entries, which meet the above requirements [4, 9].

- Space-time coding matrix for $M = 2$ and $r_s = 1$ (Alamouti scheme)

$$\mathbf{S} = \begin{pmatrix} c_1 & -c_2^* \\ c_2 & c_1^* \end{pmatrix} = \begin{pmatrix} \mathbf{s}_1 & \mathbf{s}_2 \end{pmatrix} \tag{23.52}$$

Two input symbols c_1, c_2 are mapped to two output space-time symbol vectors $\mathbf{s}_1, \mathbf{s}_2$. As there are $M = 2$ transmit antennas, $\mathbf{S}$ has two rows. The input and output clock rates are equal, thus there is no bandwidth increase. The mapping by the space-time encoder cannot start until the last QAM symbol c_2 is cached at the input. Consequently, the output signal has a delay of two time intervals.

- Space-time coding matrix for $M = 3$ and $r_s = \frac{1}{2}$

$$\mathbf{S} = \begin{pmatrix} c_1 & -c_2 & -c_3 & -c_4 & c_1^* & -c_2^* & -c_3^* & -c_4^* \\ c_2 & c_1 & c_4 & -c_3 & c_2^* & c_1^* & c_4^* & -c_3^* \\ c_3 & -c_4 & c_1 & c_2 & c_3^* & -c_4^* & c_1^* & c_2^* \end{pmatrix}$$

$$= \begin{pmatrix} \mathbf{s}_1 & \mathbf{s}_2 & \mathbf{s}_3 & \mathbf{s}_4 & \mathbf{s}_5 & \mathbf{s}_6 & \mathbf{s}_7 & \mathbf{s}_8 \end{pmatrix} \tag{23.53}$$

Four input symbols, $c_1, c_2, \ldots, c_4$ are mapped to eight output space-time symbol vectors $\mathbf{s}_1, \mathbf{s}_2, \ldots, \mathbf{s}_8$. The matrix has three rows, because of $M = 3$. The output speed is increased by factor 2, also the required bandwidth. The encoder has a delay of four input time intervals. The second half of $\mathbf{S}$ is the conjugate complex of the first half.

- Space-time coding matrix for $M = 4$ and $r_s = \frac{1}{2}$

$$
\mathbf{S} = \begin{pmatrix}
c_1 & -c_2 & -c_3 & -c_4 & c_1^* & -c_2^* & -c_3^* & -c_4^* \\
c_2 & c_1 & c_4 & -c_3 & c_2^* & c_1^* & c_4^* & -c_3^* \\
c_3 & -c_4 & c_1 & c_2 & c_3^* & -c_4^* & c_1^* & c_2^* \\
c_4 & c_3 & -c_2 & c_1 & c_4^* & c_3^* & -c_2^* & c_1^*
\end{pmatrix}
$$

$$
= \begin{pmatrix} \mathbf{s}_1 & \mathbf{s}_2 & \mathbf{s}_3 & \mathbf{s}_4 & \mathbf{s}_5 & \mathbf{s}_6 & \mathbf{s}_7 & \mathbf{s}_8 \end{pmatrix} \tag{23.54}
$$

Four input symbols, $c_1, c_2, \dots, c_4$ are mapped to eight output space-time symbol vectors $\mathbf{s}_1, \mathbf{s}_2, \dots, \mathbf{s}_8$. The matrix has four rows because of $M = 4$. The output speed is increased by factor 2 and also the required bandwidth. The output signal has a delay of four input time intervals. The second half of $\mathbf{S}$ is the conjugate complex of the first half.

- Space-time coding matrix for $M = 3$ and $r_s = \frac{3}{4}$

$$
\mathbf{S} = \begin{pmatrix}
c_1 & -c_2^* & \frac{\sqrt{2}}{2}c_3 & \frac{\sqrt{2}}{2}c_3 \\
c_2 & c_1^* & \frac{\sqrt{2}}{2}c_3^* & -\frac{\sqrt{2}}{2}c_3^* \\
\frac{\sqrt{2}}{2}c_3 & \frac{\sqrt{2}}{2}c_3 & -\mathrm{Re}[c_1] + j\mathrm{Im}[c_2] & \mathrm{Re}[c_2] + j\mathrm{Im}[c_1]
\end{pmatrix}
$$

$$
= \begin{pmatrix} \mathbf{s}_1 & \mathbf{s}_2 & \mathbf{s}_3 & \mathbf{s}_4 \end{pmatrix} \tag{23.55}
$$

Three input symbols c_1, c_2, c_3 are mapped to four output space-time symbol vectors $\mathbf{s}_1, \mathbf{s}_2, \dots, \mathbf{s}_4$. The matrix has three rows, because of $M = 3$. The output speed is increased by factor $\frac{4}{3}$, also the required bandwidth. The mapping cannot start until the last QAM symbol c_3 is cached at the input. Consequently, the output signal has a delay of three input time intervals.

- Alternative space-time coding matrix for $M = 3$ and $r_s = \frac{3}{4}$, [10].

$$
\mathbf{S} = \begin{pmatrix}
c_1 & c_2^* & c_3^* & 0 \\
-c_2 & c_1^* & 0 & -c_3^* \\
c_3 & 0 & c_1^* & c_2^*
\end{pmatrix}
$$

$$
= \begin{pmatrix} \mathbf{s}_1 & \mathbf{s}_2 & \mathbf{s}_3 & \mathbf{s}_4 \end{pmatrix} \tag{23.56}
$$

Three input symbols c_1, c_2, c_3 are mapped to four output space-time symbol vectors $\mathbf{s}_1, \mathbf{s}_2, \dots, \mathbf{s}_4$. As $M = 3$, the matrix has three rows. The output speed is increased by factor $\frac{4}{3}$, also the required bandwidth and the output signal has a minimal delay of three input time intervals.

- There exist also space-time coding matrices for real symbols c_k only, which are feasible for amplitude shift keying or real-valued phase shift keying. In case of $M = 2$, the Alamouti matrix (23.52) can be employed by setting $c_i^* = c_i$; $i = 1, 2$. For $M = 3$, the matrix (23.53) is applicable, if all conjugate complex elements are dropped resulting in a 3×4 matrix. Similarly for $M = 4$, (23.54) can provide the solution, if reduced to a 4×4 matrix by deleting the conjugate complex elements.

- For all given space-time coding matrices, it is easy to check that the rows are pairwise orthogonal with property (23.50).

Layered space-time coding is an enhancement of spatial multiplexing. We will discuss such schemes in a separate Sect. 23.5, because they paved the way for the first applications of MIMO systems in wireless communications.

23.4   Principle of Space-Time Trellis Coding

Space-time trellis codes are an extension of conventional trellis codes [11] to MIMO systems. They have been introduced by [12]. Since then a large number of codes and their design have been widely explored and the improved coding gain and spectral efficiency have been demonstrated for fading channels. However, it was also recognized that the decoder is much more complex compared to the orthogonal space-time block codes. A survey is available in [4]. Here, we are outlining the principle by showing an example. In general, the input–output relation is given by a trellis diagram which is a state transition diagram enhanced by a discrete-time coordinate k. As in the case of sequential detection of a data signal, a trellis decoder is also used here employing the Viterbi algorithm. Figure 23.6 shows an example of a space-time trellis encoder with 4-PSK, which provides transmit delay diversity. We maintain the general term QAM symbol also for the PSK symbols. Consecutive pairs of bits ($\kappa = 2$) of the input bit stream $b(n)$ are mapped to one QAM symbol $c(k)$ by the mapper. The space-time encoder allocates to each $c(k)$ a space-time symbol vector $\mathbf{s}(k)=\left(s_1(k) \ s_2(k) \right)^T$. In case of the transmit delay diversity $s_1(k) = c(k)$ and $s_2(k) = c(k - 1)$ hold. As in the same time frame one QAM symbol is going in and one space-time symbol vector is output, the spatial code rate is $r_s = 1$ and consequently for the space-time coding matrix follows $\mathbf{S}(k) = \mathbf{s}(k)$.

We describe the space-time trellis encoder by means of a trellis diagram. To this end, we determine all independent state variables, which are associated with

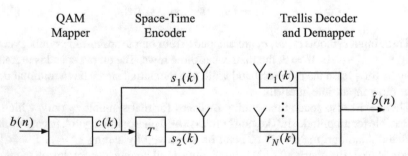

Fig. 23.6 Principle block diagram of a space-time trellis encoder and decoder for delay diversity

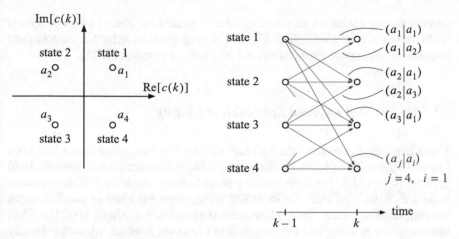

Fig. 23.7 Signal constellation diagram for 4-PSK (left), trellis diagram of space-time trellis encoder (right)

the storage elements. The encoder in Fig. 23.6 owns only one state variable $s_2(k) = c(k - 1)$. With 4-QAM, the symbol alphabet consists of $2^\kappa = 4$ distinct signal points, which we denote as $\{a_1, a_2, a_3, a_4\}$ and which are shown in the complex domain in Fig. 23.7(left). The state variable can take on these 4 values. Consequently, the system can stay in 4 different states, which can be denoted arbitrarily such as state 1, state 2, state 3, and state 4. The state transitions are shown in the trellis diagram depicted in Fig. 23.7(right). Depending on the input signal, the system can remain in the present state or change into another state. At the bottom, the time instants $k - 1$ and k are indicated enhancing the state transition diagram to a trellis diagram. The transitions are labeled as $(c(k) \mid s_2(k))$, where the first element indicates the input QAM symbol and the second element is the output symbol $s_2(k) = c(k - 1)$. Horizontal arrows reveal that the system maintains the present state, if a new symbol comes in, whereas oblique arrows illustrate a change of the state from time instants $k - 1$ to k. Please note that not all transitions are sketched in Fig. 23.7 to avoid an overload of the graph. In our notation of the states and the state variables, the transitions are indicated as $(a_j \mid a_i)$, which means, if the encoder is in state i and the new input signal is a_j, then the encoder moves to state j. The output signals are $s_2(k) = a_i$ and $s_1(k) = a_j$. As already mentioned, the receiver acts as a trellis decoder. Its complexity grows exponentially with the number of states, in our example 2^κ.

23.5 Layered Space-Time Architecture

The layered space-time architecture is based on the demultiplexing of the incoming bit stream into parallel layers, each feeding a transmit antenna. The bit stream can undergo forward error correction and mapping to symbols before or after the demultiplexing operation. In essence, layered space-time coding employs spatial multiplexing as described in Sect. 23.2 and is categorized in vertical, horizontal, and diagonal layered space-time coding, but with some ambiguity, [4, 13].

23.5.1 Vertical Layered Space-Time Coding

This scheme originates from the Bell Laboratories [14] and is also called Bell Labs Layered Space-Time (BLAST). The principle block diagram is straightforward and depicted in Fig. 23.8. The main building block is a demultiplexer $1:M$ as described in Sect. 23.2, which divides the incoming bit sequence $b(n)$ into M parallel output streams. Their bit rates are reduced by factor M compared to the input $b(n)$. The QAM mapper allocates κ−tuples of consecutive bits to QAM symbols, which are directly the output signals $s_1(k), \ldots , s_M(k)$ composing the space-time symbol vector $\mathbf{s}(k)$. In view of the vertical shape of this column vector, which allocates the QAM symbols to each transmit antenna from top to bottom, the scheme is denoted as vertical layered space-time coding or V-BLAST. First versions have even abstained from temporal encoding and have employed the same QAM signal constellations for each layer [14]. If the information bit sequence $b'(n')$ is input to a forward error correction encoder including a temporal interleaver and the code rate between $b'(n')$ and $b(n)$ is r_t, then the overall code rate of the layered space-time transmitter results in

$$r_{total} = r_t \kappa M \qquad (23.57)$$

Obviously, r_{total} can be larger than M. For example, if we take 16-QAM ($\kappa = 4$) and a typical temporal code rate of $r_t = \frac{1}{2}$ for wireless communications the overall code rate is $r_{total} = 2M$. A vertical layered space-time encoder without temporal forward error correction ($r_t = 1$) is just a spatial demultiplexer and thus exhibits no transmit diversity. We then say the transmit diversity order is one. If temporal encoding is used, i.e., $r_t < 1$, the transmit diversity order can exceed one and consequently the total diversity order of the overall system will be larger than the number of receive antennas N.

In Fig. 23.9, an equivalent structure of a V-BLAST encoder is shown. It results from the scheme in Fig. 23.8, if the QAM mapping is moved to the input of the demultiplexer. Figure 23.10 shows a receiver for a MIMO signal with vertical space-time encoding. The first building block can be a linear zero-forcing or a MMSE receiver. The M output signals $y_1, y_2, \ldots , y_M$ enter a spatial multiplexer $M : 1$, which operates as a parallel-to-serial converter providing estimates $\hat{c}(k')$ of the transmit QAM symbols. After demapping, the estimates $\hat{b}(n)$ of the encoded bit stream are available, which are fed into a forward error correction decoder yielding the estimates $\hat{b}'(n')$. Obviously, the output signal of the spatial multiplexer is processed as in case of a digital single input single output (SISO) receiver. Consequently, all advanced techniques can be applied for improved demapping and decoding, such as iterative (Turbo) decoding or soft demapping [15] using the Turbo principle indicated by

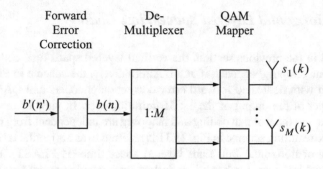

Fig. 23.8 Block diagram of the original vertical layered space-time encoder (V-BLAST)

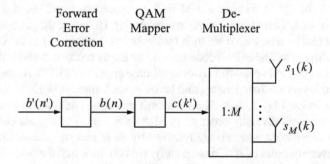

Fig. 23.9 Block diagram of an alternative vertical layered space-time encoder

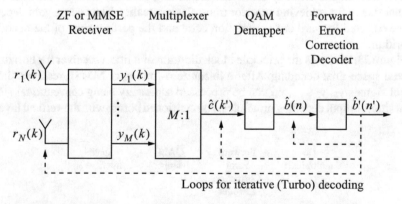

Loops for iterative (Turbo) decoding

Fig. 23.10 Block diagram of a linear receiver for vertical layered space-time decoding

the dashed feedback lines in Fig. 23.10. Alternatively, the linear receiver including the spatial multiplexer can be enhanced by an iterative a-posterior detector for the receive vector $(r_1 \; r_2 \; \ldots \; r_N)^T$. Then, the complexity of the receiver is significantly increased. The original V-BLAST receiver employed the "BLAST algorithm" [14, 16] with ordered successive interference cancellation (OSIC).

23.5.2 *Horizontal Layered Space-Time Coding*

As outlined in the previous section, the vertical layered space-time coder operates
with only one temporal encoder, if at all. Alternatively, the scheme in Fig. 23.8 can
be equipped with dedicated forward error correction encoders and QAM mappers
for each layer at the output of the $1 : M$ demultiplexer. In view of the horizontal
arrangement and the fact that coding and mapping are independent from one layer to
the other, the resulting scheme in Fig. 23.11 is referred to as horizontal layered space-
time coding or horizontal Bell Labs layered space-time (H-BLAST). Please note
that the spatial interleaver indicated by dashed lines is only present for the diagonal
layered space-time encoder described in the next section. Obviously, the temporal
encoder and the QAM mapper are M-fold compared to Fig. 23.9, if M transmit
antennas are used. This is an increase in complexity. However, the processing of the
signals is at clock rates, which are by a factor M lower compared to the input $b'(m)$
of the demultiplexer. The overall code rate can be easily calculated and is the same as
for the vertical layered space-time coding scheme given by (23.57). As the signals in
the different layers are independent and because each antenna sends no other signal
than that of its own layer, there is no transmit diversity available on symbol level.
Hence, the transmit diversity order of H-BLAST is one. The signals coming from
each of the M transmit antennas are received by all N receive antennas resulting in
a receive diversity order of N. Consequently, we conclude that the diversity order of
the horizontal layered space-time coding scheme on a symbol basis cannot be larger
than N. Besides the diversity order, also the coding gain of a transmission scheme is
of importance for achieving superior transmission quality. The coding gain depends
on the selected forward error correction code and the performance of the decoding
algorithm.

Figure 23.12 shows the principle block diagram of a linear receiver for horizontal
layered space-time decoding. After a linear zero-forcing or MMSE receiver, the M
output signals $y_1, y_2, \ldots, y_M$ can be processed separately using conventional single
input single output decoding principles. As mentioned before with the vertical layered

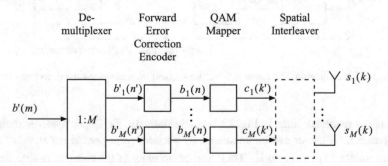

Fig. 23.11 Block diagram of horizontal layered space-time encoding. The spatial interleaver is in
operation only for diagonal layered space-time coding

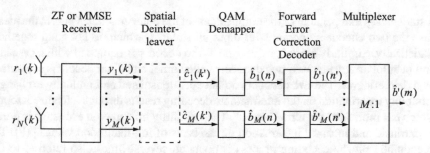

Fig. 23.12 Block diagram of a linear receiver for horizontal layered space-time decoding. The spatial deinterleaver is in operation only for diagonal layered space-time coding

system, also maximum likelihood joint decoding of the components of the receive vector $(r_1 \ r_2 \ \dots \ r_N)^T$ is an alternative here, however, with much higher complexity.

23.5.3 Diagonal Layered Space-Time Coding

This scheme builds upon the horizontal layered space-time system. By inserting a spatial interleaver, as shown by dashed lines in Fig. 23.11, symbols of different layers can be interleaved crosswise. This cross-layer operation has motivated the nomenclature diagonal layered space-time coding or diagonal (D-) BLAST. In [14], a stream rotator with periodic antenna allocation is preferred as an interleaver. If the code words are made large enough, bits or even codewords of each layer are transmitted over all M antennas and the transmit diversity order approaches M. An appropriate receiver with N antennas then can get full diversity order MN of the overall system. The total code rate is the same as in (23.57). The receiver is more complex than that of the horizontal layered system owing to the spatial interleaving. V-, H-, and D-BLAST schemes have been one of the first MIMO systems introduced in the second half of the 1990s starting with zero-forcing or MMSE receivers combined with successive interference cancellation.

23.5.4 Iterative Receivers for Layered Space-Time Systems

Iterative detection using the Turbo principle can significantly increase the performance of a receiver. Turbo codes and their iterative decoding (also referred to as Turbo decoding) have been first introduced by [17] in the framework of two concatenated codes for forward error correction. A comprehensive survey on the Turbo principle is given in [18]. A serially concatenated Turbo coding scheme consists of the cascade of a first encoder (outer encoder), an interleaver, and a second encoder (inner encoder). Ideally, the interleaver generates an input signal for the inner encoder, which is statistically independent of the output signal of the outer encoder. Thus, in the ideal case the two encoders operate with input bit streams, which do not hang together statistically. Similarly at the receiver, the Turbo decoder is equipped with a cascade or a parallel arrangement of two decoders called inner and outer decoders separated by a deinterleaver. The two decoders do not operate isolated and mutually exchange reliability information on the intermediate decoding results through a feedback loop, which has motivated the term "Turbo". The reliability information exchange is done as extrinsic and intrinsic information on the basis of log-likelihood values [19]. To decode one bit the decoding process is operated several times also referred to as decoding iterations. After each iteration, the decoding result has improved, i.e., the mean bit error rate has decreased. Depending on the convergence conditions, the iteration cycle is finally abandoned and the output of the outer decoder, which is an a-posterior probability (APP) decoder, is subject to hard decision and provides the final decoded bit. For APP decoding, the BCJR algorithm invented by [20] is optimal. Owing to its high computational complexity, several suboptimal solutions have been proposed, such as the soft-output Viterbi algorithm [21]. For code design and convergence properties, the EXIT chart has proven to be a powerful method [22–24].

An effective yet not very complex method is a MIMO receiver with soft demapping, as indicated in Fig. 23.10. Only the middle feedback loop depicted by dashed lines is employed. The multiplexed samples $\hat{c}(k')$ are entering the demapper. A conventional demapper would perform a hard decision and would output the associated binary codewords from a look-up table. To apply soft demapping at the receiver, the transmitter is equipped with a forward error correction encoder, which is named as outer encoder and the mapper plays the part of the inner encoder. Likewise for soft demapping at the receiver, the forward error correction decoder operates as outer decoder and the demapper acts as the inner decoder. In contrast to a conventional demapper, the output of the soft demapper are log-likelihood values [19]. Through a recursive loop, the outer decoder provides a-priori information for the soft demapper. After several iterations, the output signal of the outer decoder undergoes a hard decision resulting in the output bits $\hat{b}'(n')$. With an increasing number of iterations the Turbo cliff builds up, which is characterized by a steep decent of the mean bit error probability as a function of the signal-to-noise ratio.

References

1. B. Friedrichs, *Error Control Coding* (Springer, Berlin, 2017)
2. T. Richardson, R. Urbanke, *Modern Coding Theory* (Cambridge University Press, Cambridge, 2008)
3. T.K. Moon, *Error Correction Coding - Mathematical Methods and Algorithms* (Wiley Interscience, New York, 2005)
4. B. Vucetic, J. Yuan, *Space-Time Coding* (Wiley, New York, 2003)
5. S. Alamouti, A simple channel diversity technique for wireless communications. IEEE J. Sel. Areas Commun. **16**, (1998)
6. Physical channels and modulation, Technical Specifications. TS 36.211, V11.5.0, 3GPP, Technical Report, 2012, p. 61, 234

7. D. Schneider, J. Speidel, L. Stadelmeier, D. Schill, A. Schwager, MIMO for inhome power line communications, in *ITG Fachberichte International Conference on Source and Channel Coding (SCC)* (2008)
8. L.T. Berger, A. Schwager, P. Pagani, D. Schneider, *MIMO Power Line Communications - Narrow and Broadband Standards, EMC and Advanced Processing* (CRC press, Boca Raton, 2014)
9. V. Tarokh, H. Jafarkhani, A. Calderbank, Space-time block codes from orthogonal designs. IEEE Trans. Inf. Theory **45**, 1456–1467 (1999)
10. B. Hochwald, T.L. Marzetta, C.B. Papadias, A transmitter diversity scheme for wideband CDMA systems based on space-time spreading. IEEE J. Sel. Areas Commun. **19**, 48–60 (2001)
11. E. Biglieri, D. Divsalar, P. McLane, M. Simon, *Introduction to Trellis-coded Modulation with Applications* (Macmillan, New York, 1991)
12. V. Tarokh, N. Seshadri, A. Calderbank, Space-time codes for high data rate wireless communication: performance criterion and code construction. IEEE Trans. Inf. Theory **44**, 744–765 (1998)
13. A. Paulraj, R. Nabar, D. Gore, *Introduction to Space-time Wireless Communications* (Cambridge University Press, Cambridge, 2003)
14. G. Foschini, Layered space-time architecture for wireless communication in a fading environment when using multi-element antennas. Bell Syst. Tech. J. **1**, 41–59 (1996)
15. S. Ten Brink, J. Speidel, R.-H. Yan, Iterativeerative demapping for QPSK modulation. Electron. Lett. **34**, 1459–1460 (1998)
16. G. Foschini, M. Gans, On limits of wireless communications in a fading environment when using multiple antennas. IEEE Wirel. Pers. Commun. **6**, 311–335 (1998)
17. C. Berrou, A. Glavieux, P. Thitimajshima, Near Shannon limit error-correcting coding and decoding, in *International Conference on Communications ICC* (1993)
18. J. Hagenauer, The Turbo principle: tutorial introduction and state of the art, in *Proceedings of 1st International Symposium on Turbo codes* (1997)
19. J. Hagenauer, E. Offer, L. Papke, Iterative decoding of binary block and convolutional codes. IEEE Trans. Inf. Theory **42**, 429–445 (1996)
20. L. Bahl, J. Cocke, F. Jelinek, J. Raviv, Optimal decoding of linear codes for minimizing symbol error rate. IEEE Trans. Inf. Theory **20**, 284–287 (1974)
21. J. Hagenauer, P. Hoeher, A Viterbi algorithm with soft-decision outputs and its applications, in *IEEE International Conference on Global Communications (GLOBECOM)* (1989)
22. S. Ten Brink, Convergence of iterative decoding. Electron. Lett. **35**, 806–808 (1999)
23. S. Ten Brink, Designing iterative decoding schemes with the extrinsic information transfer chart. AEÜ Int. J. Electron. Commun. **54**, 389–398 (2000)
24. S. Ten Brink, Design of concatenated coding schemes based on iterative decoding convergence, Ph.D. dissertation, University of Stuttgart, Institute of Telecommunications, Shaker Publication, ISBN 3-8322-0684-1 (2001)

Chapter 24
Principles of Multi-user MIMO Transmission

24.1 Introduction

Hitherto, we have considered the MIMO transmission between a base station and one user, which we call single-user MIMO transmission. In a communications network, the base station has to serve a large number of users, e.g., in an in-house area with a wireless local area networks (WLAN) according to the standard IEEE 802.11. Also in outdoor scenarios, a multitude of users has to be addressed with the cellular networks of the type 3G (year 2004), 4G (year 2010), and 5G (year 2020) with data rates of about 8 Mbit/s, 100 Mbit/s, and up to 1 Gbit/s, respectively. In this chapter, we investigate methods for data communication between the base station and the users. Each transmitter and receiver shall be equipped with multiple antennas to benefit from the MIMO principle, which we call multi-user MIMO (MU MIMO) transmission. Also in the case that each user equipment has only one antenna, the term MIMO is used, because the antennas of all users taken together are considered as multiple input or multiple output. We differentiate between the directions from the base station to the users called the downlink and between the link from the users to the base station, which is referred to as uplink. Conventionally, without the MIMO principle, the base station allocates certain time slots or frequency bands to the users. With multiple antennas a multi-user MIMO scheme can serve all users at the same time and in the same frequency band, hence providing a higher efficiency. While the WLAN standards IEEE 802.11 a, b, g, and n do not support multi-user MIMO techniques, the later versions AC-WLAN or AC Wave 2 own these benefits. However, the transmission of different signals in the same frequency band and in the same time slots in the downlink gives rise to interference at the user terminals. Of course, a similar situation will occur in the uplink, when many users are going to address the same base station. Hence, an important question for the system designer is how to overcome or at least minimize the impact of the interference. Several methods for multi-user MIMO systems are described in [1, 2] as a survey. In the next sections, we outline some selected principles in more detail. On top of these methods, protocols of the higher system layers, such as the data link and the network layer,

© Springer Nature Switzerland AG 2019
J. Speidel, *Introduction to Digital Communications*, Signals and Communication
Technology, https://doi.org/10.1007/978-3-030-00548-1_24

can reduce residual interference by proper scheduling of the user access. However, this is beyond the scope of this chapter.

24.2 Precoding for Multi-user MIMO Downlink Transmission

24.2.1 Precoding by "Channel Inversion"

As the base station transmits signals to the various users in the same frequency band and at the same time, we have to make sure that each user gets the individually devoted signal without interference from the other users. Hence, for each user terminal the *multi-user interference* also called *multi-access interference* has to be minimized. Several schemes operate with precoding at the transmitter to keep the complexity at the cost-sensitive user equipment low. In Chap. 22, linear precoding is discussed in quite some detail as an alternative to a linear receiver for the reduction of the interchannel interference for single-user MIMO systems. The prefilter for multi-user applications, which will be outlined in the following, shall provide a reduction of the multi-user interference at all receive terminals. We will see that even the interchannel interference can be canceled by some of these methods.

The principle block diagram is depicted in Fig. 24.1. Apparently, the downlink transmission can be regarded as a broadcast or point-to-multipoint mode, in which the base station transmits signals to all users. Hence, the combined channel with matrix $\mathbf{H}(k)$ is also called broadcast channel. Let us start with the receiver, where U user terminals also denoted as user equipment and indicated by the index $u = 1, 2, ..., U$ have to be served.

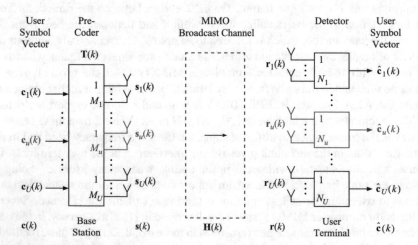

Fig. 24.1 Block diagram of a multi-user MIMO downlink transmission between a base station and various user terminals

The user terminal u shall be equipped with N_u receive antennas and its receive signal is denoted as $\mathbf{r}_u(k) \in \mathbb{C}^{N_u \times 1}$. In the following, we drop the discrete-time variable k to simplify the notation. The combined receive column vector can be written as

$$\mathbf{r} = \begin{pmatrix} \mathbf{r}_1 \\ \vdots \\ \mathbf{r}_u \\ \vdots \\ \mathbf{r}_U \end{pmatrix} \in \mathbb{C}^{N \times 1} \tag{24.1}$$

and the total number of receive antennas for all users shall be

$$N = \sum_{u=1}^{U} N_u \tag{24.2}$$

Similarly, the additive noise vector for the user terminal u is $\mathbf{n}_u \in \mathbb{C}^{N_u \times 1}$ and the combined noise vector is defined as

$$\mathbf{n} = \begin{pmatrix} \mathbf{n}_1 \\ \vdots \\ \mathbf{n}_u \\ \vdots \\ \mathbf{n}_U \end{pmatrix} \in \mathbb{C}^{N \times 1} \tag{24.3}$$

At the transmitter side the base station has structured its MIMO antennas into U groups, where the group u owns M_u antenna elements and the associated transmit signal vector is $\mathbf{s}_u \in \mathbb{C}^{M_u \times 1}$, $u = 1, 2, ..., U$. Then the total number of transmit antenna elements at the base station is

$$M = \sum_{u=1}^{U} M_u \tag{24.4}$$

We assume a fading channel with frequency flat spectrum defined by an $N \times M$ matrix $\mathbf{H} \in \mathbb{C}^{N \times M}$ and introduce the combined vector of all transmit signals

$$\mathbf{s} = \begin{pmatrix} \mathbf{s}_1 \\ \vdots \\ \mathbf{s}_u \\ \vdots \\ \mathbf{s}_U \end{pmatrix} \in \mathbb{C}^{M \times 1} \tag{24.5}$$

The input signals to the precoder are the QAM symbols devoted to the various users and are also structured as vectors $\mathbf{c}_u \in \mathbb{C}^{N_u \times 1}$, $u = 1, 2, ..., U$. Hence, $\mathbf{c}_u$ represents the vector of N_u streams of QAM symbols entering the precoder. Then the combined input signal vector of the precoder can be defined as

$$\mathbf{c} = \begin{pmatrix} \mathbf{c}_1 \\ \vdots \\ \mathbf{c}_u \\ \vdots \\ \mathbf{c}_U \end{pmatrix} \in \mathbb{C}^{N \times 1} \tag{24.6}$$

Consequently, as portrait in Fig. 24.1 the precoding matrix $\mathbf{T}$ must have the dimension $M \times N$ with in general complex entries. $\mathbf{T} \in \mathbb{C}^{M \times N}$ maps the input signal vector $\mathbf{c}$ to the output vector $\mathbf{s}$ yielding

$$\mathbf{s} = \mathbf{Tc} \tag{24.7}$$

Please note at the moment no individual mapping of an input $\mathbf{c}_u$ to an output $\mathbf{s}_j$ is required. We obtain for the receive signal vector

$$\mathbf{r} = \mathbf{Hs} + \mathbf{n} \tag{24.8}$$

Plugging (24.7) into (24.8) yields

$$\mathbf{r} = \mathbf{HTc} + \mathbf{n} \tag{24.9}$$

As described in Sect. 22.2, the zero-forcing and the MMSE precoder with matrix

$$\mathbf{T} = \mathbf{H}^{++} = \mathbf{H}^H \left(\mathbf{HH}^H \right)^{-1} ; \quad M \geq N \tag{24.10}$$

completely removes the interference, where $\mathbf{H}^{++}$ is the pseudo-inverse matrix of $\mathbf{H}$. Please note that the channel matrix $\mathbf{H}$ must have full rank N otherwise the inverse in (24.10) does not exist. Hence, the total number M of transmit antennas at the base station must be equal to or larger than the sum N of the receive antennas of all user equipment

$$\sum_{u=1}^{U} M_u \geq \sum_{u=1}^{U} N_u \tag{24.11}$$

With

$$\mathbf{HT} = \mathbf{HH}^{++} = \mathbf{I}_N \tag{24.12}$$

follows the receive signal

$$\mathbf{r} = \mathbf{I}_N \mathbf{c} + \mathbf{n} = \mathbf{c} + \mathbf{n} \tag{24.13}$$

and the multi-user interference as well as the interchannel interference are completely canceled. The right-hand side of (24.13) can be interpreted as the diagonalization of the system of equations (24.9), where the diagonal matrix is just the identity matrix $\mathbf{I}_N$. From (24.13) the individual signal for the user u can be decomposed with the help of (24.1), (24.3), and (24.6) as

$$\mathbf{r}_u = \mathbf{c}_u + \mathbf{n}_u \; ; \; u = 1, 2, ..., U \tag{24.14}$$

showing that the noise $\mathbf{n}_u$ remains as the only impairment for the receive signal. (24.14) requires that the dimension of $\mathbf{c}_u$ equals the dimension of $\mathbf{r}_u$. Consequently, the number N_u of signal components constituting the transmit symbol vector $\mathbf{c}_u \in \mathbb{C}^{N_u \times 1}$ determines the number N_u of receive antennas for the user u, which is a prerequisite of our derivation. Apparently, the base station devotes N_u parallel symbol streams to user u. Furthermore, we recognize that no individual mapping of an input $\mathbf{c}_u$ to an output $\mathbf{s}_j$ is required at the transmitter.

Now, consider an application, in which only one user u is part of the system. Then $M = M_u$ and $N = N_u$ and from the condition $M \geq N$ in (24.11) follows

$$M_u \geq N_u \tag{24.15}$$

This scenario can occur for any terminal. Therefore, (24.15) must hold for all users $u = 1, 2, ..., U$.

As shown in Fig. 24.1, the last stage is the detector at the user terminals, in many cases a maximum likelihood detector, which takes $\mathbf{r}_u$ as the input and decides for the most likely QAM symbol vector $\hat{\mathbf{c}}_u$. Of course, if the channel matrix is a square matrix $\mathbf{H} \in \mathbb{C}^{N \times N}$ with full rank N the pseudo-inverse matrix $\mathbf{H}^{++}$ turns into the inverse matrix resulting in $\mathbf{T} = \mathbf{H}^{-1}$. This is the reason why the described method for multi-user interference reduction is also called *channel inversion*.

Furthermore, for all precoding techniques discussed in this chapter, the transmitter must have the full knowledge about the channel matrix $\mathbf{H}$. Hence, channel estimation at the receiver has to be performed and the entries of the estimated matrix are then sent as channel state information (CSI) to the transmitter via a separate feedback loop. Therefore, this technique is called *closed-loop precoding*. Depending on the number of users and antennas per user terminal the dimension of $\mathbf{H}$ can be reasonably large and the calculation of the inverse or pseudo-inverse becomes expensive. In the next section, a transmitter consisting of an individual precoder per user is discussed.

Example 1

Consider the multi-user downlink with $U = 3$ user terminals. User 1 will be provided with two and the other two users with one symbol stream. The channel matrix $\mathbf{H}$ is estimated at the receivers and in a small time interval

$$\mathbf{H} = \begin{pmatrix} 1 & 0.2 & -0.1 & 0.1 & 0 \\ -0.2 & 1 & 0.1 & 0 & 0.1 \\ 0.1 & -0.1 & 1 & 0.2 & -0.1 \\ 0 & -0.1 & -0.2 & 1 & 0.2 \end{pmatrix} \tag{24.16}$$

is given. For simplicity, we assume real matrix entries. Precoding with a matrix $\mathbf{T}$ shall be applied to minimize the multi-user interference. Find $\mathbf{T}$ using the method of "channel inversion".

Solution:

From the channel matrix we conclude $M = 5$ and $N = 4$. Furthermore, $N_1 = 2$ and $N_2 = N_3 = 1$ are given. Thus, in Fig. 24.1 $\mathbf{c}_1$ and $\mathbf{r}_1$ are 2x1 vectors whereas c_2, c_3, r_2, and r_3 are scalars. Apparently, $\mathbf{H}$ is non-quadratic, and hence, no inverse matrix exists. Therefore, we calculate the precoding matrix as the pseudo-inverse (24.10) of $\mathbf{H}$. The result is

$$
\mathbf{T} = \mathbf{H}^H \left(\mathbf{H}\mathbf{H}^H \right)^{-1} = \begin{pmatrix} 0.95 & -0.19 & 0.09 & -0.10 \\ 0.20 & 0.94 & -0.07 & -0.02 \\ -0.08 & 0.11 & 0.94 & -0.16 \\ 0.01 & 0.09 & 0.20 & 0.93 \\ -0.02 & 0.11 & -0.09 & 0.19 \end{pmatrix} ; \quad \mathbf{HT} = \begin{pmatrix} 1 & 0 & 10^{-4} & 0 \\ 0 & 1 & 0 & 10^{-4} \\ 0 & 0 & 1,0001 & 0 \\ 0 & 10^{-4} & 0 & 1 \end{pmatrix}
$$

(24.17)

Apparently, $\mathbf{HT}$ is a good approximation of the identity matrix $\mathbf{I}_4$. Then follows from (24.9) with $\mathbf{c}_1 = \left(c_{11} \; c_{12} \right)^T$, $\mathbf{r}_1 = \left(r_{11} \; r_{12} \right)^T$, and $\mathbf{n}_1 = \left(n_{11} \; n_{12} \right)^T$

$$
\begin{pmatrix} r_{11} \\ r_{12} \\ r_2 \\ r_3 \end{pmatrix} = \begin{pmatrix} c_{11} \\ c_{12} \\ c_2 \\ c_3 \end{pmatrix} + \begin{pmatrix} n_{11} \\ n_{12} \\ n_2 \\ n_3 \end{pmatrix}
$$

(24.18)

where the elements with double indices belong to user 1. It is well appreciated that the multi-user as well as the interchannel interference are completely removed and the multi-user downlink turns into three independent single-user links with receive signals $\mathbf{r}_1$, r_2, and r_3.

24.2.2   Precoding with Block Diagonalization

Input–Output Relation of the Downlink

An interesting closed-form method to cancel the multi-user interference at each user terminal in the downlink is the design of a precoding matrix $\mathbf{T}$ in such a way that the interference term $\mathbf{HT}$ in (24.9) becomes a block diagonal matrix. According to linear algebra a matrix $\mathbf{G}$ is said to be block diagonal, if

$$
\mathbf{G} = \text{diag} \left(\mathbf{G}_1 \mathbf{G}_2 \ldots \mathbf{G}_U \right)
$$

(24.19)

holds, where the $\mathbf{G}_i$ are matrices with smaller dimensions than the dimension of $\mathbf{G}$. We consider the block diagram of the multi-user MIMO downlink in Fig. 24.2. The receiver side with the U user terminals is the same as in Fig. 24.1. How-

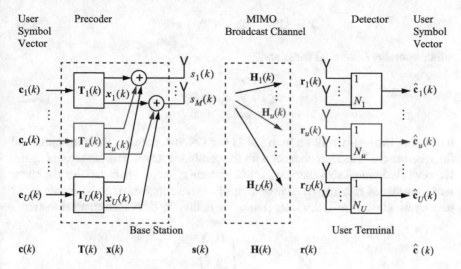

Fig. 24.2 Multi-user MIMO downlink transmission between a base station with individual precoders and various user terminals

ever, at the transmitter the base station employs individual precoders with matrices $\mathbf{T}_u \in \mathbb{C}^{M \times \zeta_u}$, $u = 1, 2, ..., U$.

The input signals at the base station are $\mathbf{c}_u \in \mathbb{C}^{\zeta_u \times 1}$, $u = 1, 2, ..., U$, where $\mathbf{c}_u$ represents the column vector composed of ζ_u QAM symbols. Obviously,

$$\zeta_u \leq N_u , \quad u = 1, 2, ..., U \tag{24.20}$$

guarantees that all transmit symbols for user u are covered by the receive antennas of this user. The precoder output vector is $\mathbf{x}_u \in \mathbb{C}^{M \times 1}$

$$\mathbf{x}_u = \mathbf{T}_u \mathbf{c}_u ; \quad u = 1, 2, ..., U \tag{24.21}$$

where M is the number of transmit antennas at the base station. The U precoder outputs are added component by component yielding the transmit signal vector

$$\mathbf{s} = \sum_{u=1}^{U} \mathbf{x}_u = \sum_{u=1}^{U} \mathbf{T}_u \mathbf{c}_u \tag{24.22}$$

with

$$\mathbf{s} = \left(s_1 \cdots s_M \right)^T \in \mathbb{C}^{M \times 1} \tag{24.23}$$

Apparently, all $\mathbf{x}_u$ must have the same number M of components; otherwise, they cannot be added. As a consequence, $\mathbf{T}_u \in \mathbb{C}^{M \times \zeta_u}$, $u = 1, 2, ..., U$ holds. Furthermore, we can define the combined precoding matrix

$$\mathbf{T} = \begin{pmatrix} \mathbf{T}_1 & \cdots & \mathbf{T}_u & \cdots & \mathbf{T}_U \end{pmatrix} \in \mathbb{C}^{M \times \zeta} \tag{24.24}$$

which contains U vertical slices and

$$\zeta = \sum_{u=1}^{U} \zeta_u \leq N \tag{24.25}$$

holds. The inequality follows from (24.2) and (24.20). Between the transmitter and the receiver the broadcast channel with the combined matrix $\mathbf{H}(k)$ is present again. However, individual sub-matrices $\mathbf{H}_u \in \mathbb{C}^{N_u \times M}$ are effective characterizing the channels from the M transmit antenna outputs of the base station to the N_u inputs of the user terminal u, $u = 1, 2, ..., U$. Hence, we define the combined channel matrix

$$\mathbf{H} = \begin{pmatrix} \mathbf{H}_1 \\ \vdots \\ \mathbf{H}_u \\ \vdots \\ \mathbf{H}_U \end{pmatrix} \tag{24.26}$$

and recognize that the sub-matrices $\mathbf{H}_u$ are horizontal slices of $\mathbf{H}$. Next, we determine the receive signals. The signal vector for user u is

$$\mathbf{r}_u = \mathbf{H}_u \mathbf{s} + \mathbf{n}_u \tag{24.27}$$

and with (24.22) follows

$$\mathbf{r}_u = \mathbf{H}_u \mathbf{T}_u \mathbf{c}_u + \mathbf{H}_u \sum_{\substack{i=1 \\ i \neq u}}^{U} \mathbf{T}_i \mathbf{c}_i + \mathbf{n}_u \ ; \ u = 1, 2, ..., U \tag{24.28}$$

where $\mathbf{n}_u$ is the additive noise at the user terminal u. We have decomposed $\mathbf{r}_u$ already into the desired receive signal $\mathbf{H}_u \mathbf{T}_u \mathbf{c}_u$ and the multi-user interference for user u

$$\mathbf{H}_u \sum_{\substack{i=1 \\ i \neq u}}^{U} \mathbf{T}_i \mathbf{c}_i \ ; \ u = 1, 2, ..., U \tag{24.29}$$

With the combined vectors (24.1), (24.6) and the matrices (24.24), (24.26) follows the matrix notation

$$\mathbf{r} = \mathbf{HTc} + \mathbf{n} = \begin{pmatrix} \mathbf{H}_1 \\ \vdots \\ \mathbf{H}_u \\ \vdots \\ \mathbf{H}_U \end{pmatrix} (\mathbf{T}_1 \cdots \mathbf{T}_u \cdots \mathbf{T}_U) \mathbf{c} + \mathbf{n} \qquad (24.30)$$

We can execute $\mathbf{HT}$ just like the multiplication of a column vector with a row vector using the block matrices as entries and obtain

$$\begin{pmatrix} \mathbf{r}_1 \\ \vdots \\ \mathbf{r}_u \\ \vdots \\ \mathbf{r}_U \end{pmatrix} = \begin{pmatrix} \mathbf{H}_1\mathbf{T}_1 & \cdots & \mathbf{H}_1\mathbf{T}_u & \cdots & \mathbf{H}_1\mathbf{T}_U \\ & \ddots & & \ddots & \\ \mathbf{H}_u\mathbf{T}_1 & \cdots & \mathbf{H}_u\mathbf{T}_u & \cdots & \mathbf{H}_u\mathbf{T}_U \\ & \ddots & & \ddots & \\ \mathbf{H}_U\mathbf{T}_1 & \cdots & \mathbf{H}_U\mathbf{T}_u & \cdots & \mathbf{H}_U\mathbf{T}_U \end{pmatrix} \begin{pmatrix} \mathbf{c}_1 \\ \vdots \\ \mathbf{c}_u \\ \vdots \\ \mathbf{c}_U \end{pmatrix} + \begin{pmatrix} \mathbf{n}_1 \\ \vdots \\ \mathbf{n}_u \\ \vdots \\ \mathbf{n}_U \end{pmatrix} \qquad (24.31)$$

Apparently, $\mathbf{HT} \in \mathbb{C}^{U \times U}$ is a block matrix composed of sub-matrices or blocks $\mathbf{H}_i\mathbf{T}_j \in \mathbb{C}^{N_i \times \zeta_j}$.

Cancelation of the Multi-user Interference

Regarding (24.31) the multi-user interference is given by the blocks $\mathbf{H}_i\mathbf{T}_j$ with unequal indices $i \neq j$. If we can find precoding matrices, which fulfill the condition

$$\mathbf{H}_i\mathbf{T}_j = \mathbf{0} \; ; \; i, j = 1, 2, ..., U \; ; \; i \neq j \qquad (24.32)$$

where $\mathbf{0}$ is the null matrix, then (24.31) results in

$$\begin{pmatrix} \mathbf{r}_1 \\ \vdots \\ \mathbf{r}_u \\ \vdots \\ \mathbf{r}_U \end{pmatrix} = \begin{pmatrix} \mathbf{H}_1\mathbf{T}_1 & \cdots & \mathbf{0} & \cdots & \mathbf{0} \\ & \ddots & & \ddots & \\ \mathbf{0} & \cdots & \mathbf{H}_u\mathbf{T}_u & \cdots & \mathbf{0} \\ & \ddots & & \ddots & \\ \mathbf{0} & \cdots & \mathbf{0} & \cdots & \mathbf{H}_U\mathbf{T}_U \end{pmatrix} \begin{pmatrix} \mathbf{c}_1 \\ \vdots \\ \mathbf{c}_u \\ \vdots \\ \mathbf{c}_U \end{pmatrix} + \begin{pmatrix} \mathbf{n}_1 \\ \vdots \\ \mathbf{n}_u \\ \vdots \\ \mathbf{n}_U \end{pmatrix} \qquad (24.33)$$

and $\mathbf{HT}$ becomes a block diagonal matrix of the type (24.19). Of course, for the precoding matrices $\mathbf{T}_j \neq \mathbf{0} \, \forall \, j$ must hold, otherwise the transmit signals will be zero. The condition (24.32) turns the multi-user downlink into a parallel structure of decoupled single-user MIMO links

$$\mathbf{r}_u = \mathbf{H}_u\mathbf{T}_u\mathbf{c}_u + \mathbf{n}_u \; , \; u = 1, 2, ..., U \qquad (24.34)$$

as depicted in Fig. 24.3. Then

$$\mathbf{r} = \text{diag} \, (\mathbf{H}_1\mathbf{T}_1, ..., \mathbf{H}_u\mathbf{T}_u, ..., \mathbf{H}_U\mathbf{T}_U) \mathbf{c} + \mathbf{n} \qquad (24.35)$$

Fig. 24.3 Multi-user MIMO
downlink decoupled into U
parallel single-user links
with block diagonalization.
The precoding matrices are
given by $\mathbf{T}_u = \mathbf{V}_u \mathbf{A}_u$, $u =$
$1, 2, ..., U$ in (24.41)

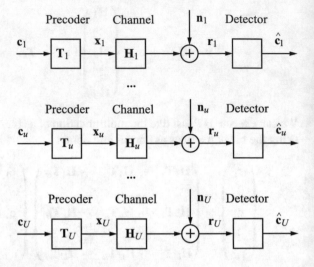

looks similar to an eigenmode transmission scheme. The interference defined in
(24.29) is zero for arbitrary transmit symbol vectors $\mathbf{c}_1, ..., \mathbf{c}_U$. The interesting question now is how the precoder matrices can be found.

Determining the Precoder Matrices for Block Diagonalization

In principle, we have to look for a linear mapping with a matrix $\mathbf{T}$, which transforms
$\mathbf{H}$ into the block diagonal form. For the multi-user MIMO downlink the solution was
first described in [1, 3–5]. We restrict ourselves to the principal computational steps.
The precoding matrix $\mathbf{T}_u$ for user u is the solution of the system of equations (24.32)
and formally given by

$$\mathbf{T}_u = \arg \begin{bmatrix} \mathbf{H}_1 \mathbf{T}_u = \mathbf{0} \\ \vdots \\ \mathbf{H}_{u-1} \mathbf{T}_u = \mathbf{0} \\ \mathbf{H}_{u+1} \mathbf{T}_u = \mathbf{0} \\ \vdots \\ \mathbf{H}_U \mathbf{T}_u = \mathbf{0} \end{bmatrix} \tag{24.36}$$

From $\mathbf{H}_i \mathbf{T}_u = \mathbf{0}$, $i = 1, 2, ..., U$, $i \neq u$ follows that each column vector of $\mathbf{T}_u$ must
be orthogonal to all row vectors of all $\mathbf{H}_i$, $i \neq u$. The solution is obtained from
a singular value decomposition of the matrix $\mathbf{H}_u^-$, which contains all sub-channel
matrices except $\mathbf{H}_u$

$$\mathbf{H}_u^- = \begin{pmatrix} \mathbf{H}_1 \\ \vdots \\ \mathbf{H}_{u-1} \\ \mathbf{H}_{u+1} \\ \vdots \\ \mathbf{H}_U \end{pmatrix} = \mathbf{U}_u \mathbf{D}_u \left(\tilde{\mathbf{V}}_u \ \mathbf{V}_u \right)^H \ ; \ \mathbf{D}_u = \begin{pmatrix} \Lambda_u^{\frac{1}{2}} & \mathbf{0} \\ \mathbf{0} & \mathbf{0} \end{pmatrix} \tag{24.37}$$

With $\mathbf{H}_u \in \mathbb{C}^{N_u \times M}$ we conclude that $\mathbf{H}_u^- \in \mathbb{C}^{(N-N_u) \times M}$ and $\mathbf{D}_u \in \mathbb{R}^{(N-N_u) \times M}$ hold. Suppose rank $\left[\mathbf{H}_u^- \right] = L_u$, then the diagonal matrix $\Lambda_u^{\frac{1}{2}}$ contains the L_u nonzero singular values of $\mathbf{H}_u^-$ and $\mathbf{V}_u$ owns the dimension $M \times (M - L_u)$. The matrix $\left(\tilde{\mathbf{V}}_u \ \mathbf{V}_u \right)$ contains the right-hand singular vectors of $\mathbf{H}_u^-$, where $\mathbf{V}_u$ assembles the $M - L_u$ vectors, which form an orthogonal basis for the null space of $\mathbf{H}_u^-$. The column vectors of $\mathbf{T}_u$ must be located in this null space. Consequently, the precoder can be determined by

$$\mathbf{T}_u = \mathbf{V}_u \tag{24.38}$$

The described procedure has to be executed for all $\mathbf{T}_u$, $u = 1, 2, ..., U$. The receive vector for the user u is obtained by inserting (24.38) into (24.34) resulting in

$$\mathbf{r}_u = \mathbf{H}_u \mathbf{V}_u \mathbf{c}_u + \mathbf{n}_u \ , \ u = 1, 2, ..., U \tag{24.39}$$

The described solution is valid for the following sufficient condition on the number of transmit and receive antennas [5]

$$M > \max \left[\sum_{\substack{i=1 \\ i \neq u}}^{U} N_i \ ; \ u = 1, 2, ..., U \right] \tag{24.40}$$

To obtain the symbol vector $\mathbf{c}_u$ from (24.39), various methods known from single-user MIMO transmission can be applied, such as the linear precoding techniques with a matrix $\mathbf{A}_u$ described in Chap. 22 yielding the total precoder matrix

$$\mathbf{T}_u = \mathbf{V}_u \mathbf{A}_u \ , \ u = 1, 2, ..., U \tag{24.41}$$

The decoupling of the single-user links is not affected, because (24.32) guarantees also $\mathbf{H}_i \mathbf{T}_j \mathbf{A}_j = \mathbf{0}$, $i \neq j$. Then we obtain the receive signal vector of user u from (24.34)

$$\mathbf{r}_u = \mathbf{H}_u \mathbf{V}_u \mathbf{A}_u \mathbf{c}_u + \mathbf{n}_u \ , \ u = 1, 2, ..., U \tag{24.42}$$

Example 3 addresses all numerical steps in quite some detail. The matrix $\mathbf{A}_u$ can be determined in different ways also to maximize the system capacity by transmit power loading using the water filling algorithm.

Finally, it should be mentioned that the channel inversion method described in the previous section can be applied as well for the multi-user MIMO scheme depicted in Fig. 24.2. The precoder is then defined by the matrix $\mathbf{T}$ in (24.10), where the channel matrix is given by (24.26).

24.2.3 Alternative Multi-user MIMO Precoding

As we have seen, block diagonalization can completely remove the multi-user interference. However, the system capacity is not taken into account with this procedure. Therefore, alternative approaches try to find a good compromise between low interference and high capacity. This could be done by just minimizing the multi-user interference rather than canceling it. Then the resulting matrix in (24.33) still shows some nonzero off-diagonal elements. Hence, the U single-user links are not anymore decoupled precisely and show some "leakage" to the other links. Several approaches have been investigated, such as precoding with minimum mean squared interference or successive interference cancelation to minimize this leakage. In addition to the precoder, the receiver terminals can also be equipped with a filter and both filters are adapted in a closed control loop between the base station and the user terminal. These methods also have the potential to relax the stringent conditions (24.11) and (24.40), which require an increase of the total number of transmit antennas M, if the number of users and/or user antennas is elevated in the system.

As outlined in Chap. 22, zero-forcing precoding is designed on the basis of low noise at the receiver. Therefore, precoders have been investigated, which use the matrix $\mathbf{H}^H \left(\mathbf{H}\mathbf{H}^H + \alpha \mathbf{I}_N \right)^{-1}$ known from the single-user minimum mean squared error (MMSE) receiver, and this modification is called "regularized channel inversion". The parameter α is used to maximize the signal-to-interference-plus-noise ratio at the receiver, [1].

In Chap. 22, we have discussed that the output of a linear precoder can generate signals $\mathbf{s}$ with large magnitudes, in particular if the channel matrix is ill conditioned. This was recognized as a general drawback of the linear precoding methods. A remedy is achieved by the introduction of nonlinear operations, such as the dirty paper precoder or the Tomlinson–Harashima precoder, which is able to reduce the transmit signal level by modulo operations on the input symbols, e.g., [1, 6]. A challenge for all precoding techniques is also the precise channel estimation combined with an effective provision of significant channel parameters for the transmitter via a feedback channel or adequate reciprocity conditions of the downlink and uplink channels.

24.3 Beamforming for Multi-user Downlink

Multi-user transmit beamforming can be considered as a special case of the previously discussed multi-user precoding in the downlink. The block diagram in Fig. 24.2 still holds in principle with some changes. The precoder matrix is reduced to just a beamforming vector

$$\mathbf{T}_u = \mathbf{t}_u \in \mathbb{C}^{M \times 1} \; ; \; u = 1, 2, ..., U \tag{24.43}$$

The base station is still equipped with M transmit antennas, but only one symbol stream per user is allocated. Thus, $\zeta_u = 1$ holds and the symbol vector per user u is

$$\mathbf{c}_u = c_u \; ; \; u = 1, 2, ..., U \tag{24.44}$$

containing only one symbol. Although the first multi-user beamforming schemes started with only one receive antenna per user, each user terminal can still maintain N_u antennas for a more general case, where the total number is given by (24.2). The individual channel from the M transmit antennas to the user terminal u is $\mathbf{H}_u$ and (24.26) holds for the total channel matrix $\mathbf{H}$. From (24.22) follows with (24.43) and (24.44) the transmit signal vector

$$\mathbf{s} = \sum_{u=1}^{U} \mathbf{t}_u c_u \tag{24.45}$$

which represents the superposition of all symbols c_u of the users weighted by the different precoding vectors. The receive signal of user u is obtained from (24.28)

$$\mathbf{r}_u = \mathbf{H}_u \mathbf{t}_u c_u + \mathbf{H}_u \sum_{\substack{i=1 \\ i \neq u}}^{U} \mathbf{t}_i c_i + \mathbf{n}_u \; ; \; u = 1, 2, ..., U \tag{24.46}$$

where the first term on the right-hand side is the desired signal $\mathbf{H}_u \mathbf{t}_u c_u$ and the second one characterizes the multi-user interference. With the same methods as discussed in Sect. 24.2.2, the multi-user interference can be completely removed. Again, the individual channel matrices $\mathbf{H}_u$ must be known at the base station via a feedback link from each user terminal, where the channel matrix is estimated.

Actually, beamforming with vectors was the first precoding technique and was later extended to full precoding matrices. Consequently, the comments given in the previous Sections apply. In particular the condition (24.40) on the number of antennas and users still holds for perfect cancelation. Again, to overcome this stringent requirement, schemes have been investigated for which the interference is only minimized rather than canceled using different minimization criteria, such as the minimal squared interference power or the maximization of the signal-to-leakage-power ratio

[7, 8]. The beamforming method allows an illustrative explanation, which is discussed in the following example.

Example 2

Consider a multi-user downlink with beamforming vectors $\mathbf{t}_u \in \mathbb{C}^{M \times 1}$; $u = 1, 2,$ ..., U. The user terminals are equipped with N_u receive antennas, $u = 1, 2, ..., U$. The noise at the receivers shall be zero. Determine the transmit signal vector, if only one user u is active and all other user symbols are zero, $c_j = 0 \,\forall\, j \neq u$. Find the receive signal of user u.

Solution:

From (24.45) follows $\mathbf{s} = \mathbf{t}_u c_u$. The user u receives the signal vector $\mathbf{r}_u = \mathbf{H}_u \mathbf{t}_u c_u$, which is obtained from (24.46). As is well known, a matrix changes the magnitude and the direction of a vector after multiplication. On the other hand, by selection of $\mathbf{t}_u$ the direction of the receive vector $\mathbf{r}_u$ can be adjusted for a given channel matrix $\mathbf{H}_u$ and we are able to steer the beam of the electromagnetic wave in the direction of the user u. As a consequence, the impact of the remaining users is significantly reduced. If a second user signal c_j is active, the corresponding beamforming vector should be aligned in such a way that the receive vector $\mathbf{r}_j$ is almost orthogonal to $\mathbf{r}_u$, which results in minimal interference. In summary, beamforming can increase the transmission quality in a multi-user downlink.

Example 3

With this exercise, we investigate the block diagonalization described in Sect. 24.2.2 in quite some detail. Consider the multi-user downlink scheme in Fig. 24.2 and Example 1 with the broadcast channel matrix (24.16). User 1 shall be served with two QAM symbols per symbol vector. Thus, $\zeta_1 = 2$ and owing to (24.20) at least $N_1 = 2$ receive antennas are required. For users 2 and 3, one symbol per symbol vector has to be transmitted, hence $\zeta_2 = \zeta_3 = 1$ and $N_2 = N_3 = 1$ are given, respectively. The base station is equipped with $M = 5$ antennas. The channel matrix in (24.16) is decomposed into horizontal stripes as in (24.26), which indicate the individual channels from the M transmit to the dedicated receive antennas of each user terminal

$$\mathbf{H}_1 = \begin{pmatrix} 1 & 0.2 & -0.1 & 0.1 & 0 \\ -0.2 & 1 & 0.1 & 0 & 0.1 \end{pmatrix} \tag{24.47}$$

$$\mathbf{H}_2 = \begin{pmatrix} 0.1 & -0.1 & 1 & 0.2 & -0.1 \end{pmatrix} \tag{24.48}$$

$$\mathbf{H}_3 = \begin{pmatrix} 0 & -0.1 & -0.2 & 1 & 0.2 \end{pmatrix} \tag{24.49}$$

The multi-user interference shall be canceled by individual precoders with the matrices $\mathbf{T}_1$, $\mathbf{T}_2$, and $\mathbf{T}_3$ using the method of block diagonalization.
(a) Find the precoder matrices.
(b) Determine the receive signals. The noise shall be neglected.

(c) Reduce the interchannel interference, if any, with an additional zero-forcing precoder.

Solution:

(a)
Calculation of $\mathbf{T}_1$
According to (24.37) we determine

$$\mathbf{H}_1^- = \begin{pmatrix} \mathbf{H}_2 \\ \mathbf{H}_3 \end{pmatrix} = \begin{pmatrix} 0.1 & -0.1 & 1 & 0.2 & -0.1 \\ 0 & -0.1 & -0.2 & 1 & 0.2 \end{pmatrix} \tag{24.50}$$

For the singular value decomposition a computer program is used yielding

$$\mathbf{H}_1^- = \mathbf{U}_1 \underbrace{\begin{pmatrix} 1.05 & 0 & 0\,0\,0 \\ 0 & 1.03 & 0\,0\,0 \end{pmatrix}}_{\mathbf{D}_1} \underbrace{\begin{pmatrix} \triangle & \triangle & -0.98 & -0.11 & 0.11 \\ \triangle & \triangle & -0.97 & 0.97 & 0.18 \\ \triangle & \triangle & 0.10 & 0.06 & 0.13 \\ \triangle & \triangle & -0.01 & 0.14 & -0.15 \\ \triangle & \triangle & 0.02 & -0.16 & 0.96 \end{pmatrix}^H}_{\left(\tilde{\mathbf{V}}_1 \; \mathbf{V}_1 \right)^H} \tag{24.51}$$

$\mathbf{D}_1$ incorporates two nonzero singular values. Therefore $\mathbf{H}_1^-$ has full rank $L_1 = 2$. The $M - L_1 = 3$ vectors on the right of $\left(\tilde{\mathbf{V}}_1 \; \mathbf{V}_1 \right)$ determine the matrix $\mathbf{V}_1$. Entries indicated by $\triangle$ belong to $\tilde{\mathbf{V}}_1$ and are of no interest. Also the matrix $\mathbf{U}_1$ is not needed in detail. Because $\zeta_1 = 2$, we only require two of the three vectors in $\mathbf{V}_1$ taken from the left to determine $\mathbf{T}_1$,

$$\mathbf{T}_1 = \begin{pmatrix} -0.98 & -0.11 \\ -0.97 & 0.97 \\ 0.10 & 0.06 \\ -0.01 & 0.14 \\ 0.02 & -0.16 \end{pmatrix} \tag{24.52}$$

We easily calculate the interference terms

$$\mathbf{H}_1 \mathbf{T}_1 = \begin{pmatrix} -1.19 & 0.09 \\ -0.76 & 0.98 \end{pmatrix} \; ; \; \mathbf{H}_2 \mathbf{T}_1 = \begin{pmatrix} 0.1 & 0 \end{pmatrix} \; ; \; \mathbf{H}_3 \mathbf{T}_1 = \begin{pmatrix} 0.07 & 0 \end{pmatrix} \tag{24.53}$$

Calculation of $\mathbf{T}_2$

$$\mathbf{H}_2^- = \begin{pmatrix} \mathbf{H}_1 \\ \mathbf{H}_3 \end{pmatrix} = \begin{pmatrix} 1 & 0.2 & -0.1 & 0.1 & 0 \\ -0.2 & 1 & 0.1 & 0 & 0.1 \\ 0 & -0.1 & -0.2 & 1 & 0.2 \end{pmatrix} \tag{24.54}$$

$$\mathbf{H}_2^- = \mathbf{U}_2 \underbrace{\begin{pmatrix} 1.11 & 0 & 0 & 0\,0 \\ 0 & 1.02 & 0 & 0\,0 \\ 0 & 0 & 0.97 & 0\,0 \end{pmatrix}}_{\mathbf{D}_2} \underbrace{\begin{pmatrix} \triangle & \triangle & \triangle & 0.09 & 0.06 \\ \triangle & \triangle & \triangle & -0.06 & -0.10 \\ \triangle & \triangle & \triangle & 0.96 & 0.20 \\ \triangle & \triangle & \triangle & 0.22 & -0.16 \\ \triangle & \triangle & \triangle & -0.17 & 0.96 \end{pmatrix}^H}_{\left(\tilde{\mathbf{V}}_2\ \mathbf{V}_2\right)^H} \tag{24.55}$$

Three singular values are unequal to zero. Therefore $\mathbf{H}_2^-$ has full rank $L_2 = 3$. Consequently, the $M - L_2 = 2$ vectors on the right of $\left(\tilde{\mathbf{V}}_2\ \mathbf{V}_2\right)$ determine the matrix $\mathbf{V}_2$. Because $\zeta_2 = 1$, we just require one vector of $\mathbf{V}_2$ taken from the left to compose $\mathbf{T}_2$,

$$\mathbf{T}_2 = \begin{pmatrix} 0.09 \\ -0.06 \\ 0.96 \\ 0.22 \\ -0.17 \end{pmatrix} \tag{24.56}$$

Then we obtain

$$\mathbf{H}_1\mathbf{T}_2 = \begin{pmatrix} 0 \\ 0 \end{pmatrix} \ ; \ \mathbf{H}_2\mathbf{T}_2 = 1.04\,; \ \mathbf{H}_3\mathbf{T}_2 = 0 \tag{24.57}$$

Calculation of $\mathbf{T}_3$

$$\mathbf{H}_3^- = \begin{pmatrix} \mathbf{H}_1 \\ \mathbf{H}_2 \end{pmatrix} = \begin{pmatrix} 1 & 0.2 & -0.1 & 0.1 & 0 \\ -0.2 & 1 & 0.1 & 0 & 0.1 \\ 0.1 & -0.1 & 1 & 0.2 & -0.1 \end{pmatrix} \tag{24.58}$$

$$\mathbf{H}_3^- = \mathbf{U}_3 \underbrace{\begin{pmatrix} 1.05 & 0 & 0 & 0\,0 \\ 0 & 1.03 & 0 & 0\,0 \\ 0 & 0 & 1.02 & 0\,0 \end{pmatrix}}_{\mathbf{D}_3} \underbrace{\begin{pmatrix} \triangle & \triangle & \triangle & -0.11 & 0.03 \\ \triangle & \triangle & \triangle & -0.01 & -0.10 \\ \triangle & \triangle & \triangle & -0.18 & 0.08 \\ \triangle & \triangle & \triangle & 0.98 & 0.01 \\ \triangle & \triangle & \triangle & 0.01 & 0.99 \end{pmatrix}^H}_{\left(\tilde{\mathbf{V}}_3\ \mathbf{V}_3\right)^H} \tag{24.59}$$

Three singular values are unequal to zero. Hence, $\mathbf{H}_3^-$ has full rank $L_3 = 3$. The $M - L_3 = 2$ vectors on the right of $\left(\tilde{\mathbf{V}}_3\ \mathbf{V}_3\right)$ determine the matrix $\mathbf{V}_3$. Because $\zeta_3 = 1$, we just require one vector of $\mathbf{V}_3$ taken from the left and get

$$\mathbf{T}_3 = \begin{pmatrix} -0.11 \\ -0.01 \\ -0.18 \\ 0.98 \\ 0.01 \end{pmatrix} \tag{24.60}$$

We easily calculate

$$\mathbf{H}_1\mathbf{T}_3 = \begin{pmatrix} 0 \\ 0 \end{pmatrix} \ ; \ \mathbf{H}_2\mathbf{T}_3 = 0 \ ; \ \mathbf{H}_3\mathbf{T}_3 = 1.02 \qquad (24.61)$$

(b) Receive vectors
We insert the numerical results into (24.28) and obtain

$$\begin{pmatrix} \mathbf{r}_1 \\ r_2 \\ r_3 \end{pmatrix} = \begin{pmatrix} \begin{pmatrix} -1.19\ 0.09 \\ -0.76\ 0.98 \end{pmatrix} & \begin{pmatrix} 0 \\ 0 \end{pmatrix} & \begin{pmatrix} 0 \\ 0 \end{pmatrix} \\ (0.1\ 0) & 1.04 & 0 \\ (0.07\ 0) & 0 & 1.02 \end{pmatrix} \begin{pmatrix} \mathbf{c}_1 \\ c_2 \\ c_3 \end{pmatrix} \qquad (24.62)$$

We recognize the block diagonal structure of the combined system matrix. For r_2 and r_3 marginal multi-user interference is present owing to numerical errors. Nevertheless, we clearly see that the precoders $\mathbf{T}_1$, $\mathbf{T}_2$, and $\mathbf{T}_3$ turn the multi-user downlink into three single-user links, which are almost completely decoupled.
(c)
The receive signal $\mathbf{r}_1$ still contains interchannel interference caused by the matrix

$$\mathbf{G}_{11} = \begin{pmatrix} -1.19\ 0.09 \\ -0.76\ 0.98 \end{pmatrix} \qquad (24.63)$$

As $\mathbf{G}_{11}$ is a square matrix with full rank 2, we can employ the inverse as an additional precoder in the zero-forcing sense

$$\mathbf{A}_1 = \mathbf{G}_{11}^{-1} = \begin{pmatrix} -0.89\ 0.08 \\ -0.69\ 1.08 \end{pmatrix} \qquad (24.64)$$

and the combined precoding matrix for link 1 then is

$$\mathbf{T}_1^{ZF} = \mathbf{T}_1\mathbf{A}_1 = \begin{pmatrix} 0.95 & -0.2 \\ 0.19 & 0.97 \\ -0.13 & 0.07 \\ -0.09 & 0.15 \\ 0.09 & -0,17 \end{pmatrix} \qquad (24.65)$$

The new multi-user interference terms associated with $\mathbf{T}_1^{ZF}$ then are

$$\mathbf{H}_1\mathbf{T}_1^{ZF} = \begin{pmatrix} 1\ 0 \\ 0\ 1 \end{pmatrix} \ ; \ \mathbf{H}_2\mathbf{T}_1^{ZF} = (0.09\ 0) \ ; \ \mathbf{H}_3\mathbf{T}_1^{ZF} = (0\ 0) \qquad (24.66)$$

Furthermore, we use $\frac{1}{1.04}\mathbf{T}_2$ instead of $\mathbf{T}_2$ as well as $\frac{1}{1.02}\mathbf{T}_3$ instead of $\mathbf{T}_3$ and obtain from (24.62) finally

$$\begin{pmatrix} \mathbf{r}_1 \\ r_2 \\ r_3 \end{pmatrix} = \left(\begin{pmatrix} 1 & 0 \\ 0 & 1 \\ (0.09 & 0) \\ (0 & 0) \end{pmatrix} \begin{pmatrix} 0 \\ 0 \\ 1 \\ 0 \end{pmatrix} \begin{pmatrix} 0 \\ 0 \\ 0 \\ 1 \end{pmatrix} \right) \begin{pmatrix} \mathbf{c}_1 \\ c_2 \\ c_3 \end{pmatrix} \tag{24.67}$$

showing that just a minor distortion of 0.09 remains in r_2.

24.4 Principles of Multi-user MIMO Uplink Transmission

24.4.1 System Model of the Uplink

In this section, we consider the uplink transmission from U user terminals to the base station. The block diagram is depicted in Fig. 24.4. The scenario is similar to Fig. 24.2; however, the transmission signals flow in the reverse direction from the right to the left. The user terminal u is equipped with M_u transmit antennas and allocates the QAM symbol vector $\mathbf{c}_u \in \mathbb{C}^{M_u \times 1}$ to the transmit signal vector $\mathbf{s}_u \in \mathbb{C}^{M_u \times 1}$, where $u = 1, 2, ..., U$. The combined transmit signal vector $\mathbf{s}$ is defined in (24.5).

The uplink channel is characterized by the channel matrix

$$\mathbf{H} = \begin{pmatrix} \mathbf{H}_1 & \cdots & \mathbf{H}_u & \cdots & \mathbf{H}_U \end{pmatrix} \in \mathbb{C}^{N \times M} \tag{24.68}$$

which is separated into horizontal blocks $\mathbf{H}_u \in \mathbb{C}^{N \times M_u}$, where $\mathbf{H}_u$ determines the uplink channel from the output of the M_u terminal antennas of the user u to the N receive antennas of the base station. Please note that we use the same notation for

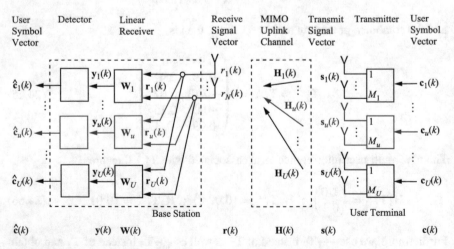

Fig. 24.4 Multi-user MIMO uplink transmission between individual user terminals and the base station

$\mathbf{H}$ and $\mathbf{H}_u$ as in the case of the downlink, although different in general. The total number of transmit antennas of all the users is M and defined in (24.4). The receive signal vector $\mathbf{r} \in \mathbb{C}^{N \times 1}$ is given by

$$\mathbf{r} = \left(r_1 \ r_2 \ \cdots \ r_N \right)^T \tag{24.69}$$

At the base station we apply linear processing using a bank of receive filters $\mathbf{W}_1, ..., \mathbf{W}_U$, where one filter is associated to each user. Thus, we abandon a pre-filter at each user terminal to keep the amount of the hardware, the software, and the power consumption of cost-conscious mobile terminals low. A higher complexity of the base station is adequate owing to the fact that its cost is shared among all users in the network. $\mathbf{r}$ is also the input of the receive filters with matrix $\mathbf{W}_u \in \mathbb{C}^{M_u \times N}$ generating the output signal vector $\mathbf{y}_u$

$$\mathbf{y}_u = \left(y_1 \ y_2 \ \cdots \ y_{M_u} \right)^T \tag{24.70}$$

where $u = 1, 2, ..., U$. Each receive filter is followed by a signal detector, which outputs an estimate $\hat{\mathbf{c}}_u \in \mathbb{C}^{M_u \times 1}$ of the transmit symbol vector $\mathbf{c}_u$.

24.4.2 Receive Signal at the Base Station

From Fig. 24.4, we determine the receive signal as

$$\mathbf{r} = \mathbf{H}_u \mathbf{s}_u + \sum_{\substack{i=1 \\ i \neq u}}^{U} \mathbf{H}_i \mathbf{s}_i + \mathbf{n} \tag{24.71}$$

and recognize the desired signal $\mathbf{H}_u \mathbf{s}_u$ for user u, which is corrupted by the multi-user interference given by the second term on the right-hand side and the noise $\mathbf{n}$ at the base station. Please note that according to Fig. 24.4 $\mathbf{r} = \mathbf{r}_1 = \mathbf{r}_2 = \cdots = \mathbf{r}_U$ holds. The similarity with the downlink in (24.28) needs no explanation. In principle, the base station has to perform a multi-user detection.

24.4.3 Zero-Forcing Receiver for Multi-user Uplink Interference Reduction

The filter output signal for user u at the base station is

$$\mathbf{y}_u = \mathbf{W}_u \mathbf{r} = \mathbf{W}_u \mathbf{H}_u \mathbf{s}_u + \mathbf{W}_u \sum_{\substack{i=1 \\ i \neq u}}^{U} \mathbf{H}_i \mathbf{s}_i + \mathbf{W}_u \mathbf{n} \qquad (24.72)$$

The multi-user interference can be written with matrix notation as

$$\mathbf{W}_u \sum_{\substack{i=1 \\ i \neq u}}^{U} \mathbf{H}_i \mathbf{s}_i = \mathbf{W}_u \mathbf{B}_u \mathbf{b}_u \qquad (24.73)$$

with

$$\mathbf{B}_u = (\mathbf{H}_1 \cdots \mathbf{H}_{u-1} \mathbf{H}_{u+1} \cdots \mathbf{H}_U) \; ; \; \mathbf{b}_u = (\mathbf{s}_1 \cdots \mathbf{s}_{u-1} \mathbf{s}_{u+1} \cdots \mathbf{s}_U)^T \qquad (24.74)$$

Below we prove that the zero-forcing receive filter

$$\mathbf{W}_u^{ZF} = \mathbf{H}_u^H \left(\mathbf{I}_N - \mathbf{B}_u \mathbf{B}_u^+ \right) \qquad (24.75)$$

can completely remove the multi-user interference independently of $\mathbf{s}$, where

$$\mathbf{B}_u^+ = \left(\mathbf{B}_u^H \mathbf{B}_u \right)^{-1} \mathbf{B}_u^H \qquad (24.76)$$

is the pseudo-inverse matrix of $\mathbf{B}_u$. Please note $\mathbf{B}_u^H \mathbf{B}_u$ must have full rank, otherwise its inverse does not exist. Under this prerequisite

$$\mathbf{W}_u^{ZF} \mathbf{B}_u = \mathbf{H}_u^H \left(\mathbf{I}_N - \mathbf{B}_u \mathbf{B}_u^+ \right) \mathbf{B}_u = \mathbf{0} \qquad (24.77)$$

holds and we obtain the filter output signal

$$\mathbf{y}_u = \mathbf{W}_u^{ZF} \mathbf{H}_u \mathbf{s}_u + \mathbf{W}_u^{ZF} \mathbf{n} \; ; \; u = 1, 2, ..., U \qquad (24.78)$$

Apparently, the multi-user MIMO uplink turns into U decoupled single-user MIMO links defined by (24.78). $\mathbf{y}_u$ still suffers some interchannel interference given by $\mathbf{W}_u^{ZF} \mathbf{H}_u$. This impairment can be reduced by conventional single-user design methods, such as an additional zero-forcing or MMSE filter, which follows the $\mathbf{W}_u^{ZF}$ filter and both can be combined to one unit. The last stage in the uplink processing in Fig. 24.4 is the decision device, which will mostly be a single-user maximum likelihood detector.

Motivated by the fact that the multi-user MIMO uplink can be turned into a parallel arrangement of single-user connections opens up several alternative methods, which try to find a good compromise between interference and noise reduction, in principle. Hence, a target is to maximize the signal-to-interference-plus-noise ratio at the receiver. Linear methods furnish a minimum mean squared error (MMSE). Then a maximum likelihood detection per user can be followed. Other approaches employ additional precoders at the transmitter, which are adapted jointly with the receive filters. Depending on the statistics of the noise and the interference an optimal solution is the joint multi-user maximum likelihood detection, yet expensive owing to the complexity, which grows exponentially as a function of the system parameters.

Proof of (24.77)

With (24.75) and (24.76) follows from the left-hand side of (24.77)

$\mathbf{W}_u^{ZF} \mathbf{B}_u = \mathbf{H}_u^H \left(\mathbf{B}_u - \mathbf{B}_u \left(\mathbf{B}_u^H \mathbf{B}_u \right)^{-1} \mathbf{B}_u^H \mathbf{B}_u \right) = \mathbf{H}_u^H \left(\mathbf{B}_u - \mathbf{B}_u \right) = \mathbf{0}$ and the proof is finished.

24.5 Outlook: Massive MIMO for Multi-user Applications

The ever increasing demand for high-speed communications has motivated the research on methods to further increase the system capacity or spectral efficiency measured in bit/s per Hz bandwidth of a wireless network. A cell is composed of a base station with M antenna elements and U user equipment with in total N antenna elements. Marzetta [9] investigated theoretically the effect of a drastic increase in the number of base station antennas, $M \to \infty$, yet his findings are also applicable for $M < \infty$. He showed the large potential, which was then verified by computer simulation and with first test beds [10–13]. With the term *massive MIMO* or *large MIMO* we characterize a multi-user MIMO scheme with a very large number of antennas at the base station and in total at the user equipment side, typically more than one hundred. Existing antenna masts for the 4G cellular networks carry already four MIMO antennas with about twenty antenna element arrays each. There are also ideas to integrate a large number of planar antennas into the facade of buildings or into the wallpaper in the in-house area. Research and development are ongoing in this field. Of course, the hardware complexity as well as the cost play an important part. In the following, we just give a survey on the principles.

First, we consider the downlink scenario in Fig. 24.2 with M transmit antennas at the base station and U user equipment. The total number N of antennas at the user side is defined in (24.2). To determine the system capacity we can apply (21.57) in Sect. 21.2 and find $C = \sum_{i=1}^{P} \log_2 \left(1 + \frac{\lambda_i E_S}{\sigma_n^2} \right)$, where $P \leq \min\{M, N\}$ is the rank of the channel matrix in (24.26). Assume $M > N$ and a channel matrix $\mathbf{H}$ with full rank N. Then follows $P = N$ and the system capacity for the cell

$$C = \sum_{i=1}^{N} \log_2 \left(1 + \frac{\lambda_i E_S}{\sigma_n^2} \right) \qquad (24.79)$$

increases with the total number N of user equipment antennas and thus with the number U of users. We observe that M must ascend as well owing to $M > N$.

Hence, with a massive MIMO antenna at the base station we can serve a large number of users and achieve a high system capacity. If the base station operates with beamforming, narrow beams between the base station an the individual user equipment can be generated showing very small overlap. This also accounts for a higher energy efficiency and transmission quality.

The situation is similar in the case of an uplink transmission depicted in principle in Fig. 24.4. Please note that M and N change their role according to our definitions, where the total number M of antennas at the user side is then given by (24.4) and N is the number of base station antennas. "Asymptotic favorable transmission" has been observed meaning that $\frac{1}{N}\mathbf{h}_u^H \mathbf{h}_j$, $u \neq j$ tends to zero for very large N in the case of a Rayleigh fading and a line of sight channel. $\mathbf{h}_u = \mathbf{H}_u \in \mathbb{C}^{N \times 1}$ are the vectors of the channel matrix $\mathbf{H}$ in (24.68), if only one antenna per user equipment is present. The property $\frac{1}{N}\mathbf{h}_u^H \mathbf{h}_j \approx 0$ can be used by a maximum ratio combiner at the base station with the matrix $\mathbf{W}_u = \mathbf{h}_u^H$, $u = 1, 2, ..., U$ to minimize the multi-user interference.

In the research and development field on massive MIMO for multi-user systems work is ongoing to circumvent some obstacles, such as pilot contamination, reduction of hardware complexity, and channel hardening. Pilot contamination can be present in a multicell system, in which the pilot signal of a user is reused in neighboring cells for other users, in cases where not sufficient orthogonal pilot signals are present. Then the users with the same pilots cannot be differentiated and thus cause impairments.

The large number of antenna elements increases the hardware complexity. Special designs for antennas and radio-frequency amplifiers are subject to intensive work. As a workaround, antenna switching techniques can be introduced and only the instantaneously active antenna is connected to an amplifier. This will reduce the number of amplifiers compared to the number of antennas.

Furthermore, asymptotic channel hardening is an observed effect, when the channel gain gets approximately constant for an increased number N of base station antennas,

$$\frac{\|\mathbf{h}_u\|^2}{\mathbf{E}\left[\|\mathbf{h}_u\|^2\right]} \to 1 \text{ for } N \to \infty \tag{24.80}$$

Thus, individual channels between the base station and the user equipment tend to become deterministic. However, many solutions for all these obstacles are in progress so that massive MIMO will become an integral part of a variety of wireless networks [14].

References

1. Q.H. Spencer, C.B. Peel, A.L. Swindlehurst, M. Haardt, An introduction to the multi-user MIMO downlink. IEEE Commun. Mag. (2004)
2. F. Khalid, J. Speidel, Advances in MIMO techniques for mobile communications - a survey. Int. J. Commun. Netw. Syst. Sci. **3** (2010)
3. L.-U. Choi, R.D. Murch, A downlink decomposition transmit preprocessing technique for multi-user MIMO systems. Proc. IST Mobile Wirel. Telecommun. Summit (2002)

4. Q. Spencer, M. Haardt, Capacity and downlink transmission algorithms for a multi-user MIMO channel, in *Proceedings of 36th Asilomar Conference on Signals, Systems, and Computers* (2002)
5. L.-U. Choi and R. D. Murch, "A transmit preprocessing technique for multiuser MIMO systems using a decomposition approach. IEEE Trans. Wirel. Commun. (2004)
6. R. Fischer, C. Windpassinger, A. Lampe, J. Huber, Space-time transmission using Tomlinson-Harashima precoding, in *4th ITG- Conference on Source and Channel Coding* (2002)
7. A. Tarighat, M. Sadek, A. Sayed, A multi-user beamforming scheme for downlink MIMO channels based on maximizing signal-to-leakage ratios, in *Proceedings IEEE International Conference on Accoustic, Speech, and Signal Processing (ICASSP)* (2005)
8. E. Bjoernson, M. Bengtsson, B. Ottersten, Optimal multi-user transmit beamforming: a difficult problem with a simple solution structure. IEEE Signal Process. Mag. (2014)
9. T.L. Marzetta, Noncooperative cellular wireless with unlimited numbers of base station antennas. IEEE Trans. Wirel. Commun. (2010)
10. J. Hoydis, S. ten Brink, M. Debbah, Massive MIMO in the UL/DL of cellular networks: how many antennas do we need? IEEE J. Select. Areas Commun. (2013)
11. E.G. Larsson, O. Edfors, F. Tufvesson, T.L. Marzetta, Massive MIMO for next generation wireless systems. IEEE Commun. Mag. **52** (2014)
12. T. Marzetta, E.G. Larsson, H. Yang, H.Q. Ngo, *Fundamentals of Massive MIMO* (Cambridge University Press, Cambridge, 2016)
13. E. Bjoernson, E.G. Larsson, T.L. Marzetta, Massive MIMO: ten myths and one critical question. IEEE Commun. Mag. (2016)
14. E. Bjoernson, J. Hoydis, L. Sanguinetti, *Massive MIMO Networks: Spectral, Energy, and Hardware Efficiency* (Now Publishers, 2017)

Correction to: Introduction to Digital Communications

Correction to:
J. Speidel, *Introduction to Digital Communications*,
Signals and Communication Technology,
https://doi.org/10.1007/978-3-030-00548-1

The book chapters 7, 8, 12, 17 and 23 of the original version were revised.
   The erratum chapter and the book have been updated with these changes.

The updated version of these chapters can be found at
https://doi.org/10.1007/978-3-030-00548-1_7
https://doi.org/10.1007/978-3-030-00548-1_8
https://doi.org/10.1007/978-3-030-00548-1_12
https://doi.org/10.1007/978-3-030-00548-1_17
https://doi.org/10.1007/978-3-030-00548-1_23

© Springer Nature Switzerland AG 2019
J. Speidel, *Introduction to Digital Communications*, Signals and Communication
Technology, https://doi.org/10.1007/978-3-030-00548-1_25

Appendix A
Some Fundamentals of Random Variables and Stochastic Processes

In the following, we give a brief overview of random variables and stochastic processes. It should be understood as a summary of the most important findings, which are needed in digital communications and signal processing. For some lemmas, the derivations and proofs are outlined. Beyond that, the reader is referred to dedicated textbooks such as [1, 2].

A.1 Continuous Random Variables

We start with some basics on random variables. Let X be a real-valued random variable and x an event.

A.1.1 Probability Density Function and Probability

$p(x)$ is denoted as probability density function (in short density function or density) with the property $p(x) \geq 0$ and $\int_{-\infty}^{\infty} p(x)dx = 1$. We call $F(x) = \int_{-\infty}^{x} p(u)du$ the (cumulative) distribution function. $F(b)$ is the probability of the event that the random variable is located in the interval $-\infty < X < b$ and we write with the probability operator P,

$$P[X < b] = P[-\infty < X < b] = \int_{-\infty}^{b} p(x)dx \tag{A.1}$$

from which we conclude

$$P[a < X < b] = \int_{a}^{b} p(x)dx = F(b) - F(a) \tag{A.2}$$

Please note, if the density function $p(x)$ contains no Dirac impulses at the boarders a and b, then also the equal sign holds for the interval. With $b \to \infty$ we obtain

© Springer Nature Switzerland AG 2019
J. Speidel, *Introduction to Digital Communications*, Signals and Communication
Technology, https://doi.org/10.1007/978-3-030-00548-1

$$P[X > a] = P[a < X < \infty] = \int_a^\infty p(x)dx \qquad (A.3)$$

A.1.2   *Two Random Variables*

Joint Probability Density Function

For the two random variables X_1 and X_2 we define the (two-dimensional) joint probability density function $p_{12}(x_1, x_2)$. The density functions of the individual random variables X_i are $p_i(x_i)$, $i = 1, 2$ and called marginal probability density functions. They are calculated as

$$p_1(x_1) = \int_{-\infty}^\infty p_{12}(x_1, x_2)dx_2 \; ; \; p_2(x_2) = \int_{-\infty}^\infty p_{12}(x_1, x_2)dx_1 \qquad (A.4)$$

Conditional Probability Density Function

We define $p_{1/2}\,(x_1 \mid X_2 = x_2)$ or with short hand notation $p_{1/2}\,(x_1 \mid x_2)$ as the conditional probability density function of X_1 under the condition $X_2 = x_2$.

$$p_{1/2}\,(x_1 \mid x_2) = \frac{p_{12}(x_1, x_2)}{p_2(x_2)} \qquad (A.5)$$

Conditional Probabilities

With the conditional densities we can calculate conditional probabilities

$$P[X_1 < x_1 \mid X_2 = x_2] = \int_{-\infty}^{x_1} p_{1/2}\,(u_1 \mid x_2)\,du_1 \qquad (A.6)$$

Bayes Theorem

The Bayes theorem relates the two conditional densities as

$$p_{1/2}\,(x_1 \mid x_2) = \frac{p_{2/1}\,(x_2 \mid x_1)\,p_1(x_1)}{p_2(x_2)}, \quad p_2(x_2) \neq 0 \qquad (A.7)$$

Statistical Independence of Random Variables

We call two random variables statistically independent, if and only if $p_{1/2}\,(x_1 \mid x_2) = p_1(x_1)$ is independent of x_2. Then follows from (A.5)

$$p_{12}(x_1, x_2) = p_1(x_1)p_2(x_2) \qquad (A.8)$$

A.2 Statistical Parameters for Random Variables

A.2.1 Expected Value

The expected value (or in short expectation) of a real-valued random variable X with density function $p(x)$ is defined as

$$\mathbf{E}[X] = \int_{-\infty}^{\infty} xp(x)dx = m_x \qquad (A.9)$$

$\mathbf{E}[X]$ is also called first moment or mean value of X. We see that $\mathbf{E}[...]$ is a linear operator.

A.2.2 Function of a Random Variable, nth Moments

Let $g(...)$ be a function of the random variable X yielding the new random variable $Y = g(X)$. Then

$$\mathbf{E}[Y] = \mathbf{E}[g(X)] = \int_{-\infty}^{\infty} g(x)p(x)dx \qquad (A.10)$$

holds. On that basis we can define the nth moment of X as

$$\mathbf{E}[X^n] = \int_{-\infty}^{\infty} x^n p(x)dx \qquad (A.11)$$

and in particular for $n = 2$ we obtain $\mathbf{E}[X^2]$, which is the quadratic mean and physically the mean power of the random variable X.

The nth central moment is defined as

$$\mathbf{E}\left[(X - m_x)^n\right] = \int_{-\infty}^{\infty} (x - m_x)^n \, p(x)dx \qquad (A.12)$$

which yields for $n = 2$ the variance of X

$$\text{var}[X] = \sigma_x^2 = \mathbf{E}\left[(X - m_x)^2\right] \qquad (A.13)$$

σ_x is known as standard deviation. It is straightforward to show that

$$\sigma_x^2 = \mathbf{E}\left[X^2\right] - m_x^2 \qquad (A.14)$$

is true. If the random variable has zero mean, then the variance equals the mean power.

A.2.3   *Covariance and Correlation of Two Random Variables*

For two real-valued random variables X_i with $\mathbf{E}[X_i] = m_i$, the marginal density functions $p_i(x_i)$; $i = 1, 2$, and the joint density function $p_{12}(x_1, x_2)$ we define the *joint central moment* of order (k, n) as

$$\mathbf{E}\left[(X_1 - m_1)^k (X_2 - m_2)^n\right] = \int_{-\infty}^{\infty} \int_{-\infty}^{\infty} (x_1 - m_1)^k (x_2 - m_2)^n p_{12}(x_1, x_2)dx_1dx_2$$

(A.15)

from which follows for $k = n = 1$ the *covariance* between X_1 and X_2 as

$$\mu_{12} = \mathbf{E}[(X_1 - m_1)(X_2 - m_2)] = \int_{-\infty}^{\infty} \int_{-\infty}^{\infty} (x_1 - m_1)(x_2 - m_2) p_{12}(x_1, x_2)dx_1dx_2$$

(A.16)

It is straightforward to show that

$$\mu_{12} = \mathbf{E}[X_1 X_2] - m_1 m_2$$

(A.17)

holds.

The *correlation coefficient* is defined as

$$\rho_{12} = \frac{\mu_{12}}{\sigma_1 \sigma_2} = \frac{\mathbf{E}[X_1 X_2] - m_1 m_2}{\sigma_1 \sigma_2}$$

(A.18)

Two random variables X_1 and X_2 are called *uncorrelated,* if and only if the covariance is zero

$$\mu_{12} = 0$$

(A.19)

which yields $\rho_{12} = 0$[1] and finally

$$\mathbf{E}[X_1 X_2] = m_1 m_2$$

(A.20)

Consequently, if we would like to see whether two random variables are uncorrelated, we have to check, if their covariance is zero.

The *Correlation* between X_1 and X_2 is defined as

$$c_{12} = \mathbf{E}[X_1 X_2] = \int_{-\infty}^{\infty} \int_{-\infty}^{\infty} x_1 x_2 p_{12}(x_1, x_2)dx_1dx_2$$

(A.21)

We call two random variables *orthogonal,* if

$$\mathbf{E}[X_1 X_2] = 0$$

(A.22)

---

[1]This is the reason why ρ_{12} is called correlation coefficient.

Let X_1 and X_2 be two random variables each with zero mean $m_1 = m_2 = 0$. If they are uncorrelated, then their covariance (A.19) is zero and from (A.20) follows for the correlation $c_{12} = \mathbf{E}[X_1 X_2] = 0$.

Let X_1 and X_2 be *statistically independent* random variables. Then (A.8) holds and from (A.21) follows

$$\mathbf{E}[X_1 X_2] = \int_{-\infty}^{\infty} \left[\int_{-\infty}^{\infty} x_1 p_1(x_1) dx_1 \right] x_2 p_2(x_2) dx_2 = \mathbf{E}[X_1]\mathbf{E}[X_2] = m_1 m_2$$

(A.23)

which yields $\rho_{12} = 0$. Hence, we conclude that statistically independent random variables are also uncorrelated. In general, the reverse is not true.

A.3 Stochastic Processes

A.3.1 Definition of a Stochastic Process

For engineers a stochastic process is best explained with the help of a physical experiment. Consider a large number $N \to \infty$ of identical resistors, each resistor i generating a random noise voltage $X_i(t)$ as a function of time t, where $i = 1, 2, \ldots, N$. The stochastic process $X(t) = \{X_1(t), X_2(t), \ldots, X_N(t)\}$ represents the family also called ensemble of all voltages and $X_i(t)$ is the ith sample function or ith realization of the process. All sample functions $(i = 1, 2, \ldots, N)$ belonging to the process have the same statistical parameters, such as probability density function, autocorrelation etc. To characterize the stochastic process statistical parameters can be defined in two ways, namely along the time axis of a dedicated sample function $X_i(t)$ or over all sample functions of $X(t)$ at a fixed time instant t_ν. Then $X(t_\nu)$ is a continuous random variable. In our measuring campaign we can further look at the stochastic process $X(t)$ at different time instants $t_1 < t_2 < t_3 < \cdots < t_M$ yielding a sequence of random variables

$$X(t_1), X(t_2), .., X(t_M) \tag{A.24}$$

and the stochastic process $X(t)$ can be regarded for each fixed time instant t as a random variable. Consequently, the corresponding definitions for random variables can be applied to describe the statistical parameters of the stochastic process. The set of random variables in (A.24) is characterized by its joint probability density function

$$p_{1M}\left(x_{t_1}, x_{t_2}, \ldots, x_{t_M}\right) \tag{A.25}$$

where $x_{t_\nu} = x(t_\nu)$ is the shorthand notation of an event of the random variable $X(t_\nu)$; $\nu = 1, 2, \ldots, M$. In general, the probability density function depends on the time instances $t_1, t_2, \ldots, t_M$, if the process is nonstationary, see Sect. A.3.4.

A.3.2   Autocovariance, Auto-, and Cross-Correlation Function

Single Stochastic Process $X(t)$

The *autocovariance function* $\mu_{xx}(t_1, t_2)$ of a stochastic process $X(t)$ is defined similar to the covariance (A.16) of two random variables $X(t_1)$ and $X(t_2)$

$$
\mu_{xx}(t_1, t_2) = \quad \mathbf{E}\left[(X(t_1) - m_x(t_1))(X(t_2) - m_x(t_2)) \right] =
$$
$$
= \int_{-\infty}^{\infty} \int_{-\infty}^{\infty} \left(x_{t_1} - m_x(t_1) \right) \left(x_{t_2} - m_x(t_2) \right) p_{12}\left(x_{t_1}, x_{t_2} \right) dx_{t_1} dx_{t_2}
$$
(A.26)

where $p_{12}\left(x_{t_1}, x_{t_2} \right)$ is the joint probability density function of $X(t_1)$ and $X(t_2)$.

The *expected values* are $m_x(t_i) = \mathbf{E}[X(t_i)]$; $i = 1, 2$. It is straightforward to show that

$$
\mu_{xx}(t_1, t_2) = \mathbf{E}[X(t_1)X(t_2)] - m_x(t_1)m_x(t_2)
$$
(A.27)

The *autocorrelation function* of the process $X(t)$ is defined similar to (A.21) as

$$
R_{xx}(t_1, t_2) = \mathbf{E}[X(t_1)X(t_2)] = \int_{-\infty}^{\infty} \int_{-\infty}^{\infty} x_{t_1} x_{t_2} p_{12}\left(x_{t_1}, x_{t_2} \right) dx_{t_1} dx_{t_2}
$$
(A.28)

Two Stochastic Processes $X(t)$ **and** $Y(t)$

We consider two stochastic processes $X(t)$ and $Y(t)$ with the corresponding random variables

$$
X(t_i) \ ; \ i = 1, 2, \ldots, M_x \ ; \ t_1 < t_2 < t_3 \ldots
$$
(A.29)

$$
Y(\tilde{t}_j) \ ; \ j = 1, 2, \ldots, M_y \ ; \ \tilde{t}_1 < \tilde{t}_2 < \tilde{t}_3 \ldots
$$
(A.30)

The joint set of random variables is characterized by the joint probability density function

$$
p_{xy}\left(x_{t_1}, x_{t_2}, \ldots, x_{t_{M_x}}; y_{\tilde{t}_1}, y_{\tilde{t}_2}, \ldots, y_{\tilde{t}_{M_y}} \right)
$$
(A.31)

where $x_{t_\nu} = x(t_\nu)$ and $y_{\tilde{t}_\nu} = y(\tilde{t}_\nu)$. Then the *cross-correlation function* can be defined as

$$
R_{xy}(t_1, t_2) = \mathbf{E}[X(t_1)Y(t_2)] = \int_{-\infty}^{\infty} \int_{-\infty}^{\infty} x_{t_1} y_{t_2} p_{xy}\left(x_{t_1}, y_{t_2} \right) dx_{t_1} dy_{t_2}
$$
(A.32)

where we have renamed $\tilde{t}_2$ as t_2.

A.3.3 Time-Domain Parameters and Ergodicity

As already alluded, we can define parameters for a stochastic process $X(t)$ along the time axis of a dedicated sample function $X_i(t)$ or over the ensemble at a fixed time instant t_ν, yielding $X(t_\nu)$. As a consequence, we differentiate between time-domain averages (or moments) on one hand and ensemble values also called expected values on the other hand. On the basis of a sample function $X_i(t)$, in general complex, we get the following time-domain parameters:

Mean Value

$$\bar{x} = \lim_{T_0 \to \infty} \frac{1}{2T_0} \int_{-T_0}^{T_0} x_i(t) dt \qquad (A.33)$$

Autocovariance Function

$$c_{xx}(\tau) = \lim_{T_0 \to \infty} \frac{1}{2T_0} \int_{-T_0}^{T_0} (x_i(t) - \bar{x})^* (x_i(t+\tau) - \bar{x})\, dt \qquad (A.34)$$

Autocorrelation Function

$$R_{xx}(\tau) = \lim_{T_0 \to \infty} \frac{1}{2T_0} \int_{-T_0}^{T_0} x_i^*(t) x_i(t+\tau) dt \qquad (A.35)$$

For $\tau = 0$ follows the **mean power**

$$R_{xx}(0) = \lim_{T_0 \to \infty} \frac{1}{2T_0} \int_{-T_0}^{T_0} |x_i(t)|^2\, dt \qquad (A.36)$$

Ergodicity

A wide sense stationary stochastic process (see Sect. A.3.4) is called ergodic, if all statistical parameters calculated on the basis of the ensemble and with respect to time of any sample function $X_j(t)$ are identical. Thus, an ergodic process can be statistically described by just one realization. In engineering, ergodicity is often assumed as a hypothesis, because the experimental proof in many cases is difficult, although important. In the following, we exclusively consider ergodic processes and focus on the ensemble values.

A.3.4   Stationary Stochastic Process

Strict Sense Stationary (SSS) Stochastic Process

- A stochastic process $X(t)$ is "strict sense stationary" (SSS), if $X(t)$ and $X(t + a)$ $\forall a$ have the same statistical parameters. In other words, their statistics do not depend on time. This holds for all probability density functions such as (A.25) and all M,

$$p_{1M}\left(x(t_1), x(t_2), .., x(t_M)\right) = p_{1M}\left(x(t_1 + a), x(t_2 + a), .., x(t_M + a)\right) \quad (A.37)$$

- Two stochastic processes $X(t)$ and $Y(t)$ are jointly strict sense stationary, if the joint statistics of $X(t)$ and $Y(t)$ are equal to the joint statistics of $X(t + a)$ and $Y(t + a)$, $\forall a$, respectively.
- A complex-valued stochastic process $Z(t) = X(t) + jY(t)$ is strict sense stationary, if this condition holds jointly for the real and imaginary part.

Wide Sense Stationary (WSS) Stochastic Process

The conditions for a "wide sense stationary" (WSS) process are much weaker than for a strict sense stationary process, as they just impose conditions on the first and second order moments. Higher moments are not touched. From (A.37) follows that $p_1\left(x_{t_1}\right) = p_1\left(x_{t_1+a}\right)$ $\forall a$ with $x_{t_\nu} = x(t_\nu)$ as before. Consequently, the expected value is constant. Furthermore (A.37) results in $p_{12}\left(x_{t_1}, x_{t_2}\right) = p_{12}\left(x_{t_1+a}, x_{t_2+a}\right)$ $\forall a$ and thus the density function, the second moments and the autocorrelation function depend only on a time difference $t_2 - t_1 = \tau$.

Definition:

- A stochastic process $X(t)$ is wide sense stationary, if its expected value is constant

$$\mathbf{E}[X] = m_x = const. \quad (A.38)$$

- and if its autocorrelation function just depends on a time difference $\tau = t_2 - t_1$. Then follows from (A.28) by using $t_2 - t_1 = \tau$ and replacing t_1 by the fixed time instant t

$$R_{xx}(t_1, t_2) = R_{xx}(t, t + \tau) = \mathbf{E}[X(t)X(t + \tau)] = R_{xx}(\tau) \quad (A.39)$$

where $R_{xx}(\tau)$ is a shorthand notation.

For a complex stochastic process, we define

$$R_{xx}(\tau) = \mathbf{E}\left[X^*(t)X(t + \tau)\right] \quad (A.40)$$

In general, the autocorrelation function exhibits the following properties:

$$R_{xx}(-\tau) = R_{xx}^*(\tau); \ R_{xx}(0) \geq |R_{xx}(\tau)| \tag{A.41}$$

$R_{xx}(0) = \mathbf{E}[|X(t)|^2]$ is always a real value and is called the mean power of $X(t)$. We also see that a SSS stochastic process is also WSS.

A.3.5 Uncorrelated WSS Stochastic Processes

A Single Process $X(t)$

To check whether a WSS stochastic process $X(t)$ is uncorrelated, we have to extend the definition of the covariance of a random variable in (A.16) to a process.

The *autocovariance function* of the WSS process $X(t)$ is thus given by

$$C_{xx}(t, t + \tau) = C_{xx}(\tau) = \mathbf{E}\left[\left(X^*(t) - m_x^*\right)(X(t + \tau) - m_x)\right] = R_{xx}(\tau) - |m_x|^2 \tag{A.42}$$

where $m_x = \mathbf{E}[X(t)]$ is the expected value of $X(t)$.

The autocovariance function $C_{xx}(\tau)$ specifies the expected value of the product of the two random variables $X^*(t) - m_x^*$ and $X(t + \tau) - m_x$ for any given time shift $\tau \neq 0$. Similar to (A.19) we can say that the two random variables are uncorrelated, if $C_{xx}(\tau) = 0$. However, we have to exclude $\tau = 0$, because in this case both random variables just differ in the sign of the imaginary part and of course are strongly correlated. Consequently, we can formulate the following meaningful definition:

A WSS stochastic process $X(t)$ is *uncorrelated*, if its autocovariance function meets the condition

$$C_{xx}(\tau) \begin{cases} \neq 0 \ ; \ \tau = 0 \\ = 0 \ ; \ \tau \neq 0 \end{cases} \tag{A.43}$$

For the *autocorrelation function* of an *uncorrelated* process then follows with (A.43) and $R_{xx}(0) = \mathbf{E}\left[|X(t)|^2\right]$

$$R_{xx}(\tau) = \begin{cases} \mathbf{E}\left[|X(t)|^2\right] \ ; \ \tau = 0 \\ m_x^2 \qquad\quad ; \ \tau \neq 0 \end{cases} \tag{A.44}$$

Now, consider a WSS process with *zero mean*, $m_x = 0$. Then we find from (A.44) that this process is *uncorrelated*, if

$$R_{xx}(\tau) = 0 \ \forall \tau \neq 0 \tag{A.45}$$

Two Processes $X(t)$ and $Y(t)$

The statistical interrelation between two WSS processes $X(t)$ and $Y(t)$ is defined by the *cross-covariance function* similar to (A.16)

$$C_{xy}(\tau) = \mathbf{E}\left[(X(t+\tau) - m_x)\left(Y^*(t) - m_y^*\right)\right] =$$

$$= R_{xy}(\tau) - m_x m_y^* \tag{A.46}$$

with $m_x = \mathbf{E}[X(t)]$, $m_y = \mathbf{E}[Y(t)]$, and the *cross-correlation function*

$$R_{xy}(\tau) = \mathbf{E}\left[X(t+\tau)Y^*(t)\right] \tag{A.47}$$

Using similar arguments as before with $C_{xx}(\tau)$, but no exception for $\tau = 0$ is required here, we define:

Two (WSS) processes $X(t)$ and $Y(t)$ are *uncorrelated* if

$$C_{xy}(\tau) = 0 \ \forall \tau \tag{A.48}$$

Then follows from (A.46)

$$R_{xy}(\tau) = m_x m_y^* \ \forall \tau \tag{A.49}$$

If at least one stochastic process has zero mean, then the processes are referred to as *orthogonal*

$$R_{xy}(\tau) = 0 \ \forall \tau \tag{A.50}$$

It is straightforward to show that the cross-correlation function has the following symmetry property:

$$R_{xy}(\tau) = R_{yx}^*(-\tau) \tag{A.51}$$

A.3.6   Statistically Independent Processes

Two stochastic processes $X(t)$ and $Y(t)$ are statistically independent, if and only if for any choice of t_i and $\tilde{t}_j$ as well as M_x and M_y

$$p_{xy}\left(x_{t_1}, x_{t_2}, \ldots, x_{t_{M_x}}; y_{\tilde{t}_1}, y_{\tilde{t}_2}, \ldots, y_{\tilde{t}_{M_y}}\right) = p_x\left(x_{t_1}, x_{t_2}, \ldots, x_{t_{M_x}}\right) p_y\left(y_{\tilde{t}_1}, y_{\tilde{t}_2}, \ldots, y_{\tilde{t}_{M_y}}\right) \tag{A.52}$$

holds, where $p_x\left(x_{t_1}, x_{t_2}, \ldots, x_{t_{M_x}}\right)$ and $p_y\left(y_{\tilde{t}_1}, y_{\tilde{t}_2}, \ldots, y_{\tilde{t}_{M_y}}\right)$ are the joint density functions of $x_{t_1}, x_{t_2}, \ldots, x_{t_{M_x}}$ and $y_{\tilde{t}_1}, y_{\tilde{t}_2}, \ldots, y_{\tilde{t}_{M_y}}$, respectively. From the statistical independence follows that the two processes are uncorrelated, but not reversely.

Two WSS stochastic processes $X(t)$ and $Y(t)$ with joint probability density function $p_{xy}(x, y)$ are *statistically independent*, if and only if

$$p_{xy}(x, y) = p_x(x)p_y(y) \tag{A.53}$$

where $p_x(x)$ and $p_y(y)$ are the marginal probability density functions of the stochastic processes $X(t)$ and $Y(t)$, respectively.

A.4 Stochastic Processes and Linear Time-Invariant Systems

A.4.1 Input–Output Relation of Linear System in Time Domain

Let $h(t)$ be the (deterministic) impulse response of a linear time-invariant system. At its input the sample function $x(t)$ of a WSS stochastic process $X(t)$ is active. Throughout the following we always consider stationary processes. Then the output process $Y(t)$ with sample function $y(t)$ is also stationary [1] and given by the convolution

$$y(t) = x(t) * h(t) = \int_{-\infty}^{\infty} x(u)h(t-u)du \qquad (A.54)$$

However, as the stochastic signals cannot be expressed by a mathematical formula, we are not in a position to explore this equation further and have to find a statistical description using autocorrelation functions and power spectral densities.

A.4.2 Wiener-Lee Theorem for Input–Output Autocorrelation functions

Let $R_{xx}(\tau)$ and $R_{yy}(\tau)$ be the autocorrelation functions of the input and output stochastic process $x(t)$ and $y(t)$, respectively. We calculate the autocorrelation function $R_{hh}(\tau)$ of the deterministic (and thus also ergodic) impulse response $h(t)$ according to (A.35)

$$R_{hh}(\tau) = \mathbf{E}\left[h^*(t)h(t+\tau)\right] = \int_{-\infty}^{\infty} h^*(t)h(t+\tau)dt \qquad (A.55)$$

where we have dropped $\lim_{T_0 \to \infty} \frac{1}{2T_0}$, because $h(t)$ is a deterministic signal with finite energy $R_{hh}(0) = \int_{-\infty}^{\infty} |h(t)|^2 \, dt$. The integral for $R_{hh}(\tau)$ can be considered as the convolution between $h(\tau)$ and $h^*(-\tau)$

$$R_{hh}(\tau) = h(\tau) * h^*(-\tau) \qquad (A.56)$$

The Wiener-Lee theorem describes the relation between the input and the output autocorrelation function of a linear time-invariant system as follows:

$$R_{yy}(\tau) = R_{hh}(\tau) * R_{xx}(\tau) = h(\tau) * h^*(-\tau) * R_{xx}(\tau) \qquad (A.57)$$

A.4.3   Wiener–Khintchine Theorem for Power Spectral Density

For communications engineers spectra of signals are important to get an idea about the required bandwidth. For deterministic signals with finite energy the Fourier spectrum exists according to the sufficient Dirichlet condition. However, a random signal $x(t)$ has infinite energy, because in general

$$\lim_{T_0 \to \infty} \int_{-T_0}^{T_0} |x(t)|^2 \, dt \to \infty \tag{A.58}$$

On the other hand an ergodic stochastic process $X(t)$ exhibits finite mean power which is

$$\mathbf{E}\left[|X|^2\right] = R_{xx}(0) = \lim_{T_0 \to \infty} \frac{1}{2T_0} \int_{-T_0}^{T_0} |x(t)|^2 \, dt < \infty \tag{A.59}$$

The Wiener–Khintchine theorem provides the power spectral density $S_{xx}(f)$ of $X(t)$ by means of the Fourier transform of the autocorrelation function of $X(t)$,

$$R_{xx}(\tau) \rightarrowtail S_{xx}(f) = \int_{-\infty}^{\infty} R_{xx}(\tau) e^{-j2\pi f \tau} d\tau \tag{A.60}$$

From the symmetry $R_{xx}(-\tau) = R_{xx}^*(\tau)$ follows the property that $S_{xx}(f)$ is real and moreover

$$S_{xx}(f) \geq 0 \tag{A.61}$$

holds. With $h(t) \rightarrowtail H(f)$ and $h^*(-t) \rightarrowtail H^*(f)$ follows from (A.56) with the Fourier transform

$$R_{hh}(\tau) \rightarrowtail S_{hh}(f) = |H(f)|^2 \tag{A.62}$$

and with $R_{yy}(\tau) \rightarrowtail S_{yy}(f)$ we obtain the power spectral density of the output process $Y(t)$ with (A.57)

$$S_{yy}(f) = |H(f)|^2 \, S_{xx}(f) \tag{A.63}$$

With the inverse Fourier transform we obtain the from (A.60)

$$R_{xx}(\tau) = \int_{-\infty}^{\infty} S_{xx}(f) e^{j2\pi f \tau} df \tag{A.64}$$

and for $\tau = 0$ the mean power of $X(t)$

$$R_{xx}(0) = \int_{-\infty}^{\infty} S_{xx}(f) df \tag{A.65}$$

Example: White noise $X(t)$ is defines by its constant power spectral density,

$$S_{xx}(f) = a = const. \ \forall \ f \tag{A.66}$$

Consequently the autocorrelation function is

$$R_{xx}(\tau) = a\delta(\tau) \tag{A.67}$$

and we see that $R_{xx}(\tau) = 0 \ \forall \ \tau \neq 0$. Thus, all pairs of random variables $X(t)$ and $X(t+\tau)$ are uncorrelated for $\tau \neq 0$. We observe that a is also the mean power of $X(t)$.

A.5   Modulation and Demodulation of a Stationary Stochastic Process

A.5.1   Modulation

We consider a WSS stationary process $X(t)$, which shall be modulated with the carrier $e^{j2\pi f_0 t}$. Then we obtain

$$Y(t) = X(t)e^{j2\pi f_0 t} \tag{A.68}$$

$X(t)$ shall have the autocorrelation function $R_{xx}(\tau) = \mathbf{E}[X^*(t)X(t+\tau)]$. For the autocorrelation function of the process $Y(t)$ we obtain

$$R_{yy}(\tau) = \mathbf{E}\left[Y^*(t)Y(t+\tau)\right] = \mathbf{E}\left[X(t+\tau)X^*(t)e^{j2\pi f_0 \tau}\right] \tag{A.69}$$

Noting that $e^{j2\pi f_0 \tau}$ is deterministic yields the final result

$$R_{yy}(\tau) = R_{xx}(\tau)e^{j2\pi f_0 \tau} \tag{A.70}$$

This shows that the modulation of a stationary stochastic process translates into the modulation of its autocorrelation function. With the frequency shifting property of the Fourier transform we obtain

$$R_{yy}(\tau) \rightarrowtail S_{yy}(f) = S_{xx}(f - f_0) \tag{A.71}$$

outlining that the modulation of $X(t)$ results in a frequency shift of its power spectral density $S_{xx}(f)$ by the carrier frequency f_0.

A.5.2 Demodulation

We consider the modulated stationary process $Y(t)$ and apply the synchronous demodulation with the carrier $e^{-j2\pi f_0 t}$ resulting in the demodulated stochastic process $Z(t)$,

$$Z(t) = Y(t)e^{-j2\pi f_0 t} = X(t) \tag{A.72}$$

Consequently, we obtain

$$R_{zz}(\tau) = R_{yy}(\tau)e^{-j2\pi f_0 \tau} = R_{xx}(\tau) \tag{A.73}$$

and

$$S_{zz}(f) = S_{yy}(f + f_0) = S_{xx}(f) \tag{A.74}$$

We also see that modulation and demodulation does not change the mean power of the processes, because with $\mathbf{E}[|X|^2] = R_{xx}(0)$, $\mathbf{E}[|Y|^2] = R_{yy}(0)$, and $\mathbf{E}[|Z|^2] = R_{zz}(0)$ follows

$$\mathbf{E}\left[|X|^2\right] = \mathbf{E}\left[|Y|^2\right] = \mathbf{E}\left[|Z|^2\right]. \tag{A.75}$$

A.6 Stationary, Real-Valued Bandpass Process

As is well known [3] any real-valued bandpass signal can be written in general as

$$n(t) = x(t)\cos(2\pi f_0 t) - y(t)\sin(2\pi f_0 t) \tag{A.76}$$

where $x(t)$ and $y(t)$ are real-valued lowpass signals with cut-off frequency f_c. This model shall be adopted to a stochastic bandpass process $N(t)$ with power spectral density $S_{nn}(f)$ and with the passband in the range of $f_0 - f_c \leq |f| \leq f_0 + f_c$, where $f_0 > f_c$ is the center frequency. $X(t)$ and $Y(t)$ shall be WSS lowpass processes with the power spectral densities

$$S_{xx}(f) \; ; \; S_{yy}(f) \begin{cases} \neq 0 \; ; \; |f| \leq f_c \\ = 0 \; ; \quad \text{else} \end{cases} \tag{A.77}$$

A.6.1 Condition for Stationarity

We would like to know under which conditions this bandpass process $N(t)$ is WSS stationary. Therefore we have to check whether $\mathbf{E}[N(t)] = const.$ holds and whether the autocorrelation function

$$R_{nn}(\tau) = \mathbf{E}\left[N(t)N(t+\tau)\right] \tag{A.78}$$

is independent of t.

Expected Value

With(A.76) follows

$$\mathbf{E}[N(t)] = \mathbf{E}[X(t)]\cos(2\pi f_0 t) - \mathbf{E}[Y(t)]\sin(2\pi f_0 t) \tag{A.79}$$

$\mathbf{E}[N(t)] = const.$ $\forall t$ holds, if

$$\mathbf{E}[X(t)] = \mathbf{E}[Y(t)] = 0 \tag{A.80}$$

Consequently from (A.79) also follows

$$\mathbf{E}[N(t)] = 0 \tag{A.81}$$

Autocorrelation Function

Next, please consider (A.78). By applying basic trigonometric formulas and using the fact, that terms with $\sin()$, $\cos()$, and the arguments $2\pi f_0 t$ and $2\pi f_0(t+\tau)$ are nonrandom and therefore can be taken out from the expectation operator, we obtain finally with (A.76)

$$
\begin{aligned}
R_{nn}(\tau) = \quad & \tfrac{1}{2}\left[R_{xx}(\tau) + R_{yy}(\tau)\right]\cos(2\pi f_0\tau)+ \\
& +\tfrac{1}{2}\left[R_{xx}(\tau) - R_{yy}(\tau)\right]\cos(4\pi f_0 t + 2\pi f_0\tau)- \\
& -\tfrac{1}{2}\left[R_{xy}(\tau) - R_{yx}(\tau)\right]\sin(2\pi f_0\tau)- \\
& -\tfrac{1}{2}\left[R_{xy}(\tau) + R_{yx}(\tau)\right]\sin(4\pi f_0 t + 2\pi f_0\tau)
\end{aligned} \tag{A.82}
$$

To get $R_{nn}(\tau)$ independent of t the following conditions must hold:

$$R_{xx}(\tau) = R_{yy}(\tau) \tag{A.83}$$

and

$$R_{xy}(\tau) = -R_{yx}(\tau) \tag{A.84}$$

Then we obtain from (A.82)

$$R_{nn}(\tau) = R_{xx}(\tau)\cos(2\pi f_0\tau) - R_{xy}(\tau)\sin(2\pi f_0\tau) \tag{A.85}$$

Knowing that an autocorrelation function provides the mean power of the process for $\tau = 0$ we conclude from (A.85) and (A.83)

$$\mathbf{E}\left[|N(t)|^2\right] = \mathbf{E}\left[|X(t)|^2\right] = \mathbf{E}\left[|Y(t)|^2\right] \tag{A.86}$$

Furthermore, we can find another property by applying (A.51) on (A.84) yielding

$$R_{yx}(-\tau) = -R_{yx}(\tau) \tag{A.87}$$

which indicates that $R_{yx}(\tau)$ is an odd function. Consequently, $R_{yx}(0) = 0$ must be true and the property (A.84) yields

$$R_{xy}(0) = 0 \tag{A.88}$$

This means that the random variables $X(t)$ and $Y(t)$ for any given t are not correlated. Please note that $R_{xy}(0) = 0$ does not require $X(t + \tau)$ and $Y(t)$ to be uncorrelated for any τ. However, if the zero mean processes $X(t)$ and $Y(t)$ are assumed to be uncorrelated for any τ, $R_{xy}(\tau) = 0 \; \forall \tau$ holds and from (A.85) follows

$$R_{nn}(\tau) = R_{xx}(\tau) \cos(2\pi f_0 \tau) \tag{A.89}$$

Using the Wiener–Khintchine theorem (A.60) and the frequency shifting property of the Fourier transform we obtain from (A.89) the power spectral density of the process $N(t)$

$$S_{nn}(f) = \frac{1}{2}\left[S_{xx}(f - f_0) + S_{xx}(f + f_0)\right] \tag{A.90}$$

which clearly exhibits a bandpass shape.

A.6.2 Summary on Stationary Bandpass Process

A WSS bandpass process $N(t)$ exhibits zero mean and is composed of the in-phase component $X(t)$ and the quadrature component $Y(t)$, which are zero mean WSS lowpass processes. Moreover, $N(t)$, $X(t)$, and $Y(t)$ have the same mean power. For the cross-correlation holds $R_{xy}(0) = \mathbf{E}[X(t)Y(t)] = 0$, which means that the random variables $X(t)$ and $Y(t)$ are assumed to be uncorrelated for any fixed t.

A.6.3 Complex Envelope of a Bandpass Process

It is straightforward to show that (A.76) can be written as

$$N(t) = \mathrm{Re}\left[Z(t)e^{j2\pi f_0 t}\right] \tag{A.91}$$

where

$$Z(t) = X(t) + jY(t) \tag{A.92}$$

is called the complex envelope. If $X(t)$ and $Y(t)$ are WSS lowpass processes, then $Z(t)$ is a WSS complex lowpass process. It is easy to show that for the autocorrelation function of $Z(t)$ follows with (A.83) and (A.84)

$$R_{zz}(\tau) = \mathbf{E}\left[Z^*(t)Z(t+\tau)\right] = 2\left[R_{xx}(\tau) - jR_{xy}(\tau)\right]. \qquad (A.93)$$

A.7 Two-Dimensional Gaussian Process

A.7.1 Joint Gaussian Probability Density Function

We consider now two real-valued SSS Gaussian processes $X(t)$ and $Y(t)$. For any fixed t they represent random variables [1] with the Gaussian joint probability density function

$$p_{xy}(x, y) = \frac{1}{2\pi\sigma_x\sigma_y\sqrt{1-\rho^2}} e^{-\frac{(x-m_x)^2\sigma_y^2 - 2(x-m_x)(y-m_y)\rho\sigma_x\sigma_y + (y-m_y)^2\sigma_x^2}{2\sigma_x^2\sigma_y^2(1-\rho^2)}} \qquad (A.94)$$

with the mean values m_x and m_y,
the variances

$$\sigma_x^2 = \mathbf{E}[X^2] - m_x^2 \; ; \; \sigma_y^2 = \mathbf{E}[Y^2] - m_y^2 \qquad (A.95)$$

the normalized cross-covariance (correlation coefficient)

$$\rho = \frac{R_{xy}(0) - m_x m_y}{\sigma_x \sigma_y} \qquad (A.96)$$

and the marginal probability density function of $X(t)$

$$p_x(x) = \frac{1}{\sqrt{2\pi}\sigma_x} e^{-\frac{(x-m_x)^2}{2\sigma_x^2}} \qquad (A.97)$$

as well as of $Y(t)$

$$p_y(y) = \frac{1}{\sqrt{2\pi}\sigma_y} e^{-\frac{(y-m_y)^2}{2\sigma_y^2}} \qquad (A.98)$$

A.7.2 Uncorrelated Gaussian Random Processes

Let $X(t)$ and $Y(t)$ be two WSS, real and uncorrelated Gaussian processes. Then $R_{xy}(\tau) = m_x m_y$ holds according to (A.49). This is valid for any τ, including $\tau = 0$. With $R_{xy}(0) = m_x m_y$ follows from (A.96)

$$\rho = 0 \qquad\qquad\qquad\qquad (A.99)$$

and consequently from (A.94) with (A.97) and (A.98) follows

$$p_{xy}(x, y) = p_x(x)p_y(y) \qquad\qquad\qquad (A.100)$$

Hence, we conclude that uncorrelated Gaussian processes $X(t)$ and $Y(t)$ are even statistically independent.

A.7.3   Complex Gaussian Random Process

Let $X(t)$ and $Y(t)$ be WSS real-valued Gaussian lowpass processes with properties given in Sect. A.7.1. Then they constitute a complex Gaussian random lowpass process

$$Z(t) = X(t) + jY(t) \qquad\qquad\qquad (A.101)$$

A.7.4   Gaussian Bandpass Process

Any real-valued bandpass process in general is given by (A.76)

$$n(t) = x(t) \cos(2\pi f_0 t) - y(t) \sin(2\pi f_0 t)$$

If $X(t)$ and $Y(t)$ are stationary Gaussian lowpass processes with properties given in Sect. A.7.1, we denote $N(t)$ as a Gaussian bandpass process. We know from Sect. A.6 that $N(t)$, $X(t)$, and $Y(t)$ have zero mean and the same mean power. Furthermore, if the two Gaussian lowpass processes $X(t)$ and $Y(t)$ are uncorrelated then they are even statistically independent.

A.8   Sampling of a Stochastic Process

The received signal in a digital communication system is a random process. Before detection, the signal is sampled. In this section, the main basics for sampling of a stochastic process are summarized.

A.8.1   Prerequisites

$X(t)$ shall be a WSS stochastic process. Then $X(t)$ and $X(t + a)$ $\forall a$ have the same density function, because $p_x(x)$ is independent of t. Consequently, after sampling with a sampling frequency $f_S = \frac{1}{T}$ the resulting samples $X(kT)$ exhibit the same probability density function $p_x(x)$ as $X(t)$. Furthermore, $X(kT) = X_S(k)$ can be considered as a sequence of equidistant random variables of the process $X(t)$, which constitute a stationary discrete-time stochastic process, where $k \in \mathbb{Z}$ is the discrete-time.

A.8.2   Auto- and Cross-Correlation Function of a Discrete-Time Stochastic Process

$X(kT)$ for every fixed $k \in \mathbb{Z}$ can be considered as a random variable. Consequently, we apply (A.40) to get the autocorrelation function as $R_{xx}(\tau) = \mathbf{E}[X^*(kT)X(kT + \tau)]$. Obviously, $X(kT) = X_S(k)$ is only defined for discrete-time arguments. Therefore $R_{xx}(\tau)$ can take on defined values only for $\tau = lT$ with $l \in \mathbb{Z}$ and the autocorrelation function (autocorrelation sequence) of $X_S(k)$ will become a function of a discrete variable to be be written as

$$R_{x_S x_S}(l) = \mathbf{E}[X_S^*(k)X_S(k + l)] \qquad (A.102)$$

Interestingly

$$R_{x_S x_S}(l) = R_{xx}(lT) \qquad (A.103)$$

can also be considered as the sampled version of the continuous-time autocorrelation function $R_{xx}(\tau)$. It is straightforward to show that the properties of (A.41) hold similarly as

$$R_{x_S x_S}(-l) = R_{x_S x_S}^*(l) \; ; \; R_{x_S x_S}(0) \geq |R_{x_S x_S}(l)| \qquad (A.104)$$

With the same arguments as for (A.47) we can define the *cross-correlation function* of two discrete-time WSS stochastic processes $X_S(k)$ and $Y_S(k)$ as

$$R_{x_S y_S}(l) = \mathbf{E}[X_S(k + l)Y_S^*(k)] \qquad (A.105)$$

Symmetry properties and conditions for uncorrelated processes are similar as for the continuous-time processes.

A.8.3   Power Spectral Density

According to the Wiener–Khintchine theorem we find the power density spectrum $S_{x_S x_S}(f)$ of the WSS stochastic process $X_S(k)$ by the Fourier transform of the sampled autocorrelation function. Applying ideal sampling on $R_{xx}(\tau)$ yields[2]

---

[2]We multiply the Dirac impulses by T to ensure that the autocorrelation functions on both sides of the equation have the same physical dimension.

$$R_{xx,S}(\tau) = R_{xx}(\tau) \sum_{l=-\infty}^{\infty} T\delta(\tau - lT) \tag{A.106}$$

With the Fourier correspondence

$$\sum_{l=-\infty}^{\infty} T\delta(\tau - lT) \;\longmapsto\; \sum_{m=-\infty}^{\infty} \delta\left(f - m\frac{1}{T}\right) \tag{A.107}$$

and with $R_{xx}(\tau) \longmapsto S_{xx}(f)$ we obtain from (A.106)

$$R_{xx,S}(\tau) \;\longmapsto\; S_{x_Sx_S}(f) = S_{xx}(f) * \sum_{m=-\infty}^{\infty} \delta\left(f - m\frac{1}{T}\right)$$

which results after executing the convolution integral in

$$S_{x_Sx_S}(f) = \sum_{m=-\infty}^{\infty} S_{xx}\left(f - m\frac{1}{T}\right) \tag{A.108}$$

We see that the spectrum is a periodic repetition of the baseband power spectral density $S_{xx}(f)$, where the period is given by the sampling frequency $\frac{1}{T}$.

Appendix B
Some Fundamentals of Linear Algebra

B.1 Eigenvalue Decomposition

In this Section, we review some properties of the eigenvalue decomposition of a matrix $\mathbf{A}$, assuming for the moment that such a decomposition shall exist for the given matrix. The eigenvalue–eigenvector problem of linear algebra can be stated as follows: Given a NxN matrix $\mathbf{A} \epsilon \mathbb{C}^{N \times N}$ with in general complex entries a_{ij}, a column vector $\mathbf{v}_i \epsilon \mathbb{C}^{N \times 1}$, and a scalar factor λ_i. We are looking for the vector $\mathbf{A}\mathbf{v}_i$, which is equal to the vector $\lambda_i \mathbf{v}_i$

$$\mathbf{A}\mathbf{v}_i = \lambda_i \mathbf{v}_i \; ; \; i = 1, \ldots, N \tag{B.1}$$

with $\mathbf{v}_i \neq \mathbf{0}$, otherwise we would have the trivial solution, which is of no interest. A matrix can change the length (by its determinant) and the direction of the vector after multiplication. Thus, we are looking for the vector $\mathbf{A}\mathbf{v}_i$ with the same direction as $\mathbf{v}_i$ but with the length changed by λ_i. The nontrivial solutions λ_i and $\mathbf{v}_i$ are called eigenvalues and eigenvectors of the matrix, respectively. The set of all eigenvalues is denoted as spectrum and the absolute value of the largest eigenvalue is referred to as spectral radius. We can rewrite (B.1) with matrix notation as

$$\mathbf{A}\mathbf{V} = \mathbf{V}\mathbf{\Lambda} \tag{B.2}$$

where

$$\mathbf{V} = \begin{pmatrix} \mathbf{v}_1 & \mathbf{v}_2 & \ldots & \mathbf{v}_N \end{pmatrix} \epsilon \mathbb{C}^{N \times N} \tag{B.3}$$

is the matrix of eigenvectors and

$$\mathbf{\Lambda} = \text{diag}\,(\lambda_1, \lambda_2, \ldots, \lambda_N) \tag{B.4}$$

is a diagonal matrix composed of the eigenvalues of $\mathbf{A}$. To solve the eigenvalue–eigenvector problem (B.1) can be written as

© Springer Nature Switzerland AG 2019
J. Speidel, *Introduction to Digital Communications*, Signals and Communication Technology, https://doi.org/10.1007/978-3-030-00548-1

$$\mathbf{Av}_i - \lambda_i \mathbf{v}_i = \mathbf{0} \iff (\mathbf{A} - \lambda_i \mathbf{I}_N)\, \mathbf{v}_i = \mathbf{0}\;;\;\; i = 1, \ldots, N \qquad (B.5)$$

This system of homogeneous equations has a nontrivial solution only if

$$\det(\mathbf{A} - \lambda_i \mathbf{I}_N) = 0\;;\;\; i = 1, \ldots, N \qquad (B.6)$$

with the $N \times N$ identity matrix $\mathbf{I}_N$. Equation (B.6) is called characteristic equation for the matrix $\mathbf{A}$ and the left-hand side is the characteristic polynomial with degree N as a function of λ_i. We can conclude that the eigenvalues of the matrix $\mathbf{A}$ are the roots of the characteristic polynomial. After all λ_i are calculated from (B.6) we can insert each into (B.5) and find the corresponding eigenvectors $\mathbf{v}_i$. Note that the solution for each $\mathbf{v}_i$ contains at least one free parameter, because (B.5) is a homogeneous system of equation and thus the rank of the matrix $\mathbf{A} - \lambda_i \mathbf{I}_N$ is

$$\mathrm{rank}\,(\mathbf{A} - \lambda_i \mathbf{I}_N) \le N - 1 \qquad (B.7)$$

The free parameters have to be used to normalize all eigenvectors such that

$$\mathbf{v}_i^H \mathbf{v}_i = \|\mathbf{v}_i\|^2 = 1\;;\;\; i = 1, \ldots, N \qquad (B.8)$$

Please note that the resulting eigenvectors associated to different eigenvalues are non-orthogonal in general, i.e.,

$$\mathbf{v}_i^H \mathbf{v}_j = 0\;;\;\; i = 1, \ldots, N\;;\;\; i \ne j \qquad (B.9)$$

does not hold. Equation (B.9) is true for Hermiteian and symmetric matrices with real entries, see Sect. B.3. On the other hand we will see later that the singular value decomposition of a matrix yields pairwise orthogonal eigenvectors. The components of the eigenvectors are in general complex, where $\mathbf{v}_i^H = \left(\mathbf{v}_i^*\right)^T = \left(\mathbf{v}_i^T\right)^*$ is the conjugate transpose vector (Hermiteian vector) to $\mathbf{v}_i$. The inverse matrix $\mathbf{V}^{-1}$ is obtained from the relation

$$\mathbf{V}^{-1}\mathbf{V} = \mathbf{V}\mathbf{V}^{-1} = \mathbf{I}_N \qquad (B.10)$$

Taking $\mathbf{V}$ we can now transform $\mathbf{A}$ into diagonal form $\mathbf{\Lambda}$. The procedure is also called principal axis transformation or eigenvalue decomposition of $\mathbf{A}$. For that purpose we multiply (B.2) from the left by $\mathbf{V}^{-1}$ and obtain with (B.10)

$$\mathbf{V}^{-1}\mathbf{A}\mathbf{V} = \mathbf{\Lambda} \qquad (B.11)$$

$\mathbf{V}$ is therefore called transform matrix. We see that there is no need for orthogonal eigenvectors. If we multiply (B.11) in a first step from the left with $\mathbf{V}$ and secondly from the right with $\mathbf{V}^{-1}$ we get

$$\mathbf{A} = \mathbf{V}\mathbf{\Lambda}\mathbf{V}^{-1} \qquad (B.12)$$

which is another form of the eigenvalue decomposition or diagonalization of $\mathbf{A}$. Such an eigenvalue decomposition will not exist for all square matrices, in other words not all square matrices are diagonalizable. In any case the inverse matrix $\mathbf{V}^{-1}$ must exist. Alternative formulation are: $\mathbf{V}$ must be a non-singular matrix or all eigenvectors $\mathbf{v}_i$ are linearly independent of each other. In the following, let us pinpoint some special $N{\times}N$ matrices, which are diagonalizable, and which are of importance for MIMO systems.

B.2   Normal Matrices

By definition $\mathbf{A}$ is a normal matrix, if and only if

$$\mathbf{A}^H\mathbf{A} = \mathbf{A}\mathbf{A}^H \tag{B.13}$$

$\mathbf{A}^H$ is called the Hermiteian[3] or conjugate transpose matrix with respect to $\mathbf{A}$. The Hermiteian operator $(...)^H$ is defined as

$$\mathbf{A}^H = \left(\mathbf{A}^*\right)^T = (\mathbf{A}^T)^* \tag{B.14}$$

It can be shown [4] that every normal matrix is unitarily diagonalizable, i.e., an eigenvalue decomposition exists, where the transform matrix $\mathbf{V}$ is a unitary matrix defined by

$$\mathbf{V}^H\mathbf{V} = \mathbf{V}\mathbf{V}^H = \mathbf{I}_N \iff \mathbf{V}^H = \mathbf{V}^{-1} \tag{B.15}$$

Then from (B.11) follows

$$\mathbf{\Lambda} = \mathbf{V}^H\mathbf{A}\mathbf{V} \tag{B.16}$$

It should be noted that (B.13) is only a sufficient and not a necessary condition for diagonalizable matrices. That means there are diagonalizable matrices which are not normal. An example is the matrix $\begin{pmatrix} 0 & 1 \\ 4 & 0 \end{pmatrix}$. Please note that the Hermiteian operator can also be applied to non-square matrices $\mathbf{A}\epsilon\,\mathbb{C}^{M{\times}N}$ and $\mathbf{B}\epsilon\,\mathbb{C}^{N{\times}M}$ with the property similar to the transposition operation

$$(\mathbf{AB})^H = \mathbf{B}^H\mathbf{A}^H \tag{B.17}$$

---

[3] Charles Hermite, French mathematician.

B.3   Hermiteian Matrices

Definition of a Hermiteian Matrix

By definition, $\mathbf{A} \in \mathbb{C}^{N \times N}$ is called a Hermiteian matrix, if and only if

$$\mathbf{A}^H = \mathbf{A} \tag{B.18}$$

It is easy to show that a Hermiteian matrix is also a normal matrix. For the proof, we insert (B.18) into (B.13) and obtain

$$\mathbf{A}^H \mathbf{A} = \mathbf{A}\mathbf{A} = \mathbf{A}\mathbf{A}^H \tag{B.19}$$

As a consequence, Hermiteian matrices are also unitarily diagonalizable given by (B.16).

Quadratic Form

Let $\mathbf{A} \in \mathbb{C}^{N \times N}$ be a Hermiteian matrix with eigenvalues λ_i ; $i = 1, \ldots, N$. Then this matrix can be defined as

$$\mathbf{A} = \mathbf{a}\mathbf{a}^H \tag{B.20}$$

with the column vector $\mathbf{a} \in \mathbb{C}^{N \times 1}$. With the column vector $\mathbf{z} \in \mathbb{C}^{N \times 1}$ we define the quadratic form

$$\mathbf{z}^H \mathbf{A} \mathbf{z} \tag{B.21}$$

which has the property

$$\mathbf{z}^H \mathbf{A} \mathbf{z} \geq 0 \ \forall \mathbf{z} \neq 0 \tag{B.22}$$

For the proof we insert (B.20) into (B.22) and find

$$\mathbf{z}^H \mathbf{A} \mathbf{z} = \mathbf{z}^H \mathbf{a}\mathbf{a}^H \mathbf{z} = \mathbf{z}^H \mathbf{a} \left(\mathbf{z}^H \mathbf{a}\right)^H = \left|\mathbf{z}^H \mathbf{a}\right|^2 \geq 0 \ \forall \mathbf{z} \neq 0 \tag{B.23}$$

Eigenvalues of a Hermiteian Matrix

All eigenvalues λ_i of a Hermiteian matrix are positive, i.e.,

$$\lambda_i \geq 0 \ ; \ i = 1, \ldots, N \tag{B.24}$$

This implies that all eigenvalues of a Hermiteian matrix are real. For the proof let $\mathbf{v}_i$ be the eigenvector associated with the eigenvalue λ_i. Then

$$\mathbf{A}\mathbf{v}_i = \lambda_i \mathbf{v}_i \tag{B.25}$$

The corresponding positive semi-definite quadratic form according to (B.23) then is

$$\mathbf{v}_i^H \mathbf{A} \mathbf{v}_i = \mathbf{v}_i^H \lambda_i \mathbf{v}_i = \lambda_i \, \|\mathbf{v}_i\|^2 \geq 0 \tag{B.26}$$

from which we conclude the proposition (B.24).

Eigenvectors of a Hermiteian Matrix

Lemma

The eigenvectors $\mathbf{v}_i \in \mathbb{C}^{N \times 1}$ and $\mathbf{v}_j \in \mathbb{C}^{N \times 1}$ of a Hermiteian matrix $\mathbf{A}$ associated with two different nonzero eigenvalues $\lambda_i \neq \lambda_j \neq 0$ are (pairwise) orthogonal, i.e.,

$$\mathbf{v}_i^H \mathbf{v}_j = \mathbf{0} \; ; \; i = 1, \ldots, N \; ; \; i \neq j \tag{B.27}$$

For the proof we use the definition of the eigenvectors $\mathbf{A}\mathbf{v}_i = \lambda_i \mathbf{v}_i$ and $\mathbf{A}\mathbf{v}_j = \lambda_j \mathbf{v}_j$. Then we calculate

$$\mathbf{v}_i^H \mathbf{v}_j \lambda_i = (\lambda_i \mathbf{v}_i)^H \mathbf{v}_j = (\mathbf{A}\mathbf{v}_i)^H \mathbf{v}_j = \mathbf{v}_i^H \mathbf{A}^H \mathbf{v}_j = \mathbf{v}_i^H \mathbf{A} \mathbf{v}_j = \mathbf{v}_i^H \mathbf{v}_j \lambda_j \tag{B.28}$$

and the result is $\mathbf{v}_i^H \mathbf{v}_j \lambda_i = \mathbf{v}_i^H \mathbf{v}_j \lambda_j$. As the eigenvalues are unequal and unequal to zero, proposition (B.27) follows and the proof is finished.

Please note that (B.27) also holds, if the eigenvectors are not normalized, which can be easily proven by checking (B.27) with the vectors $\alpha_i \mathbf{v}_i$ and $\alpha_j \mathbf{v}_j$.

Orthogonal and normalized vectors are called orthonormal. For the matrix $\mathbf{V}$ of eigenvectors in (B.3) then follows

$$\mathbf{V}^H \mathbf{V} = \begin{pmatrix} \mathbf{v}_1^H \\ \mathbf{v}_2^H \\ \vdots \\ \mathbf{v}_N^H \end{pmatrix} \begin{pmatrix} \mathbf{v}_1 & \mathbf{v}_2 & \ldots & \mathbf{v}_N \end{pmatrix} = \mathbf{I}_N \tag{B.29}$$

and with (B.10) we conclude that $\mathbf{V}$ is unitary, $\mathbf{V}^H = \mathbf{V}^{-1}$.

B.4 Unitary Matrices

Definition

$\mathbf{V} \in \mathbb{C}^{N \times N}$ is called a unitary matrix, if and only if

$$\mathbf{V}^{-1} = \mathbf{V}^H \tag{B.30}$$

The inverse matrix $\mathbf{V}^{-1}$ is the solution of

$$\mathbf{V}^{-1} \mathbf{V} = \mathbf{V} \mathbf{V}^{-1} = \mathbf{I}_N \tag{B.31}$$

Consequently, with (B.30) follows

$$\mathbf{V}^H \mathbf{V} = \mathbf{V} \mathbf{V}^H = \mathbf{I}_N \tag{B.32}$$

$\mathbf{V}$ is composed of orthonormal column vectors.

$$\mathbf{V} = \begin{bmatrix} \mathbf{v}_1 & \mathbf{v}_2 & \dots & \mathbf{v}_N \end{bmatrix} \tag{B.33}$$

satisfying (B.8) and (B.9), i.e.,

$$\mathbf{v}_i^H \mathbf{v}_j = \begin{cases} 1 \; ; \; i = j = 1, \dots, N \\ 0 \; ; \; i, \; j = 1, \dots, N \; ; \; i \neq j \end{cases} \tag{B.34}$$

Properties

- An inverse matrix is defined only for a square matrix. Therefore, all unitary matrices are square matrices.
- Equation (B.32) is also the property of a normal matrix. Consequently, unitary matrices are a subset of normal matrices and thus unitarily diagonalizable. With a unitary transform matrix $\mathbf{V}$ the eigenvalue decomposition of (B.12) can be written as

$$\mathbf{A} = \mathbf{V} \mathbf{\Lambda} \mathbf{V}^H \tag{B.35}$$

- All eigenvalues ϱ_i of a unitary matrix $\mathbf{V} \epsilon \mathbb{C}^{N \times N}$ have absolute values equal to 1

$$|\varrho_i| = 1 \; ; \; i = 1, \dots, N \tag{B.36}$$

Proof: We calculate the scalar product of the two vectors and obtain with (B.32)

$$(\mathbf{V} \mathbf{v}_i)^H \, \mathbf{V} \mathbf{v}_i = \mathbf{v}_i^H \mathbf{V}^H \mathbf{V} \mathbf{v}_i = \|\mathbf{v}_i\|^2 \tag{B.37}$$

On the other hand the eigenvalue–eigenvector condition

$$\mathbf{V} \mathbf{v}_i = \varrho_i \mathbf{v}_i \tag{B.38}$$

holds. The left-hand side of (B.37) yields with (B.38)

$$(\mathbf{V} \mathbf{v}_i)^H \, \mathbf{V} \mathbf{v}_i = \varrho_i^* \varrho_i \mathbf{v}_i^H \mathbf{v}_i = |\varrho_i|^2 \, \|\mathbf{v}_i\|^2 \tag{B.39}$$

The left-hand sides of (B.37) and (B.39) are identical. Consequently, this must also hold for the right-hand sides

$$\|\mathbf{v}_i\|^2 = |\varrho_i|^2 \, \|\mathbf{v}_i\|^2 \tag{B.40}$$

from which the proposition (B.36) directly follows and the proof is finalized.

- The input signal $\mathbf{s}$ and output signal $\mathbf{y} = \mathbf{Vs}$ of a system described by a unitary matrix $\mathbf{V}$ have the same mean power

$$\mathbf{E}\left[\|\mathbf{y}\|^2\right] = \mathbf{E}\left[\|\mathbf{s}\|^2\right] \tag{B.41}$$

Proof:

$$\mathbf{E}\left[\|\mathbf{y}\|^2\right] = \mathbf{E}\left[(\mathbf{Vs})^H \mathbf{Vs}\right] = \mathbf{E}\left[\mathbf{s}^H \mathbf{V}^H \mathbf{Vs}\right] = \mathbf{E}\left[\mathbf{s}^H \mathbf{I}_N \mathbf{s}\right] = \mathbf{E}\left[\|\mathbf{s}\|^2\right]. \tag{B.42}$$

B.5 Norm of a Vector, Norm of a Matrix

The squared norm of a vector $\mathbf{v} = \left(v_1 \ v_2 \ \cdots \ v_N \right)^T$ with complex components is given by the sum of the squared absolute values of the components

$$\|\mathbf{v}\|^2 = \sum_{i=1}^{N} |v_i|^2 \tag{B.43}$$

The squared norm (Frobenius norm) of a matrix $A \in \mathbb{C}^{M \times N}$ with complex entries a_{ij} is given by the sum of the squared absolute values of the entries

$$\|\mathbf{A}\|_F^2 = \sum_{i=1}^{M} \sum_{j=1}^{N} |a_{ij}|^2 = \sum_{j=1}^{N} \|\mathbf{a}_j\|^2 \tag{B.44}$$

Alternatively, the Frobenius norm can be calculated as the sum of the squared norms of the column vectors $\mathbf{a}_j$ or row vectors of $\mathbf{A}$, respectively.

B.6 Singular Value Decomposition

The Procedure

The Singular Value Decomposition (SVD) of a matrix $\mathbf{H} \in \mathbb{C}^{N \times M}$ is given by

$$\mathbf{H} = \mathbf{UDV}^H \tag{B.45}$$

with

$$\mathbf{D} = \begin{pmatrix} \sqrt{\lambda_1} & 0 & 0 & \cdots & 0 & 0 & \cdots & 0 \\ 0 & \sqrt{\lambda_2} & 0 & \cdots & 0 & 0 & \cdots & 0 \\ & & \ddots & & & & & \\ 0 & 0 & 0 & \cdots & \sqrt{\lambda_P} & 0 & \cdots & 0 \\ 0 & 0 & 0 & \cdots & 0 & 0 & \cdots & 0 \\ & & \vdots & \cdots & & & \ddots & \\ 0 & 0 & 0 & \cdots & 0 & 0 & \cdots & 0 \\ 0 & 0 & 0 & \cdots & 0 & 0 & \cdots & 0 \end{pmatrix} = \begin{pmatrix} \mathbf{\Lambda}_P^{\frac{1}{2}} & 0 \cdots 0 \\ 0 & 0 \cdots 0 \\ \vdots & \vdots \cdots \vdots \\ 0 & 0 \cdots 0 \end{pmatrix} \epsilon \, \mathbb{R}^{N \times M}$$

$$\tag{B.46}$$

$\lambda_1 \geq \lambda_2 \geq \cdots \lambda_P > 0$, and $\lambda_{P+1} = \lambda_{P+2} = \cdots = \lambda_N = 0$ are the N eigenvalues of the Hermiteian matrix

$$\mathbf{Q}_N = \mathbf{H}\mathbf{H}^H \; \epsilon \, \mathbb{C}^{N \times N} \tag{B.47}$$

and

$$P = \text{rank}\,(\mathbf{Q}_N) \tag{B.48}$$

is the rank of the matrix $\mathbf{Q}_N$. In general the rank of a matrix $\mathbf{H} \epsilon \, \mathbb{C}^{N \times M}$ is defined as the number of linearly independent rows or columns of the matrix, thus

$$\text{rank}\,(\mathbf{H}) \leq \min\{M, N\} \tag{B.49}$$

From this definition follows for (B.48)

$$P \leq N \tag{B.50}$$

$\sqrt{\lambda_i}$; $i = 1, \ldots, P$ are called the singular values of the matrix $\mathbf{H}$.
 $\mathbf{U} \epsilon \, \mathbb{C}^{N \times N}$ and $\mathbf{V} \epsilon \, \mathbb{C}^{M \times M}$ are unitary matrices, thus

$$\mathbf{U}^{-1} = \mathbf{U}^H \; ; \; \mathbf{V}^{-1} = \mathbf{V}^H \tag{B.51}$$

hold. Furthermore, $\mathbf{U}$ is the matrix of the normalized eigenvectors with respect to the eigenvalues $\lambda_1, \lambda_2, \cdots, \lambda_N$. Let

$$\mathbf{\Lambda}_N = \text{diag}\,(\lambda_1, \lambda_2, \ldots, \lambda_P, 0, \ldots, 0) \; \epsilon \, \mathbb{R}^{N \times N} \tag{B.52}$$

be a diagonal matrix composed of the eigenvalues of $\mathbf{Q}_N$. Then the eigenvalue decomposition of $\mathbf{Q}_N$ is

$$\mathbf{U}^H \mathbf{Q}_N \mathbf{U} = \mathbf{\Lambda}_N \tag{B.53}$$

One method to find the matrix $\mathbf{V}$ in (B.45) is the eigenvalue decomposition of the matrix

$$\mathbf{Q}_M = \mathbf{H}^H \mathbf{H} \; \epsilon \, \mathbb{C}^{M \times M} \tag{B.54}$$

which is

$$\mathbf{V}^H \mathbf{Q}_M \mathbf{V} = \mathbf{\Lambda}_M \tag{B.55}$$

with the diagonal matrix

$$\mathbf{\Lambda}_M = \text{diag} \; (\lambda_1, \lambda_2, \ldots, \lambda_P, 0, \ldots, 0) \; \epsilon \, \mathbb{R}^{M \times M} \tag{B.56}$$

$\mathbf{V}$ is the matrix of eigenvectors of $\mathbf{Q}_M$ with respect to the eigenvalues $\lambda_1, \lambda_2, \cdots, \lambda_M$. Note that the eigenvalues $\lambda_1, \lambda_2, \cdots, \lambda_P$ are the same as for the matrix $\mathbf{Q}_N$. Furthermore

$$\text{rank} \, (\mathbf{Q}_M) = \text{rank} \, (\mathbf{Q}_N) = P \tag{B.57}$$

holds and $\mathbf{\Lambda}_M$ as well as $\mathbf{\Lambda}_N$ contain the same diagonal matrix

$$\mathbf{\Lambda}_P = \text{diag} \; (\lambda_1, \lambda_2, \ldots, \lambda_P) \; \epsilon \, \mathbb{R}^{P \times P} \tag{B.58}$$

of the P eigenvalues, which are unequal to zero. Note that in (B.46)

$$\mathbf{\Lambda}_P^{\frac{1}{2}} = \text{diag} \; \left(\sqrt{\lambda_1}, \sqrt{\lambda_2}, \ldots, \sqrt{\lambda_P} \right) \tag{B.59}$$

holds.

Notes

- In contrast to the eigenvalue decomposition, which is only feasible for square matrices ($M = N$), the SVD in (B.45) can be done for any matrix $\mathbf{H} \epsilon \, \mathbb{C}^{N \times M}$ with arbitrary M and N.
- Exercise: Consider the SVD of a square matrix $\mathbf{H} \epsilon \, \mathbb{C}^{M \times M}$.
- The matrix $\mathbf{D}$ in (B.46) contains the square matrix $\mathbf{\Lambda}_P^{\frac{1}{2}} \epsilon \, \mathbb{R}^{P \times P}$ defined in (B.59). As the remaining elements in $\mathbf{D}$ are zero, the SVD can also be formulated with $\mathbf{D} = \mathbf{\Lambda}_P^{\frac{1}{2}}$, a non-square matrices $\mathbf{U} \epsilon \, \mathbb{C}^{N \times P}$, and $\mathbf{V} \epsilon \, \mathbb{C}^{M \times P}$.

Proof of SVD Lemma

Proof of (B.45) and (B.46)

We prove that with the eigenvalue decomposition (B.53) of $\mathbf{Q}_N$ and with a unitary matrix $\mathbf{V} \epsilon \, \mathbb{C}^{M \times M}$ the proposition (B.45) with (B.46) follows. First, we easily see that $\mathbf{Q}_N$ in (B.47) is a Hermiteian matrix, i.e., $\mathbf{Q}_N = \mathbf{Q}_N^H$ holds. We know from sections (B.2) and (B.3) that for any Hermiteian matrix an eigenvalue decomposition exists according to (B.53). By inserting (B.47) we obtain

$$\mathbf{U}^H \mathbf{H} \mathbf{H}^H \mathbf{U} = \mathbf{\Lambda}_N \tag{B.60}$$

Now we introduce the following identity matrix $\mathbf{I} = \mathbf{V} \mathbf{V}^H$ making use of the prerequisite (B.51) that $\mathbf{V}$ is a unitary matrix, which yields

$$\mathbf{U}^H \mathbf{H} \mathbf{V} \mathbf{V}^H \mathbf{H}^H \mathbf{U} = \mathbf{U}^H \mathbf{H} \mathbf{V} \left(\mathbf{U}^H \mathbf{H} \mathbf{V} \right)^H = \mathbf{\Lambda}_N \tag{B.61}$$

Next we decompose the diagonal matrix on the right-hand side into the product of two matrices

$$\mathbf{\Lambda}_N = \mathbf{D} \mathbf{D}^H \tag{B.62}$$

Inserting (B.62) into (B.61) results in

$$\mathbf{U}^H \mathbf{H} \mathbf{V} \left(\mathbf{U}^H \mathbf{H} \mathbf{V} \right)^H = \mathbf{D} \mathbf{D}^H \tag{B.63}$$

and by comparison of the left and right-hand part we obtain

$$\mathbf{U}^H \mathbf{H} \mathbf{V} = \mathbf{D} \tag{B.64}$$

from which we conclude the proposition $\mathbf{H} = \mathbf{U} \mathbf{D} \mathbf{V}^H$ and the proof ends. The only condition we have imposed so far on $\mathbf{V} \in \mathbb{C}^{M \times M}$ is the requirement that $\mathbf{V}$ is a unitary matrix. Moreover, we see that the derived SVD holds for arbitrary matrices $\mathbf{H} \in \mathbb{C}^{N \times M}$.

Proof of (B.55)

We now prove that $\mathbf{V} \in \mathbb{C}^{M \times M}$ can be obtained by the eigenvalue decomposition of $\mathbf{Q}_M = \mathbf{H}^H \mathbf{H}$. Assume that the singular value decomposition of $\mathbf{H}$ is given by (B.45). From this equation follows by applying the Hermiteian operation on both sides

$$\mathbf{H}^H = \mathbf{V} \mathbf{D}^H \mathbf{U}^H \tag{B.65}$$

Multiplying (B.65) from the right-hand side with (B.45) and knowing that $\mathbf{U}^H \mathbf{U} = \mathbf{I}_N$ results in

$$\mathbf{H}^H \mathbf{H} = \mathbf{V} \mathbf{D}^H \mathbf{U}^H \mathbf{U} \mathbf{D} \mathbf{V}^H = \mathbf{V} \mathbf{D}^H \mathbf{D} \mathbf{V}^H \tag{B.66}$$

From (B.46) follows

$$\mathbf{D}^H \mathbf{D} = \mathbf{\Lambda}_M \tag{B.67}$$

and we obtain from (B.66)

$$\mathbf{Q}_M = \mathbf{H}^H \mathbf{H} = \mathbf{V} \mathbf{\Lambda}_M \mathbf{V}^H \tag{B.68}$$

From (B.68) follows by multiplication with $\mathbf{V}^H$ and $\mathbf{V}$ directly the eigenvalue decomposition $\mathbf{V}^H \mathbf{Q}_M \mathbf{V} = \mathbf{\Lambda}_M$ of $\mathbf{Q}_M$ in (B.55). Consequently, $\mathbf{V}$ must be the matrix of eigenvectors associated to the eigenvalues given in $\mathbf{\Lambda}_M$. As $\mathbf{Q}_M$ is a Hermiteian matrix, we know from Sect. B.3 that $\mathbf{V}$ is unitary. Furthermore, we see from (B.68) and (B.53) together with (B.52) and (B.56) that $\mathbf{Q}_M$ and $\mathbf{Q}_N$ have the same positive eigenvalues $\lambda_1, \lambda_2, \ldots, \lambda_P$ and that their remaining eigenvalues are zero. This finalizes the proof.

B.7 Some Lemmas of Determinants

The proofs can be found in [4].

In the following, we assume "compatible" matrices $\mathbf{A}$, $\mathbf{B}$, $\mathbf{C}$, and $\mathbf{I}$, which means that their dimensions allow multiplication and addition.

- Determinant of the product of two matrices

$$\det(\mathbf{AB}) = \det(\mathbf{A})\det(\mathbf{B}) \tag{B.69}$$

- Determinant of the sum of matrices

$$\det(\mathbf{A} + \mathbf{BC}) = \det(\mathbf{A} + \mathbf{CB}) \ ; \ \text{if } \mathbf{AB} = \mathbf{BA} \tag{B.70}$$

- Determinant of the sum of matrices with cyclic permutation

$$\det(\mathbf{I} + \mathbf{ABC}) = \det(\mathbf{I} + \mathbf{BCA}) = \det(\mathbf{I} + \mathbf{CAB}) \tag{B.71}$$

- Let $\lambda_1, \lambda_2, \ldots, \lambda_N$ be the eigenvalues of the matrix $\mathbf{A} \in \mathbb{C}^{N\times N}$. Then

$$\det(\mathbf{A}) = \lambda_1\lambda_2, \ldots, \lambda_N. \tag{B.72}$$

B.8 Trace of a Matrix

Definition of Trace

Given the square matrix
$$\mathbf{A} = (a_{ik}) \in \mathbb{C}^{N\times N} \tag{B.73}$$

The trace of $\mathbf{A}$ is defined as

$$\text{tr}(\mathbf{A}) = a_{11} + a_{22} + \ldots + a_{NN} \tag{B.74}$$

With a scalar factor α follows

$$\text{tr}(\alpha\mathbf{A}) = \alpha\,\text{tr}(\mathbf{A}) \tag{B.75}$$

Note, for a non-square matrix the trace does not exist, because there is no main diagonal. The proof of the following Lemmas is straightforward.

Cyclic Permutation

Let $\mathbf{A} \in \mathbb{C}^{N\times M}$; $\mathbf{B} \in \mathbb{C}^{M\times N}$ and $\mathbf{C} \in \mathbb{C}^{N\times N}$. Consequently, the product $\mathbf{ABC}$ is an $N\times N$ square matrix. Then

$$\mathrm{tr}\,(\mathbf{ABC}) = \mathrm{tr}\,(\mathbf{BCA}) = \mathrm{tr}\,(\mathbf{CAB}) \neq \mathrm{tr}\,(\mathbf{ACB}) \tag{B.76}$$

In particular

$$\mathrm{tr}\,(\mathbf{AB}) = \mathrm{tr}\,(\mathbf{BA}) \tag{B.77}$$

This also holds for $M = N$.

Trace of the Sum of Matrices

Let $\mathbf{A} \in \mathbb{C}^{N \times N}$ and $\mathbf{B} \in \mathbb{C}^{N \times N}$ be square matrices of the same dimension. Then

$$\mathrm{tr}\,(\mathbf{A} + \mathbf{B}) = \mathrm{tr}\,(\mathbf{A}) + \mathrm{tr}\,(\mathbf{B}) \tag{B.78}$$

With (B.77) follows

$$\mathrm{tr}\,(\mathbf{AB} - \mathbf{BA}) = 0 \tag{B.79}$$

Trace and Eigenvalues

Let $\lambda_1, \lambda_2, \ldots, \lambda_N$ be the eigenvalues of the matrix $\mathbf{A} \in \mathbb{C}^{N \times N}$. Then

$$\mathrm{tr}\,(\mathbf{A}) = \sum_{i=1}^{N} \lambda_i \tag{B.80}$$

and

$$\mathrm{tr}\,(\mathbf{A}^{-1}) = \sum_{i=1}^{N} \lambda_i^{-1} \tag{B.81}$$

For the latter, the eigenvalues must be unequal to zero.

B.9 Differentiation of a Scalar Function $f(\mathbf{Z})$ with Respect to a Matrix $\mathbf{Z}$

For the proof see [5].

Definition

Differentiation of a scalar function with respect to a matrix is a shorthand notation meaning that the scalar function is partially differentiated with respect to all matrix elements of $\mathbf{Z}$ and arranged in a matrix. Example:

$$\mathbf{Z} = \begin{pmatrix} z_{11} & z_{12} & z_{13} \\ z_{21} & z_{22} & z_{23} \end{pmatrix}; \; f(\mathbf{Z}) = f(z_{11}, z_{12}, z_{13}, z_{21}, z_{22}, z_{23}); \; z_{ik} \in \mathbb{C} \tag{B.82}$$

Obviously, $f(\mathbf{Z})$ is a multivariate scalar function of $z_{11}, \ldots, z_{23}$. Then we define

$$\frac{\partial f}{\partial \mathbf{Z}^*} = \begin{pmatrix} \frac{\partial f}{\partial z_{11}^*} & \frac{\partial f}{\partial z_{12}^*} & \frac{\partial f}{\partial z_{13}^*} \\ \frac{\partial f}{\partial z_{21}^*} & \frac{\partial f}{\partial z_{22}^*} & \frac{\partial f}{\partial z_{23}^*} \end{pmatrix} \tag{B.83}$$

For complex variables z_{ik} we define

$$\frac{\partial f}{\partial z_{ik}^*} = \frac{1}{2}\left(\frac{\partial f}{\partial \mathrm{Re}\,(z_{ik})} + j\frac{\partial f}{\partial \mathrm{Im}\,(z_{ik})}\right) \tag{B.84}$$

Differentiation of the Trace of a Matrix with Respect to a Matrix

We start with the differentiation of a constant α

$$\frac{\partial \alpha}{\partial \mathbf{Z}^*} = 0 \ ; \ \mathbf{Z} \in \mathbb{C}^{N \times M} \ ; \ \alpha \in \mathbb{C} \tag{B.85}$$

Please note in the following that the argument of tr (...) has to be a square matrix.

$$\frac{\partial \mathrm{tr}\,(\mathbf{Z}^H)}{\partial \mathbf{Z}^*} = \mathbf{I}_N \ ; \ \mathbf{Z} \in C^{N \times N} \tag{B.86}$$

$$\frac{\partial \mathrm{tr}\,(\mathbf{Z})}{\partial \mathbf{Z}^*} = 0 \ ; \ \mathbf{Z} \in \mathbb{C}^{N \times N} \tag{B.87}$$

$$\frac{\partial \mathrm{tr}\,(\mathbf{A}\mathbf{Z}^H)}{\partial \mathbf{Z}^*} = \mathbf{A} \ ; \ \mathbf{A} \in \mathbb{C}^{N \times M} \ ; \ \mathbf{Z} \in \mathbb{C}^{N \times M} \ ; \ \mathbf{A}\mathbf{Z}^H \in \mathbb{C}^{N \times N} \tag{B.88}$$

$$\frac{\partial \mathrm{tr}\,(\mathbf{A}\mathbf{Z})}{\partial \mathbf{Z}^*} = 0 \ ; \ \mathbf{A} \in \mathbb{C}^{M \times N} \ ; \ \mathbf{Z} \in \mathbb{C}^{N \times M} \ ; \ \mathbf{A}\mathbf{Z} \in \mathbb{C}^{M \times M} \tag{B.89}$$

$$\frac{\partial \mathrm{tr}\,(\mathbf{Z}\mathbf{Z}^H)}{\partial \mathbf{Z}^*} = \mathbf{Z} \ ; \ \mathbf{Z} \in \mathbb{C}^{N \times M} \ ; \ \mathbf{Z}\mathbf{Z}^H \in \mathbb{C}^{N \times N} \tag{B.90}$$

$$\frac{\partial \mathrm{tr}\,(\mathbf{A}\mathbf{Z}\mathbf{Z}^H)}{\partial \mathbf{Z}^*} = \mathbf{A}\mathbf{Z} \ ; \ \mathbf{A} \in \mathbb{C}^{N \times N} \ ; \ \mathbf{Z} \in \mathbb{C}^{N \times M} \ ; \ \mathbf{A}\mathbf{Z}\mathbf{Z}^H \in \mathbb{C}^{N \times N} \tag{B.91}$$

With cyclic permutation Lemma (B.76) we obtain from (B.91)

$$\frac{\partial \mathrm{tr}\,(\mathbf{Z}\mathbf{Z}^H\mathbf{A})}{\partial \mathbf{Z}^*} = \frac{\partial \mathrm{tr}\,(\mathbf{A}\mathbf{Z}\mathbf{Z}^H)}{\partial \mathbf{Z}^*} = \mathbf{A}\mathbf{Z} \ ; \ \mathbf{A} \in \mathbb{C}^{N \times N} \ ; \ \mathbf{Z} \in \mathbb{C}^{N \times M} \ ; \ \mathbf{Z}\mathbf{Z}^H\mathbf{A} \in \mathbb{C}^{N \times N} \tag{B.92}$$

$$\frac{\partial \mathrm{tr} \left(\mathbf{Z A Z}^H \mathbf{B} \right)}{\partial \mathbf{Z}^*} = \mathbf{B Z A} \ ; \ \ \mathbf{A} \in \mathbb{C}^{M \times M} \ ; \ \ \mathbf{Z} \in \mathbb{C}^{N \times M} \ ; \ \ \mathbf{B} \in \mathbb{C}^{N \times N} \ ; \ \ \mathbf{Z A Z}^H \mathbf{B} \in \mathbb{C}^{N \times N}$$

$$\text{(B.93)}$$

References

1. A. Papoulis, S.U. Pillai, *Probability, Random Variables, and Stochastic Processes*, 4th edn. (McGraw-Hill, Boston, 2002)
2. A. Papoulis, *Probability, Random Variables, and Stochastic Processes* (McGraw-Hill, Boston, 1965)
3. J.G. Proakis, M. Salehi, *Digital Communications* (McGraw-Hill, Boston, 2007)
4. R.A. Horn, C.R. Johnson, *Matrix Analysis* (Cambridge University Press, Cambridge, 2013)
5. A. Hjorungnes, *Complex-Valued Matrix Derivatives with Applications in Signal Processing and Communications* (Cambridge University Press, Cambridge, 2011)

Index

A

AC-WLAN, 269
Age variable, 93
Alamouti encoder, 251
A-posterior probability, 36
A-priori probabilities, 36
Associativity, 107
Asymptotic favorable transmission, 290
Autocorrelation, 128
Autocorrelation function, 27, 298
Autocovariance function, 298
AWGN channel, 211

B

Bandpass noise, 23, 25
Bandpass process, 306
Beamforming, 167, 282
Bessel function, 85
Bitrate, 8
Block diagonalization, 274, 278
Block diagonal matrix, 274
Block fading, 248
Block-wise operation, 246
Branch metric, 41
Broadcast channel, 270

C

Capacity, 209
Capacity of eigenmodes, 215
Cascade, 116–118
Causal time-variant system, 99
Channel capacity, 210
Channel hardening, 290
Channel impulse response, 14
Channel input, 7

Channel inversion, 270, 273
Channel matrix, 219
Channel state information, 273
Clarke and Jakes model, 84
Closed-loop precoding, 273
Code rate, 5
Coherence bandwidth, 135
Coherence time, 135
Commutativity, 106
Conditional probab. density function, 294
Conditional probabilities, 294
Constellation diagrams, 8
Continuous-time equivalent baseband system, 28
Convolution, 6, 104, 105
Correlated transmit signal, 220
Correlation, 296
Correlation coefficient, 296
Correlation functions, 127
Covariance, 296
Covariance matrix, 154
Cross-correlation function, 298, 311
Cross-covariance function, 301
Cumulative distribution function, 293
Cut-off frequency, 19
Cyclic permutation, 323

D

Delay, 68
Delay cross power spectral density, 132, 133
Delay Doppler spread function, 114, 151
Delay spread function, 47, 48, 70, 75, 96, 145
Delay time, 93
Demodulator, 10
Detection methods, 31

© Springer Nature Switzerland AG 2019
J. Speidel, *Introduction to Digital Communications*, Signals and Communication
Technology, https://doi.org/10.1007/978-3-030-00548-1

Determinant, 323
Diagonalizable matrix, 315
Differentiation of the trace, 325
Differentiation with respect to a matrix, 324
Dirac impulse, 48
Dirty paper precoder, 280
Discrete-time equivalent baseband system, 29
Discrete-time, time-variant system, 98
Distributivity, 108
Doppler effect, 47
Doppler power spectrum, 134
Doppler shift, 74
Doppler spread function, 71, 115
Doppler-variant impulse response, 114
Doppler-variant transfer function, 115

E
Eigenmode decomposition, 203, 216
Eigenmode system, 206
Eigenvalue, 313
Eigenvalue decomposition, 313
Eigenvalues, 219, 323, 324
Eigenvector, 313
Equal gain combiner, 188
Equalizer, 45
Equivalent baseband, 11, 13
Equivalent baseband system model, 145
Equivalent time-variant baseband system, 57
Ergodic capacity, 231
Ergodicity, 299
EXIT chart, 266
Expected value, 295
Exponential covariance model, 161

F
Finite impulse response channel, 149
Finite Impulse Response (FIR), 149
Fixed networks, 91
Forward error correction, 4
Frequency flat channel, 152
Frequency selective fading, 77
Frobenius norm, 319
Function of random variable, 295

G
Gaussian multipath model, 81
Gaussian noise, 27
Gaussian process, 309

H
Hermiteian matrix, 316
Hermiteian operator, 315

I
IEEE 802.11, 269
I.i.d. Gaussian MIMO channel, 152
Ill conditioned, 280
Impulse response, 48, 95
Inter-channel interference, 166, 176, 236, 239
Intersymbol interference, 14
Intersymbol interference, time-variant, 60
Iterative detection, 265

J
Joint central moment, 296
Joint probability density function, 294

L
Lagrange method, 225
Layered space-time (BLAST), 261
Layered space-time (D-BLAST), 265
Layered space-time (H-BLAST), 264
Layered space-time (V-BLAST), 263
Leakage, 280
Likelihood function, 31
Likelihood probability density function, 40
Linear combiner, 185
Linear MIMO receivers, 165
Linear time-variant systems, 95
Linearity, 104

M
Mapper, 5
Massive MIMO, 289
Matrix of eigenvectors, 313
Maximum likelihood detection, 31
Maximum likelihood detector, 194
Maximum likelihood MIMO receiver, 193
Maximum likelihood sequence detection, 38
Maximum ratio combiner, 186, 187
Mean noise power, 24
Mean power, 295
MIMO Kronecker model, 158
MIMO operation modes, 166
MIMO precoding, 233
MIMO receivers, 165
MIMO soft demapping, 266
2x2 MIMO system, 255

MMSE precoder, 238
MMSE receive matrix, 183
MMSE receiver, 181
Mobile receiver, 73
Modified impulse response, 92, 93
Modified time-variant impulse response, 97
Modulation of stochastic process, 305
Modulator, 6
Moore–Penrose inverse, 170
Multi-access interference, 270
Multi-user detection, 287
Multi-user interference, 270, 276
Multi-user maximum likelihood detection, 289
Multi-user MIMO, 269, 270
Multi-user MIMO downlink, 270
Multi-user MIMO uplink, 286
Multi-user transmit beamforming, 281
Multipath model, 67
Multipath propagation, 67
Multiple Input Multiple Output (MIMO), 143, 147
Multiple Input Single Output (MISO), 147, 251
Mutual information, 209

N
Nakagami-m fading, 83
Noise, 23, 26
Noise after sampling, 26
Normalized channel matrix, 179
Normal matrix, 315
Null space, 279
Nyquist criterion, 17
Nyquist frequency, 18
Nyquist lowpass, 19, 20

O
Ordered successive interference cancellation, 196
Orthogonal space-time coding, 250
Outage capacity, 231
Output time, 93
Overall code rate, 248
Overall delay spread function, 51
Overall Doppler spread function, 52
Overall time-variant impulse response, 49, 50
Over-determined, 170

P
Path loss, 63, 68
Path metric, 41
Power allocation, 224
Power spectral density, 24, 130
Precoder, 233
Precoder matrix, 272
Precoding, 233
Precoding multi-user MIMO, 270
Prefilter, 213, 233
Probability, 293
Probability density function, 24, 293
Pseudo inverse, 170, 175
PSK, 8
Pulse shaper, 5, 6

Q
QAM transmitter, 7
Q-function, 33
Quadratic form, 316
Quadrature Amplitude Modulation (QAM), 3, 9

R
Raised cosine, 20, 21
Rayleigh fading, 82
Receive diversity, 167
Receive lowpass, 10
Receiver, 10
Redundancy, 5
Reflections, 68
Regularized channel inversion, 280
Rician fading, 83
Roll-off factor, 19

S
Sampling stochastic process, 310
Scattering, 67
Sequence detector, 41
Sequential detection, 31
Serial-to-parallel converter, 247, 249
Shadowing, 64
Signal-to-interference-plus-noise ratio, 288
Signal-to-leakage-power ratio, 281
Signal-to-noise ratio, 177, 178, 180, 186, 187, 189, 235
Single Input Multiple Output (SIMO), 185
Single Input Single Output (SISO), 146
Single-user MIMO, 277
Singular Value Decomposition (SVD), 203, 278, 283, 319

Singular values, 204
Space-time codes, 258
Space-time coding, 245
Space-time coding matrix, 246
Space-time encoder, 245
Space-time trellis coding, 260
Spatial code rate, 247
Spatial demultiplexer, 249
Spatial diversity, 167
Spatial interleaver, 265
Spatial multiplexing, 166, 249
Squared norm of matrix, 319
Squared norm vector, 319
Standard deviation, 295
State space trellis diagram, 39
Statistically independent, 294, 297
Statistically independent processes, 302
Stochastic process, 297
Strict sense stationary process, 300
SVD-based precoder, 240
SVD-based receiver, 240
Symbol alphabet, 8
Symbol-by-symbol detection, 31
Symbol error probability, 33, 34
Symbol rate, 5, 8
Symbol sequence, 5
System capacity, 210
System functions, 113

T
Threshold detection, 32
Time-invariant convolution, 103
Time-invariant system, 95
Time-variant channel, 47, 48
Time-variant convolution, 103
Time-variant finite impulse response filter,
    98
Time-variant impulse response, 50
Time-variant transfer function, 113, 150
Tomlinson–Harashima precoder, 280
Tomlinson–Harashima scheme, 236
Trace of a matrix, 323
Transfer function, 10

Transmission matrix, 246
Transmission system, 4, 48
Transmit diversity, 167, 251
Transmit lowpass filter, 5
Transmit prefilter, 175
Trellis diagram, 39, 40
Turbo principle, 265
Two-dimensional convolution, 125
Two-dimensional impulse response, 47

U
Uncorrelated scattering, 85, 132
Uncorrelated WSS processes, 301
Under-determined, 174
Uniform circular array, 148
Uniform linear array, 148
Unit impulse, 99
Unitary matrix, 317
User equipment, 270
User terminals, 270

V
Variance, 295
Viterbi algorithm, 42
Viterbi equalizer, 45

W
Water filling algorithm, 222
Wide sense stationarity, 129
Wide sense stationary process, 300
Wide-sense stationary, 84
Wiener–Khintchine theorem, 304
Wiener-Lee theorem, 303
WLAN, 269

Z
Zero-Forcing precoder, 236
Zero-forcing receiver, 168–170, 174, 287
Z-transform, 5

Printed in the United States
By Bookmasters